The Wind Power Story

The Wind Power Story

A Century of Innovation that Reshaped
the Global Energy Landscape

Brandon N. Owens

IEEE PRESS

WILEY

Published by John Wiley & Sons, Inc., Hoboken, New Jersey.
Published simultaneously in Canada.

For general information on our other products and services or for technical support, please contact our Customer Care Department within the United States at (800) 762-2974, outside the United States at (317) 572-3993 or fax (317) 572-4002.

Wiley also publishes its books in a variety of electronic formats. Some content that appears in print may not be available in electronic formats. For more information about Wiley products, visit our web site at www.wiley.com.

Library of Congress Cataloging-in-Publication Data

Paperback: 9781118794180

Illustrations credit by Maciej K. Kozak

Set in 10/12pt Warnock by SPi Global, Pondicherry, India

Printed in the United States of America

V10012434_073119

To Colleen, Lauren, Cameron, and Logan

Contents

Preface

When I joined the National Renewable Energy Laboratory in 1994, wind power appeared to have reached its nadir. The California Wind Rush of the 1980s had crashed, global energy prices had fallen from their peaks, and wind turbine manufacturers and developers were struggling to stay afloat. Wind power technology itself appeared unrefined and sometimes unreliable. This was before policy makers in the United States and Europe began enacting support programs that would ultimately jumpstart the rise of wind power in the twenty-first century. When US manufacturer Kenetech filed for bankruptcy in the spring of 1996, it seemed to confirm our worst fears about the future of wind energy.

Yet—to those who studied the technology—it was abundantly clear that wind power still had enormous potential to transform the global energy landscape. Both public and private research efforts had advanced the technology to the point where it had become evident that wind power could successfully integrate into power networks across the globe and—with continued effort—we believed that wind power would eventually achieve the cost and reliability targets needed to compete head-to-head with conventional generation options in a commercial environment. The technology just needed more time and additional policy support in order to get the ball rolling again.

Today, as we look back from 2019, what's happened since has exceeded the expectations of even the most optimistic wind power believers. From a global installed base of 3.5 gigawatts (GW) in 1994, wind power grew to nearly 600 GW by the end of 2018. In other words, the installed base of wind power increased over one hundred and fifty times. Wind power is now being added to power networks around the world at a rate of over fifty gigawatts per year. Here in the United States, there's more wind and solar being added to the grid than all other sources combined. The rise of wind power has been breathtaking.

Wind power's rise occurred just in time. The growing threat of climate change is the greatest challenge facing humankind today, and zero-greenhouse gas emitting wind power technologies represent one of the great technological solutions at our disposal. Wind power is now playing a critical role in ushering much-needed transformation into the global energy system. Given the urgency of the climate crisis, it is more important than ever to hasten the adoption of wind power and all other low carbon technologies. Given the success to date, it's now time to hit the accelerator, not the brake. Our future depends upon it.

Because of the important role that wind power is playing on the global stage, I believe it is important to convey the full story of wind power's arrival on the global energy scene. As I dug deeper into the history of wind power, it became clear to me that the nadir that I witnessed in 1994 was just another cycle in wind power's long history. Most modern observers believe that the development of wind power started in earnest in Denmark or California in 1980s. More astute observers trace the beginning of wind power back to the global energy crisis in 1973 and the national research programs that were initiated in the wake of that crisis.

However—as it turns out—wind power's origins trace back to the late nineteenth century. In fact, to truly appreciate the wind power story, we need to go all the way back to Scotland in the winter of 1887. This is when university professor James Blyth affixed an electric generator to a windmill in order to produce electricity for his cottage. That was the birth of the wind turbine—a combination of the centuries-old windmill and the nineteenth century electric generator.

Wind power's journey across the twentieth century is a colorful tale. The journey was filled with starry-eyed dreamers such as Germany's Herman Honnef who envisioned massive direct drive wind turbines in the 1930s that were sixty years ahead of his time, and pragmatists—such as Americans Joe and Marcellus Jacobs—who built their first small wind turbines in the 1920s using spare car parts. The story is also full of interesting illuminations and novel innovations, like Ulrich Hütter's use of fiberglass composite wind turbine blades for the first time in the 1950s, and Johannes Juul's invisible solution to taming Mother Nature using a passive stall wind turbine design for the first time. Let us not forget about the craftsmen in Denmark who reinvented wind power in the 1970s, and the cowboys in California who presided over the California Wind Rush in the 1980s. My goal in writing this book is to capture all of these stories in one place, while providing the broader context of the surrounding energy technology, policy, and market landscape. Along the way, I trace the evolution of wind power technology innovations and highlight the people that made them possible.

My hope is that with this material in one place, we can reflect upon wind power's long struggle and recent emergence, and use the insights gained to help guide our collective energy future. There will be new energy breakthroughs on the horizon and we need to learn to better cultivate those opportunities, lest we allow one of our most promising technologies to languish at the doorstep of tomorrow. In lieu of this broader aspiration, my hope is that readers will simply find the wind power story educational and interesting. I hope you have as much fun reading it as I did writing it.

Golden, Colorado   *Brandon N. Owens*

1

The Wind Power Pioneers

When we see a new form or structure with new qualities, we are really seeing new arrangements of what already existed.
—David Christian (2018)

1.1 Work of the Devil

In 1887, eleven years after Scotsman Alexander Graham Bell called "Mr. Watson, come here, I want to see you." into his newly invented telephone, another Scot developed a new way to generate electricity—the wind turbine. James Blyth—a professor at Anderson's College in Glasgow—combined a traditional vertical-axis windmill with an electric generator.[*]

Of course, windmills themselves were centuries old in 1887. But up until this time, they were used primarily to pump water, or as a mechanical power source in brickyards, spinning mills, and wood workshops.[†] So until Blyth did it, no one had coupled the centuries-old windmill with the new nineteenth century electric generator—or "dynamo"—as it was called at the time. The first rudimentary dynamo is credited to Frenchman Hippolyte Pixii in 1832. Antonio Pacinotti enabled it to provide continuous direct current (DC) power by 1860; and in 1867, Werner von Siemens, Charles Wheatstone, and S.A. Varley nearly simultaneously devised the "self-exciting dynamo-electric generator."[1]

---

[*] Vertical-axis windmills are those with the main shaft of the rotor mounted vertically and horizontal-axis windmills are those with the main shaft of the rotor mounted horizontally. Europeans began using horizontal-axis windmills in the twelfth century. Despite occasional experimentation with vertical-axis designs and some experimental promise, the horizontal-axis design has become the preferred orientation for modern wind turbines.

[†] The first established use of windmills occurred 900 years earlier, in the tenth century CE, in what is now Iran (Shepard 1990).

The Wind Power Story: A Century of Innovation that Reshaped the Global Energy Landscape, First Edition. Brandon N. Owens.
© 2019 by The Institute of Electrical and Electronics Engineers, Inc.
Published 2019 by John Wiley & Sons, Inc.

Blyth's contribution was to combine the electric generator and the windmill. He fused two technologies from two different eras to create something new.[‡]

This was the beginning of a century long journey that would eventually lead to the development of the modern horizontal-axis wind turbine, the most common commercially available wind turbine today. In the twenty-first century, wind power is a big deal. Wind turbines are now generating electricity in ninety countries around the world and the total installed base is nearly 600 GW.[2] That is a lot of electricity. In fact, total global installed wind capacity today is enough to meet the electricity needs of every household in the US.[§] Furthermore, investment in wind, solar, and other renewable resources reached US$300 billion in 2018. It was the largest source of global electricity spending.[3] Together with solar power, wind energy investments account for two-thirds of all global power plant additions. Wind power is not just a passing fad either. Looking ahead, the International Energy Agency (IEA)—which keeps track of these things—expects the world to add another 300 GW of wind turbines by 2022.[4]

Wind power is such an important part of the global energy landscape today in the twenty-first century that we almost take it for granted. It wasn't always so. Wind power had very humble beginnings, starting with Mr. Blyth's fusion of a vertical-axis windmill and the electric dynamo in 1887. Born in 1839 in the village of Marykirk, James Blyth displayed a proclivity toward science from an early age. He was educated at the local parish school before winning a scholarship to the General Assembly Normal School in Edinburgh. After receiving a BA from the University of Edinburgh, he taught mathematics for a while before earning an MA and being appointed Professor of Natural Philosophy at Anderson's College in 1880, which is now the University of Strathclyde. It is here that he began his research into wind power.[5]

After putting his wind turbine in operation in July 1887, Blyth delivered a paper to the Philosophical Society of Glasgow the following May. In the paper, Blyth described his wind turbine as being "of a tripod design, with a 33-feet windshaft, four arms of 13 feet with canvas sails, and a Burgin dynamo driven from the flywheel using a rope."[6] Blyth used the electrical output from the generator to charge lead-acid batteries that powered the lights in his holiday home in Marykirk. Once it was up and running, he soon realized that his wind turbine provided more electricity than his cottage lights needed, so he offered to light up the local main street. Apparently, not knowing what to make of his

---

‡ A turbine is a machine that extracts energy from a fluid and converts it to useful work. Here, a machine that converts wind energy into electric power is referred to as a wind turbine.
§ If 591 GW of wind turbines generate power at a level that is—on average—a quarter of their rated power throughout the year, then the annual level of wind generated electricity would be about 1,300 billion kWh. Total US residential electricity consumption is about 1,450 billion kWh.

offer—and wind power in general—the townspeople labeled it "the work of the devil."[7] Blyth, it seems, would be the first in a long list of distinguished wind power innovators that were a step ahead of their contemporaries.

Blyth was awarded a patent in the United Kingdom for his "wind engine" in 1891.[8] Four years later, he licensed the Glasgow engineering company—Mavor and Coulson—to build a second, improved wind turbine. The improved wind turbine was eventually used to supply emergency power to the Lunatic Asylum, Infirmary and Dispensary of Montrose, which is known today as Sunnyside Royal Hospital.[9] Blyth's promising research came to an end with his death in 1906. Scotland thus relinquished its first—and only—lead in wind power technology.

As it turns out, Professor Blyth was not the only innovator tinkering with wind power at the end of the nineteenth century. During the same winter, another inventor was busy building a wind turbine prototype. To provide power to his mansion and basement laboratory, Mr. Charles F. Brush built a wind turbine in his backyard in Cleveland, Ohio. Mr. Brush's wind turbine began generating electricity just a few months after Blyth's started.

At the time, Brush was one of America's most distinguished inventors. He received his mining engineering degree from the University of Michigan in 1869. He worked four years in Cleveland as a chemist before forming an iron dealing partnership. In 1876, Brush received one of his fifty career patents for the open coil-type dynamo and subsequently began selling them commercially. His next invention—an improvement to the arc light—made Brush world famous. Arc lights preceded Edison's incandescent light bulb in commercial use and they were suited to applications where very bright light was needed, such as street lights and lighting in commercial and public buildings.

Brush's experience as an electric industry pioneer put him in a unique position to invent a new electricity generation technology like wind power. He had already improved both the arc light and the electric generator. In fact, for a time, his improved generator was the largest in the world. Cities across the country began using Mr. Brush's arc light. In Wabash, Indiana, the city fathers hired Brush to set up four 3,000-arc light displays at the local courthouse. By 1881, Brush's invention was lighting the streets of New York, Boston, Philadelphia, Baltimore, and San Francisco.[10]

In 1880, Mr. Brush formed the Brush Electric Company, which was later bought by Thomson-Houston Electric Company in 1889 and eventually merged with the Edison General Electric Company in 1891 to help form General Electric (GE).[11] After the success of the Brush Electric Company, Brush was financially secure enough to focus on his personal interests, which apparently included creating the first wind turbine in North America in the backyard of his mansion on Euclid Avenue—or "Millionaire's Row"—in Cleveland.

Electricity from Brush's wind turbine was used to charge twelve batteries, which were the power source for 350 incandescent lamps, two arc lamps, and three motors.[12] Like Blyth, Brush had combined a traditional windmill with a DC generator. Brush used the more common horizontal-axis windmill, a machine with its main shaft positioned horizontally—or parallel—to the ground, as opposed to Blyth's vertical-axis windmill, which had its main shaft oriented vertically—or perpendicular—to the ground. Although vertical-axis wind turbines (VAWT) have undergone development and innovation through the twentieth century, the horizontal-axis wind turbine (HAWT) has now become the preferred orientation in modern wind turbines, although VAWT innovators are still trying to change the tide.

Brush used 144 cedar blades, each seventeen meters in length. Mr. Brush's wind turbine stood 18 m tall and weighed forty tons.[13] Thus, not only was Brush's wind turbine larger than Blyth's, it was also absolutely enormous compared to the small windmills that were used throughout the United States in the nineteenth century to pump water. These windmills typically had a blade length of 3 m, less than one-fifth the size of Brush's wind turbine blades. The larger blade length of Brush's wind turbine allowed his machine to capture more energy from the wind.

Power output from a wind turbine is a function of the air density, the "power coefficient" or efficiency of the wind turbine, the rotor's "swept area," and the cube of the wind speed. The "swept area" of a wind turbine is the size of the circular disc that is created when the blades are spinning around the hub. The greater the blade length, the larger the swept area. The swept area of the Brush turbine was 908 m^2, compared to just 28 m^2 for the smaller mechanical windmills that were prevalent during the late nineteenth century. Mr. Brush was aware—on an intuitive level at least—that when it came to harnessing power from the wind, bigger is better.

The most common windmill in the United States during Brush's time was the Halladay Windmill, named for its creator, engineer, and businessman Daniel Halladay. Americans used Halladay Windmills for grinding and pumping water. The use of the Halladay Windmills was so widespread that it became known as the "American Windmill." Halladay Windmills became a fixture of the American West. Even today, many still dot the landscape of the United States.

With its fan of wooden blades, the American Windmill was a "high-solidity" machine, which means that the ratio of the blade area to the rotor area was high. High-solidity windmills are very good at producing torque, or rotational force, which is ideal for lifting the pistons of water pumps. To avoid damage in high winds, American Windmills were equipped with automatic speed regulators that adjusted the angle of the blades based on the wind speed. They also had tail vanes that automatically rotated—or "yawed"—the windmill so that it

Figure 1.1 Mr. Brush's Windmill Dynamo. Built in 1887, Charles Brush's 12-kW wind turbine was the first operational wind turbine in the United States. Situated in the backyard of Brush's mansion in Cleveland, the wind turbine was used to charge twelve batteries that were the power source for 350 incandescent lamps, two arc lamps, and three motors in Brush's mansion.

faced the direction of the oncoming wind. It was an elegant technology, embodying the collective knowledge that had been developed over the course of centuries, and it was ideally suited to meet the needs of nineteenth century America.

American Windmills worked so well that when Mr. Brush created his wind turbine, he didn't trouble himself with developing an entirely new design—he simply built a super-sized version of the American Windmill and connected it to an electric generator. The rotor of Mr. Brush's wind turbine was turned out of the wind by a wind flag that was positioned vertically to the rotor. Although they were an ocean apart, two men, Blyth and Brush, simultaneously developed similar solutions to turn wind energy into electric power. Along the way, they helped give birth to a technology that would one day transform the world's electricity landscape.

To be fair though, at the beginning of the wind power journey, we must not give full credit to Blyth and Brush for the creation of wind turbine. The key components were available, and indications are that other wind turbines had been built prior to Blyth and Brush's, including a small multibladed wind turbine with sheet metal blades displayed at the World Exhibition in

Philadelphia in 1876.[††] Further, there is some evidence that wind turbines were used on board ships to power on-board batteries. Two examples are the *Fram*, which sailed to the Antarctic, and the *Chance*, which sailed out of New Zealand. Another example is Charles-Marie-Michel de Goyon's wind generator near Normandy in France in 1887. However, Blyth and Brush developed the most complete and well-documented solutions during this period; so, for our purposes, the wind power story starts here.

Brush's wind turbine ran for twelve years without requiring any major repair and ran for twenty years before being put out of commission in 1909. Henry Ford, a close friend of Brush, attempted to purchase it—first from Brush, who verbally agreed, and later from the Brush estate after Mr. Brush's death from bronchitis in June 1929. However, Ford's effort was eventually blocked by the City of Cleveland, which had intended to preserve the machine for public display. Unfortunately, the money needed to restore the machine was not forthcoming in depression-era Cleveland. The components of Mr. Brush's wind turbine were put into storage and later destroyed.[14]

Despite the technical success of Mr. Brush's wind turbine, there was one problem with the machine: the blades just didn't turn fast enough to generate much electricity. The maximum speed of the rotor on Bush's wind dynamo was 10 rpm and, with the assistance of a 50-to-1 step-up gearbox, the main shaft of the wind turbine made fifty revolutions for every one revolution of the wind turbine rotor. At its top speed, the electrical output was just 12 kW.[‡‡] A rotational speed of 10 rpm was fine for a windmill, where torque—not rotational speed—matters most. But a wind turbine is different. Rotational speed is more important than torque because higher speeds correspond to greater quantities of electrical output. Clearly, this dilemma would have to be solved if the wind turbine was going to be useful for generating electricity.

Thus, the key technical question facing Brush was how to increase the rotational speed of the blades. There are no records to indicate whether Mr. Brush asked, or attempted to answer this question. He moved on to other projects. His curious nature, which made him an ideal wind turbine inventor, drove him to explore other research areas. He started research in the kinetic theory of

---

†† Ball (1908, 322) notes that he received numerous letters about wind power experimentations and that one machine had been in operation for thirteen years. Ulrich Hütter (1954, 2) notes that "the many-blade sheet metal turbine first shown at the World Exhibition in Philadelphia in 1876 had relatively small areas swept out by the turbine wheels."
‡‡ The maximum output of a generator is known as the machine's "rated capacity" and is measured in kilowatts (kW). The power output from a generator is measured in kilowatt-hours (kWh). A wind turbine with a 12 kW rated capacity like Mr. Brush's, could produce 1,752 kWh per year if it operated during 20 percent of the hours in the year (since 12 kW × 8,760 hours/ year × 20 percent = 1,752 kWh).

gravitation, a theory that holds that gravitational force can be explained in terms of tiny particles that impact all objects. This theory was studied by scientists until the beginning of the twentieth century, when it was eclipsed by Albert Einstein's theory of general relativity. Mr. Brush might have been better sticking with wind power research.

1.2 The Danish Edison

In Denmark, meteorologist Poul la Cour was poised to make an even greater contribution to the development of wind power than either Blyth or Brush. With university degrees in Physics and Meteorology, la Cour was already an accomplished scientist. While Brush was developing his arc light in 1878, la Cour was busy perfecting his improvements to the telegraph. La Cour's enhancement allowed one hundred telegraphs to be sent on the same wire at the same time. Thomas Edison had only managed to send four telegraphs simultaneously. This—and other successful inventions—earned la Cour the well-deserved nickname "The Danish Edison."[15]

Although la Cour was just as intellectually gifted as America's Thomas Edison, he never gained the notoriety, nor the wealth, of his American counterpart. Quite the opposite actually. The expenses associated with his experiments nearly ruined him financially, and friction with his employer over his "inventive spirit" forced him to change careers in 1878. This apparently wasn't a matter of great concern to la Cour because his primary aim was to invent, not to make money. By all accounts, he wanted to create technologies that helped make people's lives easier. La Cour's passion for inventing was driven by his philanthropic nature and utopian vision of society.

In 1891, four years after Brush and Blyth built their wind turbines, la Cour built Denmark's first wind turbine. He built it during the same year that Denmark's first power plant opened in Odense.[16] A teacher at the Danish Folk High School in the rural town of Askov, la Cour understood the positive impact that electricity could have on people's lives. He envisioned a unique way to generate and store electricity and he applied for funding from the Finance Committee of the Danish Parliament.[17] He was awarded a research grant from Copenhagen and successfully constructed his first wind turbine in 1891. In light of the important role that wind power would play in Denmark's history, la Cour's breakthrough in 1891 was a key achievement.

In the same way that Bush designed his wind turbine after the American Windmill, which was ubiquitous in the United States during the time, la Cour designed his wind turbine after Tower Mills that were found scattered throughout Denmark. Tower Mills were typically constructed of a stone or brick tower and had a rotating cap that held the rotor and could be turned to face the direction of the oncoming wind. Tower Mills had been part of the European landscape since 1390 CE.[18] In contrast to the high-solidity multibladed American Windmill, Tower Mills were "low-solidity" machines with just four blades. The blades used by la Cour were originally invented in 1772 by Andrew Meikle in Great Britain; they replaced sail cloths with rectangular plate-like structures that automatically opened as the blades turned.[19] As it turns out, the blades of la Cour's wind turbine spun faster, but provided less torque than those on Charles Brush's wind turbine. The difference in design and performance between Brush's American Windmill design and la Cour's Tower Mill design would turn out to have important implications for la Cour's invention and for the future of wind power.

The blades in la Cour's wind turbine were 7.4 m in length and yielded a swept area of 172 m^2. The turbine was connected to two 9-kW DC electric generators. In the process of developing his wind turbine, la Cour also developed the first rudimentary solution to convert the natural speed fluctuations of the wind—what we now call "variability"—into a constant rotational speed for his wind turbine blades. This was useful because variations in the wind turbine's rotational speed created changes in the frequency and voltage of the electrical output. The importance of constant rotational speed would grow in future years when wind turbines began to be connected to transmission systems. La Cour's goal was to ensure that a steady voltage was provided to his battery in order to avoid premature wear and tear.

In order to try and maintain a constant rotational speed, la Cour introduced a second shaft to the wind turbine and connected it to the main shaft via a pulley system. The electric generator was then attached to the second shaft rather than the main shaft. By adjusting the tension of the pulley belts based on the wind speed, la Cour could create just the right amount of slippage in the pulley belts to keep the second shaft rotating at a constant speed. La Cour called his system "Kratostate."[20] It was a crude solution, but suitable for his purposes.

In addition to constructing Europe's first wind turbine, la Cour also devised a new method to store wind power. He recognized that one of the drawbacks of wind energy is that it is a variable energy resource, which means that power is generated intermittently when the wind blows, which may not be when people actually need electricity. This issue was particularly problematic for la Cour because his wind turbines were intended to provide electricity to Denmark's rural farmers who needed light for working and reading on long, dark—and often windless—winter nights.[21] In America, Mr. Brush had used batteries to store wind power for later use. However, la Cour felt that batteries were too expensive and wanted a different approach. He needed a new solution—and if one wasn't readily available, he come up with his own.

Inspired by Italian professor Pompeo Garuti, la Cour developed a unique solution for the storage problem: he would direct the electricity generated by the wind turbine into a tub of water where oxygen would be separated from

Figure 1.2 Poul La Cour's experimental wind turbine in Askov in 1891. The turbine was connected to two 9-kW DC electric generators. In the process of developing his wind turbine, la Cour also developed the first rudimentary solution to convert the natural speed fluctuations of the wind into a constant rotational speed for his wind turbine.

hydrogen through electrolysis.[22] The hydrogen and oxygen gas was then collected in tanks and burned later for lighting. He built his storage concept in the summer of 1891. On the windiest days, his test wind turbine could produce over 264 gallons of hydrogen and 132 gallons of oxygen.[23] It worked as planned. However, la Cour's solution was not seriously reconsidered as a complement to wind power until the twenty-first century.

This was an elegant solution to address the variability of wind power, but there was one problem with La Cour's approach to electrolysis: it did not separate the hydrogen and oxygen molecules completely. This led to the inconvenient problem of an occasional hydrogen gas explosion. One explosion blew out the windows of the Askov High School.[24] There were no casualties as far as we know, but this problem forced la Cour to remove his hydrogen system and replace it with batteries. Batteries may not have been the cheapest option, but they were not prone to spontaneous combustion, which was important if humans were nearby.

La Cour's next challenge was to determine how to build the most efficient wind turbine. Was it better to have many blades or just a few? Was it better to have rectangular blades or curved blades? How should the blades be positioned toward the wind? La Cour wanted to know the answers to these questions so that he could develop a set of rules to help guide wind turbine design. To answer these questions, la Cour designed a series of wind tunnel experiments between 1896 and 1900. He used a wind tunnel to test wind turbines with a variety of different blade designs and positions.

As it turns out, similar blade research had been conducted previously by John Smeaton who documented his findings in his concisely titled 1760 book, *An Experimental Enquiry Concerning the Natural Powers of Water and Wind to Turn Mills, and Other Machines, Depending on a Circular Motion*. Smeaton built and tested a small windmill to derive a set of laws that describe the operation and performance of windmill blades. Although he did not recommend a specific number of blades, Smeaton found that—beyond a certain point—the maximum speed of the blades will diminish if too many blades are added so that "when the whole cylinder of wind is intercepted, it does not then produce the greatest effect."[25] This is an argument in favor of low-solidity wind turbines. He also found that the windmill performed the best when the blade's angle to the oncoming wind—known as the "angle-of-attack"—was 15°–18°. He recommended the use of twisted blades over flat sail blades, so that the blade angle varies from 15° to 7.5° from root to tip. Finally, Smeaton observed that the maximum energy of the windmill is directly proportional to the cube of the wind speed.

La Cour's blade research provided a nineteenth century refresh of Smeaton's original seventeenth century research. When he concluded his work, la Cour developed guiding principles for the construction of wind turbines. In his book *Forsøgsmøllen*—published in 1900—he made recommendations about wind turbine blades. First, he recommended a low-solidity, four-blade design as the most efficient configuration for a wind turbine. Next, he recommended a twisted blade, positioned at a 20° angle-of-attack at the base, declining to a 10° angle at the tip. While Smeaton was only able to get his windmill blades to turn at a speed equal to the oncoming wind, la Cour's experimental wind turbine reached a speed equal to 2.4 times the speed of the oncoming wind. Finally, la Cour confirmed Smeaton's finding that the maximum power of the wind turbine was directly proportional to the cube of the wind speed, but he also added that it was directly proportional to the swept area of the blade. Thus, longer wind turbine blades generate more power.

Based on these findings, it became clear that there are three avenues for increasing the power output of a wind turbine: (1) configure the blades properly; (2) increase the blade length; and (3) increase the wind speed. Thus, the way to increase the output of Mr. Brush's wind turbine back in Cleveland would have been to use fewer blades, give them an aerodynamic twist, and position them with a very small angle-of-attack. Word was not—apparently—sent to Mr. Brush back in Cleveland.

In 1897, la Cour received additional funding from Copenhagen for a larger wind turbine. He ordered a six-bladed rotor for his new windmill from Christian Sørensen, an inventor who had established the Skanderborg Wind Motor Factory in 1889 to build small windmills for industrial use. However, the Sørensen rotor was plagued with problems that forced la Cour to eventually replace it with a traditional four-bladed Danish-style rotor.[26]

By 1902, la Cour was done with experimentation and was ready to achieve his more practical objective of electrifying rural Denmark using his newly fashioned wind turbines. During this time, Denmark's electricity consumption was 16.7 million kWh and just 1.5 percent was from rural areas.[27] La Cour believed that his wind turbine design was the key to electrifying rural Denmark. However, he was an inventor—not an electrical equipment manufacturer—so he gave his designs to the Lykkegaard Company of Ferritslev. Lykkegaard built wind turbines ranging from 10 to 35 kW each. The Lykkegaard rotors had a diameter of up to 20 m and they used four shutter-type sails as blades.[28] Yawing was carried out by two fantail side wheels. A DC generator was installed at the base of the tower and the generator was driven by the rotor through a long main shaft and gearbox.[29]

Being a teacher by trade, la Cour was also determined to spread the gospel of wind power through education. In 1903, he established the Danish Wind Electricity Society (DVES). DVES instructed approximately twenty electricians a year in Askov. They learned wind power theory for three months and studied the maintenance of the Askov turbine along with their classroom work. DVES published a bimonthly journal on wind electricity with la Cour as the author of most of the articles. Through his education programs, la Cour planted the seeds for future Danish wind power development by teaching the next generation of wind energy innovators.

In order to gain admittance to la Cour's wind power course, one of his students—Johannes Juul—falsified his registration documents indicating that he was eighteen years old, when he was only seventeen and too young for the course. When this was discovered, he was forced to wait until his eighteenth birthday before he was granted a diploma. Juul spent the time working as an apprentice. He captured his experience with his camera and created the only photographic documentation of la Cour's early wind power practitioners at work.[30] Mr. Juul wasn't quite done with wind power, but his contributions to wind power history wouldn't occur for another fifty years.

La Cour's efforts produced tangible results. By 1908, Lykkegaard had built seventy-two wind turbines that supplied power to rural Denmark. By 1910, more than one hundred wind turbines were installed throughout Denmark for agricultural use. Ten years later, 105 small wind turbines generated electricity in combination with diesel generators.[31]

La Cour died the same year, leaving Denmark with a legacy of wind power research, education, and manufacturing. The Danish government's funding of la Cour's wind research was the first documented instance of government support for wind energy research. At the time, la Cour recognized the significance of government funding for his research. In 1900, he considered the research and wrote: "One may find it surprising that a small country like Denmark, taking a bold step ahead of any others, is willing to devote such large sums of money to research on the use of wind power. Nothing could be more natural, however, considering that this country, which has no waterfalls or coal, is exposed to the risk of paying increasingly high rates for the purchase of fuel from foreign countries...."[32]

La Cour—for one—had done his part to ensure that Denmark embarked upon a path that encouraged domestic energy independence—the rest would be up to future generations.

1.3 The War of the Currents

Around the same time that Blyth, Brush, and la Cour were experimenting with wind power, electric power systems were beginning to take shape in the United States and Europe. Thermal power plants would become the workhorse of twentieth century power systems. Whereas wind turbines convert wind energy to electric power, thermal power plants convert heat energy to electric power. The world's first thermal power plant—Thomas Edison's Pearl Street Station in New York City—began supplying power in September 1882, five years before Blyth's wind turbine began operation.[§§] Pearl Street Station used reciprocating steam engines that used the heat created in a coal boiler. As the twentieth century progressed, the reciprocating engine would soon be supplanted by the steam turbine, which itself would be complimented by the gas turbine in ultraefficient "combined cycle" configurations toward the end of the twentieth century.

The Pearl Street Station was composed of six reciprocating engines. Each engine was connected to a 12-kW electric generator that operated at 700 rpm. If it operated every hour of the year, it could produce over 680,000 kWh. The plant initially was used to provide power to fifty-nine customers. Demand for power from Pearl Street Station grew tenfold in the first three years, and within fourteen months the Pearl Street Station had 508 customers.[33]

After the Pearl Street Station, the amount of electricity that could be produced by a single power plant—the plant's "rated capacity" measured in kilowatts—grew quickly. The development of ever larger power plants was first facilitated by Charles Curtis's steam turbine. Mr. Curtis presented the concept

---

§§ The world's first power plant was the old Schoelkopf Power Station No. 1 near Niagara Falls on the US side of the US-Canadian border, which began to produce electricity in 1881.

of a steam turbine to drive a generator to GE management in 1896. By 1897, he was directing steam turbine development for the company. GE placed its first turbine into operation in November 1901. It was a 500-kW turbine that operated at 1,200 rpm. By 1902, GE offered turbines with rated capacities of 500, 1,500, and 5,000 kW. Wind power turbines wouldn't reach these size ranges until the twenty-first century.

Although larger power plants were more expensive to build, the average cost of power declined rapidly as the rated capacity of the plants increased. That is—the plants themselves were more expensive, but they could produce a lot more output, which turned out to be cheaper per unit if the total cost of the plant was divided by the total electrical output. Large power plants were profitable only if all the output was sold to customers. However, even the most densely populated cities like New York didn't have enough homes and businesses located nearby to support megawatt-scale power plants.*** Thus, to support large power plant development and drive down costs, electricity would have to be distributed to far-flung customers through long distance transmission lines.

At the time, however, DC transmission lines were unable to transmit power over more than one-half of a mile without large voltage losses and, nonetheless, transmission lines were very expensive because of the high price of copper.[34] This meant that power plants sizes were limited by the amount of electricity demand within a half-mile radius. Although steam turbine technology enabled large power plants to be built, the transmission system couldn't yet support them.

At this moment in time, the future path of the electric power system was uncertain. If the transmission system could be modified to enable long distance power transmission without as much voltage loss, then large steam turbines could be built and would serve as the backbone to a centralized, or "central station," electric power system that followed a hub-and-spoke model. In this design, large power plants serve as the hub and the transmission lines act as the spokes that delivered power to outlying customers on the edge. On the other hand, if the limitations of the transmission system could not be overcome, then many small electric generators could be placed right where electricity was needed most at the point of use. Power systems designed using this approach are known as "distributed power" systems.

Distributed power systems are fully decentralized and do not exploit the economies of scale gained through the construction of large power plants. However, they allow homes, businesses, and factories to own and operate their own power plants instead of relying upon remote, corporate-owned electric utilities to meet their power needs. Because of its small size and

---

*** 1 megawatt (MW) = 1,000 kilowatts (kW) and 1 gigawatt (GW) = 1,000 MW.

distributed nature, wind power had the potential to thrive within a distributed power system, but it was unlikely to survive as a viable option within the central station approach, which increasingly relied upon large power plants that could provide power when called upon. The ability of a power plant to provide power when needed is called "dispatch." The rudimentary wind turbines of the day were neither large, nor dispatchable, so they were unsuitable for the central station model.

The obvious solution to the transmission problem was to increase the voltage of the electricity before transmitting it. This would work because a doubling of the voltage would reduce the losses by three-quarters.[†††] Transformers—which had the ability to increase or decrease the voltage of an electric current—were widely available and easily manufactured in the late 1800s. However, they could be used only with alternating current (AC) equipment. In AC, the flow of the electric charge periodically reverses direction, whereas in DC current the flow of electric charge is only in one direction. The entire system—generators, transmission and distribution lines, and end-use equipment—would have to be switched over from DC to AC in order to make this solution work. But given the heterogeneous nature of the emerging power system, a sea change would be required to make the transition to AC.

1.4   The Colorado Connection

No history of the electric power industry would be complete without mentioning George Westinghouse. Westinghouse was an inventor and engineer who gained his first patent at the age of nineteen. Westinghouse understood the potential of AC as a high-voltage, long-distance electricity distribution system. In 1885, Westinghouse installed a Siemens AC generator into his workshop in Pittsburgh to begin experimenting. He eventually purchased rights to all of the equipment patents required to assemble a complete AC system. The last piece of the puzzle was an AC motor invented by Nikola Tesla, another electric power industry giant.

The tide turned in Westinghouse's favor in 1891 when a 75-kW AC power system was constructed in Telluride, Colorado. The Telluride system had a 2.65 mile transmission line and demonstrated the feasibility of a complete AC power system—from AC generator through high-voltage transmission to the AC electric motor. The Telluride system was built by the Gold King Mine to support its mining operations. The mine had suffered financially because of the high cost of transporting coal to its steam generator located 12,000 feet

---

††† For DC, $L = P^2R/V^2$, where L = transmission line losses (watts), P = electric power delivered (watts), R = resistance (ohms), and V = voltage (volts). By this equation, if V is doubled, then losses are reduced by three-quarters.

above sea level. Attorney Lucien Nunn was hired to explore alternative sources of power for the mine. Nunn couldn't help but notice the potential energy of the South Fork of the San Miguel River just three miles way and 200 feet below the mine and instructed his brother Paul to investigate "the subject of transmission by power of electricity."[35] Despite the technical uncertainty, Nunn moved forward with the idea of transmitting hydropower from the San Miguel to the Gold King Mine and contacted Westinghouse to supply the equipment. The 75-kW Westinghouse AC generator and the Tesla-designed single-phase AC motor arrive in the winter of 1890. The Telluride system was up and running by early 1891.

With the completion of the Telluride system, Westinghouse was ready to fight to ensure that the high-voltage AC system supplanted the existing low-voltage DC network, which was then championed by Thomas Edison and GE. The outcome of the AC–DC battle would determine the future electric power industry as well as the future of wind power: either wind would become a valuable part of a DC distributed power system, or it would be challenged to find a place in the AC central station model that favored large power plants.

With battle lines drawn, the infamous "War of the Currents" played out. Edison went to extreme lengths to discredit Westinghouse's AC system. His efforts included arranging demonstrations where AC power was used to electrocute animals live on stage in order to demonstrate the inherent dangers of AC power. In 1893, Westinghouse won the contract to light up the Chicago World's Fair with a bid that came in at half that of Edison's GE. The fair's "City of Light" was illuminated by Westinghouse and Tesla, and it provided a compelling and successful demonstration of AC power. Westinghouse had won. After the Chicago fair, 80 percent of all electrical devices sold in the United States were for the AC network.[36]

Westinghouse's victory, and the subsequent movement from DC to AC, solidified the central station model of electricity production and delivery. This resulted in the development of increasingly large power plants. By 1903, Chicago Edison's 5,000-kW steam turbine—the Fisk Street Station—was operational. Just ten years later, the largest generator in the United States was 35,000 kW—over 1,000 times the size of the wind turbines being built in Denmark at the time. By 1922, 175,000-kW power plants were being constructed. The largest wind turbines of this era were Lykegaard's 35 kW units.[37]

Beyond their small size, wind turbines possessed another characteristic that made it difficult to integrate them into the AC system: they were "variable speed" machines which meant that voltage and frequency of the electrical output fluctuated with the speed of the wind. Fluctuating power output itself didn't make wind incompatible with the AC system, but it did mean that other power plants would have had to constantly adjust their output to accommodate fluctuations in wind power output, a capability they didn't have in the early twentieth century. The bigger problem though was the fluctuating frequency.

This type of power is known as "wild AC" and it was incompatible with the AC system, which requires power at a constant frequency.

Although frequency standards had not yet been established in the United States or Europe—by 1918, London still used ten different frequencies—one thing was clear: power plants operating on the AC transmission system were required to run at the preferred local network frequency, and this requirement was at odds with the variable speed operation of a wind turbine.[38] While it is true that wind turbines could be made to produce electricity at a constant frequency—la Cour had accomplished as much with his Kratostate system— the existing technology could not be depended upon to regulate wind turbine frequency well enough to maintain synchronism with the AC transmission network. Nineteenth century wind power technology was simply unable to successfully integrate into the AC transmission network.

Thus, despite the technical success achieved by Blyth, Brush, and la Cour, wind power was largely swept aside as electric power systems grew in the United States and Europe. Wind power's small size and its variable nature were not compatible with the central station power plant model and the newly dominant AC transmission system. Instead of becoming a key part of a distributed electricity network, wind power's contribution was limited to meeting the needs of rural customers.

Therefore, in the first decades of the twentieth century, wind power innovators would have to find ways to provide electricity to rural consumers that were out of reach of the electric transmission system. Given the ever-expanding reach of the transmission network, even this modest objective would turn out to be a formidable task.

Notes

1 Harvey, Larson, and Patel, 2017.
2 GWEC, 2018.
3 IEA, 2017a, 43.
4 IEA, 2017b, 29.
5 Science on the Streets, n.d.
6 Price, 2017.
7 Hardy, 2010.
8 Price, 2017.
9 Hardy, 2010.
10 Green Energy Ohio, n.d.
11 *Enclyclopedia of Cleveland History*, n.d.
12 Green Energy Ohio, n.d.
13 *Mr. Brush's Windmill Dynamo* 1890.
14 Green Energy Ohio, n.d.

15 Nissen, 2009.
16 Pedersen, 2010.
17 Nissen, 2009.
18 Shepard, 1990.
19 Beurskens, 2014.
20 Nissen, 2009.
21 Ibid.
22 Pedersen, 2010.
23 Nissen, 2009.
24 Ibid.
25 Smeaton, 1760.
26 Christensen, 2013.
27 Pedersen, 2010.
28 Nissen, 2009.
29 Hau, 2013.
30 Christensen, 2013.
31 Pedersen, 2010.
32 Champly, 1933.
33 Nye, 1992.
34 Ibid.
35 Britton, 1972.
36 PBS, 2004.
37 Christensen, 2013.
38 Nye, 1992.

Bibliography

Ball, Robert S. *Natural Sources of Power*. London, UK: Archibald Constable & CO., 1908.

Beurskens, Jos. "The History of Wind Energy." In *Understanding Wind Power Technology: Theory, Deployment and Optimization*, edited by Alois Schaffarczyk. Translated by Gunther Roth. West Sussex, UK: John Wiley, 2014.

Britton, Charles C. "An Electric Power Facility in Colorado." *The Colorado Magazine* 49, no. 3 (1972): 185–95.

CASE. n.d. "Encyclopedia of Cleveland History." *Case Wastern University*. Accessed August 21, 2017. https://case.edu/ech/articles/b/brush-charles-francis/.

Champly, R. *Wind Motors: Theory, Construction, Assembly and Use in Drawing Water and Generating Electricity*. Translated by Leo Kanner Associates Sponsored by NASA. Paris, France: Dunod Publishers, 1933.

Christensen, Benny. "History of Danish Wind Power." In *The Rise of Modern Wind Energy: Wind Power for the World*, edited by Preben Maegaard, Anna Krenz, and Wolfgang Palz, 642. Singapore: Pan Stanford Publishing, 2013.

Christian, David. (2018). *Origin Story: A Big History of Everything*. New York, NY: Little, Brown and Company.

la Cour, Poul. *Forsøgsmøllen*. Copenhagen, Denmark: Ernst Bojesen, 1900.

Green Energy Ohio. n.d. *Charles F. Brush*. Accessed August 21, 2017. http://www. greenenergyohio.org/page.cfm?pageId=341.

GWEC. *Global Wind Report 2016*. Brussels, Belgium: Global Wind Energy Council, 2017.

———. *Global Wind Statistics 2017*. Brussels, Belgium: Global Wind Energy Council, 2018.

Hardy, Chris. 2010. "Renewable Energy and Role of Marykirk's James Blyth." *The Courier*, July 6, 2010. Dundee: D.C. Thomson.

Harvey, Abby, Aaron Larson, and Sonal Patel. 2017. "History of Power: The Evolution of the Electric Generation Industry." *Power Magazine*, October 1, 2017.

Hau, Erich. *Wind Turbines: Fundamentals, Technologies, Appications, Economics*. Berlin, Heidelberg: Springer-Verlag, 2013.

Hütter, Ulrich. "The Development of Wind Power Installations for Electrical Power Generation in Germany." *Zeitschrift BWK* 6, no. 7 (1954): 270–8.

IEA. *World Energy Investment 2017*. Paris, France: International Energy Agency, 2017a.

———. *Renewables 2017: Analysis and Forecasts to 2022*. Paris, France: International Energy Agency, 2017b.

"Mr. Brush's Windmill Dynamo." *Scientific American* 63, no. 25 (1890): 389.

Nissen, Povl-Otto. "The Scientist, Inventor and Teacher Poul la Cour." In *Wind Power-The Danish Way*, edited by The Poul la Cour Foundation, 6–11. Vejen, Denmark: The Poul la Cour Foundation, 2009. Accessed August 22, 2017. http://www.poullacour.dk.

Nye, David E. *Electrifying America*. Cambridge, MA: MIT Press, 1992.

Owens, Brandon N. *The Rise of Distributed Power*. Corporate Report. Boston, MA: General Electric, 2014.

PBS. 2004. *Tesla Master of Lighting*. Accessed August 22, 2017, https://www.pbs. org/tesla/index.html.

Pedersen, Jørgen Lindgaard. *Science, Engeering and People with a Mission: Danish Wind Energy in Context 1891–2010*. Schumpeter Conference Report, 1–22. Copenhagen, Denmark: Technical University of Denmark (DTU), 2010.

Price, Trevor J. "Blyth, James (1839-1906)." In *Oxford Dictionary of National Biography*. Oxford: Oxford University Press, 2017.

Science on the Streets. n.d. *James Blyth*. Accessed August 21, 2017. http:// scienceonstreets.phys.strath.ac.uk/new/James_Blyth.html.

Shepard, Dennis G. *Historical Development of the Windmill*. Government Report. Cleveland, OH: Lewis Research Center, National Aeronautics and Space Administration (NASA), 1990.

Smeaton, John. *An Experimental Enquiry Concerning the Natural Powers of Water and Wind to Turn Mills, and other Machines, depending on a circular Motion*. London, UK, 1760.

2

The Age of Small Wind

The Rural Electrification Administration pretty well killed the demand for self-contained DC wind power systems. AC was just too readily available everywhere.

—Marcellus Jacobs (1973)

2.1 Networks of Power

At the turn of the twentieth century, the growing AC power generation and transmission system facilitated the adoption of the central station model of electricity production and distribution, which favored increasingly large power generators.* This was followed by the rapid electrification of the United States and Europe. The electricity displays at the 1893 World's Fair in Chicago captured the American imagination and left the public clamoring for more electric power. Electric power technology development accelerated, and so did power plant construction and electricity consumption.

In America, starting with the first power production at Pearl Street Station in 1882, electricity consumption increased from about 6 million kWh in 1900 to approximately 115 million kWh in 1930. Annual electricity use per residential customer rose from about 200 kWh per year around the turn of the century to 550 kWh per year by 1930. The percentage of homes and farms with electricity service soared from less than 5 percent in 1900 to nearly 70 percent by 1930.[1]

---

* The central station model of power generation and delivery is basically a hub and spoke model where a large power plant is centrally located, and transmission lines are built to distribute power to customers near and far. This is opposed to a distributed power model where smaller power plants are located at or near the point of consumption.

The Wind Power Story: A Century of Innovation that Reshaped the Global Energy Landscape, First Edition. Brandon N. Owens.
© 2019 by The Institute of Electrical and Electronics Engineers, Inc.
Published 2019 by John Wiley & Sons, Inc.

Electricity production and distribution companies were formed to serve new customers. Because the largest companies could realize the greatest cost reductions through economies of scale in the central station model, electricity production and distribution became the exclusive domain of large "electric holding companies," each of which owned and operated a portfolio of electric utilities. The nascent electric power industry was generally viewed as a natural monopoly.[†] Monopolies had been outlawed by the Sherman Antitrust Act of 1890; so cities—and eventually states—regulated the electricity industry through the formation of Public Utility Commissions. In 1907, Wisconsin and New York became the first states to regulate electric utilities. By 1914, forty-three additional states had followed suit.[2] The twentieth century structure of the power industry was solidified. Wind power was nowhere in sight.

This was also the beginning of a political bargain between the electric holding companies and regulators. Electric utilities were protected from competition and provided with a guaranteed rate of return in exchange for providing electricity at "reasonable" prices and an obligation to serve all customers with minimal interruptions. Reasonable prices were defined as prices that covered the cost of electricity production and delivery plus a pre-determined rate of return.[3]

This bargain established electric utilities as the gatekeepers and operators of the national electricity system. It enabled them to select only those generation technologies that best suited their purposes—such as large steam turbines—while erecting barriers to technologies that didn't fit with the central station model, such as wind power. In the United States, the electric utilities' grip on generation technology would not be loosened until the Public Utilities Regulatory Policies Act (PURPA) of 1978, which opened the door to smaller, so-called "nonutility" generators. PURPA—plus federal and state incentives—would eventually lead to a California wind power boom in the early 1980s. Thus, the structure of the electric power industry—which was settled in the first decades of the twentieth century—would impede the entry of wind power until the last decades of the twentieth century.

The electric power industry growth pattern in the United States was mirrored in Europe. Europeans shared America's newfound passion for electricity and electrified in tandem in the first two decades of the twentieth century. Immediately after Pearl Street, power plants were built in London, Berlin, and Saint Petersburg.[4] Inspired by the International Electrical Exhibition in Frankfurt, Germany in 1891, Germany led the way in electrification, followed by France, Switzerland, and the United Kingdom.

---

† A natural monopoly is a firm or industry where high up-front costs and other barriers to entry give the largest supplier a natural cost advantage. This certainly applied to electricity suppliers, where the largest power plants generally had the lowest average cost of electricity production and those who first installed power lines had exclusive access to customers.

Interestingly, Denmark electrified as well but did so in a manner that accommodated both distributed and central station generators because it had a much more distributed population. Development of the transmission system in Europe mirrored growth in the United States.[5] Importantly for wind power, transmission system growth in Denmark was largely focused on urban areas during the initial phase of development through the 1930s. This would become a key factor for wind power later as small wind energy manufacturers looked for niches to fill with their machines.

Increasingly large steam turbines were the favored technology for the central station model because they could be built at the megawatt-scale, and they could be turned on and off at will—a feature known as "dispatch" in electric industry parlance—to respond to changes in electricity demand. Wind turbines possessed none of these characteristics and were therefore of little use to electric utilities at the time. The central station model may not have been the only way to produce and deliver electricity to meet the needs of homes and businesses, but it was the dominant approach, and wind power had a limited role to play within this system.

As the centerpiece of the central station model, steam turbines experienced a high degree of innovation. Between 1903 and 1907, GE alone obtained forty-nine steam-turbine related patents. Fourteen of these led to improvements in steam turbine performance.[‡] In contrast, little innovation occurred around wind power in the first decade of the twentieth century. Nothing was done to follow up on Brush's wind turbine. Instead of a model for the future, it quickly became a relic of the past, representing one failed attempted to transform an ancient technology into a modern one.

A handful of small companies—such as the Lewis Electric Company, Fairbanks, Morse and Company, and the Aermotor Company—manufactured wind turbines in the United States in the first decade of the century. These models were short lived and met with limited success. In Europe, several companies adopted the wind turbine design of American manufacturers and began production. In Britain, R.A. Fessenden built an experimental wind turbine in 1894 and started the Rollasen Wind Motor Company to manufacture wind turbines.[6] In Germany, Stahlwind, Dresden, Köster, and Heide/Holstein manufactured wind turbines. All these manufacturers ceased production by 1912.[7,8]

As a result of the efforts of Poul la Cour, Denmark was the only country where wind power gained any ground in the first decade of the twentieth century. Through the end of the World War I, la Cour's research efforts resulted

---

‡ Zink (1996) provides some details on GE's steam turbine innovations: "In 1903, the automatic governor appeared. In 1905 and 1906, patents were issued for reheating steam during expansion. A mechanism to adjust for longitudinal expansion of the rotor was patented in 1905. In 1907, GE patented an impulse machine with full peripheral admission and increasing bucket lengths from stage to stage."

in advances in wind power in Denmark. By 1908, Lykkegaard had built seventy-two small wind turbines, which supplied power to rural Denmark. By 1920, 105 small wind turbines generated electricity in combination with diesel generators.[9] However, this would turn out to be the apex for wind power in Denmark during this era. The fall in diesel prices after World War I pushed Denmark toward the adoption of diesel generators in rural areas, which led to a decline of wind power even in la Cour's Denmark.

2.2   An Interesting Twist

The original windmill designs that Blythe, Brush, and la Cour used to build the first wind turbines were not hastily constructed machines. On the contrary—by the late nineteenth century—windmills had been under development for hundreds of years. In fact, the nineteenth century windmill represented the state of the art that embodied the innovation of the world's most talented minds for generations. The windmill blade evolved in Europe over a 500-year period, between the fifteenth and nineteenth centuries. Over this time, it was transformed from a flat board to a cloth sail with an aerodynamic twist. These innovations occurred incrementally because of gradual learning and knowledge transferred from the use of other technologies, such as the sails on boats.[§]

Yet, despite its deep technological lineage, something was still missing. The transformation of the windmill into the wind turbine was not yet complete. Windmills were designed to provide torque for grinding and water pumping, whereas wind turbines required high rotational speed to enable increased electrical output. In essence, a design change was needed to further transform slowly rotating nineteenth-century windmills into fast rotating twentieth-century wind turbines.

Thanks to Wilbur and Orville Wright's first flight in 1903, the solution was—quite literally—in the air. The development of the airplane and its application in World War I provided spillover benefits for wind turbine technology. Airplane propeller blades—in some cases without any modification whatsoever—were applied to wind turbines after the war. The integration of aerodynamic propellers and wind power turned the tide for small wind power and initiated a period of technology progress that enabled it to successfully compete in remote areas without access to the power system through the transmission network.

American researchers Harve Stuart and Elisha Fales first applied a twisted-airfoil airplane propeller to wind turbines in 1917.[10] They conducted their work as part of the United States government's aircraft research program in Dayton, Ohio.[11] The twisted-airfoil shape is highly effective at increasing a

---

§ See Shepard (1990) for a discussion of windmill technology evolution.

wind turbine's rotational speed and raising its maximum efficiency, thereby increasing the power output of a wind turbine. In Denmark—in 1919—Poul Vinding and Johannes Larsen succeeded in creating a 30-kW test turbine that produced AC electricity using an asynchronous generator and delivered power directly to the transmission grid from a test station in Buddinge near Copenhagen.[12] Professor Larsen also proposed the use of a DC-to-AC converter to send DC output from a wind turbine to the AC transmission network. This approach was too complicated and expensive at the time, but it represented an advance in wind turbine theory and practice that wouldn't be revisited until German Hermann Honnef's large wind turbine designs of the 1930s.

In the eighteenth century, Smeaton already taught the world that blade position matters when he showed that the smaller the angle-of-attack the better. La Cour concurred and built upon Smeaton's work at the end of the nineteenth century. La Cour found that to be most effective, wind turbine blades should be aerodynamically twisted. But it wasn't until researchers affixed a twisted airfoil propeller blade to a wind turbine that the importance of blade shape was fully appreciated. This was the innovation of Stuart and Fales.

Why does blade shape even matter so much to a wind turbine anyway? The answer has to do with how much "lift" can be created by the wind. When wind strikes a horizontal-axis wind turbine blade, it is broken into two forces: lift and drag. Lift is the perpendicular force that pushes the blade up, and drag is the parallel force that pushes the blade back. The rotational speed of the wind turbine blade, and the wind turbine's efficiency, is determined by the strength of the lift force relative to the drag force, which is measured by the "lift-to-drag coefficient."

Strong lift is generated several ways. The first is by positioning the blade at the correct angle-of-attack, or the angle of the blade relative to the oncoming wind. The smaller the angle-of-attack, the greater the lift-to-drag coefficient. This is true up to the point where an increasingly small angle-of-attack creates excessive turbulence on the backside of the wind turbine blade. The optimal angle of attack is around 10°–15°. This was first discovered by Smeaton, and later confirmed by la Cour.

The second way to generate a strong lift is to use an airfoil-shaped blade. The easiest way to understand the importance of airfoil-shaped blades is in the context of Bernoulli's principle. Airfoils come in many different shapes, but generally have a convex top and a flat bottom so that air must travel a greater distance along the top side than the bottom side. The difference in travel distance means that air travels faster along the top side of the blade than the bottom side. According to Bernoulli's principle, differences in the velocity of the wind will be accompanied by a pressure difference between the top and bottom sides of the blade. The pressure difference generates a lift that forces the blade up. The combined impact of a small angle-of-attack and an airfoil shape generates a much stronger lift compared to a flat blade positioned at a less acute angle-of-attack.

That's enough reason to use an airfoil-shaped blade on a wind turbine, but there is a third factor: if the blade is twisted from root to tip, then even more lift will be created. To understand this, consider the concept of "apparent" wind speed and direction. This is the wind speed and direction as seen from a point on the moving wind turbine blade. In mathematical language, it is the vector sum of the actual wind speed and direction and the blade's wind speed and direction. The faster a blade is moving, the faster the apparent wind speed and the more acute the apparent direction.

As it turns out, the apparent speed of a wind turbine blade varies at different positions on the blade. The apparent wind speed is much faster for points close to the blade root, and the apparent speed is much slower for points close to the blade tip. This means that the optimal angle-of-attack varies when moving up from the blade root, where the blade is spinning faster and the apparent wind direction is more acute, to the blade tip, where the blade is spinning slower and the apparent wind direction is less acute. Twisting the blade just so ensures the optimal angle-of-attack matches the wind speed along the entire length of the moving blade. This increases the lift that is generated across the entire blade length.

All of this means that when Harve Stuart and Elisha Fales strapped an airplane propeller to a wind turbine, they added an innovation that was almost as important to the modern wind turbine as the initial fusion of the windmill and electric generator by Blyth and Brush back in 1887. Because they had the right shape, Stuart and Fales's airfoil blades increased the rotation speed of wind turbines.

The positive impact that an airfoil blade can have on a wind turbine can be measured by the wind turbine's "tip-speed ratio." The tip-speed ratio is the ratio of the rotational speed of the wind turbine blades at the tip of the blade, to the speed of the oncoming wind. In a horizontal-axis wind turbine, the tip-speed ratio increases from zero to some maximum value as the wind speed increases. The tip-speed ratio then declines at higher wind speeds due to effects of turbulence. A wind turbine's optimal tip-speed ratio is the ratio that maximizes the wind turbine's power output. A similar concept—"rated wind speed"—is the wind speed that produces a wind turbine's optimal tip-speed ratio.

The optimal tip-speed ratio of a wind turbine depends upon the number of blades—or the solidity of the wind turbine—the blade shape (airfoil or straight), and position (i.e., the angle-of-attack). Brush's multibladed machine based on the American Windmill design had an optimum tip-speed ratio of about two. La Cour's four-bladed designs had an optimum tip-speed ratio of about four. Stuart and Fales showed that their blades increase the tip-speed ratio of a wind turbine to twelve. This change alone could increase the power output form a wind turbine from less than 10 to greater than 60 kW.[13]

In 1926, Stuart and Fales applied a patent for what later became known as the "Stuart Airfoil" with apologies to Mr. Fales. In the patent application, they note that their invention enables "a high-speed wind turbine that is free from the disadvantages of the so-called 'American,' the 'Dutch' or other types of windmills." Indeed, Stuart and Fales understood that airfoil-shaped blades were the key improvement needed to transform slowly rotating windmills into high speed wind turbines.[14] With the addition of the airfoil-shaped blade, the transition from the windmill to the wind turbine was complete.

2.3 The Savior of the Windmills

Like Stuart and Fales, German Kurt Bilau was passionate about wind turbine blades. During World War I, he saw many of Europe's windmills up close when he used them as an observation post as an artilleryman in the German army. After the war, he refurbished the surviving windmills, upgrading their blades so that they would not be torn down. To do this, he covered the blades with airfoil-shaped sheet metal coverings.[15] When he did this, he found that he could increase the optimal tip-speed ratio to 4.5.[16]

Mr. Bilau became recognized for his efforts to improve European windmills. He noted that he was always greeted as the "savior of the windmills" when arriving at the site of an old windmill to work his magic.[17] After his efforts refurbishing windmills across Europe, Bilau decided to apply his knowledge to wind turbines. In 1920, he constructed a 45-kW four-bladed test wind turbine with airfoil blades at the Göttingham Institute of Aerodynamics. It weighed 3 tons and stood 12 m tall. The blade length was 8 m. After several successful tests, Bilau claimed that his wind turbine had twice the efficiency of other designs.[18] He patented his wind turbine in October 1921, and opened a business to begin manufacturing his designs.

Bilau—like his contemporaries Stuart and Fales—was successful in increasing the rotational speed of wind turbines using airfoil blades. However, for Mr. Bilau, dealing with high speed wind turbines—as opposed to slowly rotating windmills—posed some new challenges. Bilau needed some way to keep the fast moving turbine blades from spinning out of control in high winds. In addition, airfoil blades were so thin that it was impossible to get the wind turbines to start spinning from a stopped position.

Ever the innovator, Bilau invented a hinged blade that he called "Ventikanten" that was modeled after an airplane wing. Bilau's blade had a front and back section connected by a hinge that could be adjusted depending upon the wind conditions. To start from a stopped position, the blade could be adjusted to provide a greater surface area to catch the wind. To stop the blade in high wind conditions, the rear component could be positioned to act as a brake. Bilau's Ventikanten is still used in windmills throughout Europe, including East

Prussia and Southern England.[19] By 1940, 160 windmills had been retrofitted with Bilau's blade.[20] Bilau's contributions to wind power development didn't end there; he also published three books related to wind power: *Wind Power in Theory and Practice* (1927), *Windmill Construction* (1933), and *Wind Utilization for Power Generation* (1942).[21]

Despite the success of Bilau's blade design on windmills, there was never any real demand for his wind turbines in Germany in the 1920s. Diesel prices had declined after World War I and the German economy was too depressed and fragile during the interwar years to support any new innovations like Bilau's wind turbines. In 1926, he closed shop. In his own words "due to the onset of the economic crisis there will be no production of wind engines."[22] Bilau held out hopes for the commercial production of his wind turbines through the 1930s and served as a consultant to German small wind turbine manufacturer Ventimotor GmbH until his death in Berlin in 1941.[23]

Bilau's interests were not limited to windmills and wind turbines. Starting in the 1930s he lectured and wrote on a wide range of subjects including astronomy, history, humanities, and western civilization. During this time, he became involved in research to prove Plato's Theory of Atlantis. He published what he believed to be the first exact map of Atlantis in 1932.[24] Despite his significant contributions to wind power, this intellectual diversion—along with his sympathies with the National Socialist German Workers' Party—have diminished his standing in the history of wind power.

During the time that Bilau was hard at work making wind turbines work better in the real world, one of his collaborators at the Göttingham Aerodynamic Research Institute was busy working on wind turbine theory. Physicist Albert Betz didn't build a single wind turbine, but his contribution to wind power would ultimately turn out to be as important as the efforts of early wind power practitioners. Betz—who had originally focused his intellectual talents by studying aircraft aerodynamics—turned his sights to understanding wind turbine efficiency. In a 1919 publication, Betz derived what is now known as the "Betz Limit," which defines the theoretical maximum value of the efficiency of a wind turbine.[25] In his book *Wind Energy and Their Use by Windmills* (1926), he summarized his findings and established a theoretical basis for wind turbines.

Betz started his work by empirically testing the performance of different airfoils in the wind tunnel at the Aerodynamic test station in Göttingen. From these tests, he derived the physical principles of the conversion of wind energy. He used the kinetic energy equation from classical mechanics to calculate the energy contained in the wind passing through a wind turbine. Next, to determine how much of this energy could be captured by the wind turbine, he considered the change in the speed of the wind as it passed through the plane of the rotor. By setting the power absorbed from the wind equal to the change in kinetic energy, Betz computed the maximum power that can be extracted from the wind. His calculations indicated that an ideal wind turbine can extract 16/27— or 59.3 percent—of the power in the passing wind.[**] This is the Betz Limit, and it represents the upper bound of a wind turbine's efficiency. No wind turbine ever made will extract more than 59.3 percent of the power from the oncoming wind.

In this manner, Betz laid the foundation for the modern theoretical understanding of wind turbine. The efficiency of a wind turbine in extracting power from the wind is what is known as the wind turbine's "power coefficient," often abbreviated as C_p. C_p is the ratio of actual electric power produced by a wind turbine divided by the total wind power flowing into the turbine blades at a specific wind speed.

A wind turbine's power coefficient varies depending upon the wind speed. At very low wind speeds, C_p is low. It rises until it reaches its maximum level, which occurs at the wind turbine's rated wind speed and optimum tip-speed ratio. The maximum C_p of Brush's slowly rotating high-solidity wind turbine was around 15 percent. La Cour's low-solidity wind turbines had a maximum power coefficient of around 20 percent. The C_p of wind turbines equipped with Stuart and Fales' propeller blades is as high as 40 percent.[26] The best wind turbines in the twenty-first century can achieve efficiencies approaching 50 percent.[27] Betz also provided a common vernacular to describe wind turbines—power coefficient, rated wind speed, and tip-speed ratio can be found in any wind energy handbook today.

2.4 From the Arctic to Antarctica

Back in the United States after 1920—in the wake of Fales and Stuart's blade research and the work of Bilau and Betz in Germany—the production of small-scale wind turbines ramped up. HEBCO—founded by Herbert E. Bucklen—was the first company to manufacture wind turbines using airplane propeller blades. Others followed suit. The Aerolite Wind Electric Company, Fritchle Wind Power Electric Manufacturing Company, and the Perkins Corporation, all had varying degrees of success manufacturing and selling small wind turbines sporting airfoil-shaped blades to farmers throughout the American West.[††]

As small wind turbine production increased in the United States, the government of the Union of Soviet Socialist Republics (USSR) was also focused on finding practical solutions to meeting its growing energy needs. As early as

---

[**] While performing his calculations, Betz assumed that the wind turbine did not possess a hub and had an infinite number of rotor blades that did not result in any drag resistance to the wind flowing through them.

[††] Righter (1996, chapter 4, 73–104) provides a good review of the history of small wind turbines in America during this period.

1918—in the aftermath of the Russian Revolution—Vladimir Lenin instructed the USSR Academy of Sciences to include wind research in its reorganization plan. In exchange for continued financial support from the new Soviet regime, the Academy refocused to conduct research on issues related to state construction.[28] The Academy's research would ultimately lead to the development of the unheralded 100-kW Balaclava wind turbine in Crimea 1931—the first large, grid-connected wind turbine in the world.

As the 1920s progressed, increasing momentum gathered around small wind turbines both in the United States and the Soviet Union. Geography was the common denominator behind the rise of small wind power in these countries. Both the United States and the Soviet Union contained expansive geographies with extensive rural areas that didn't have access to transmission lines. Transmission development in the early decades of the twentieth century had largely focused on urban areas. As a result, there was a built-in opportunity for small wind power systems in remote areas.

Wind turbines were also used to charge radio batteries in the 1920s and 1930s. The most popular wind turbine of the period was Wincharger's 6 V machine. Although the voltage was too small to power electrical appliances—which required 32 V—the 6 V machine was ideal for charging vacuum-tube radios. It was marketed jointly by Wincharger and Zenith as the "Radio Charger" and, as advertised, saved radio listeners a trip to the nearest town to charge their radio battery using a diesel generator. Zenith later added a larger 32 V wind turbine to their lineup, which was able to power complete farms.[29]

The most noteworthy manufacturer of wind turbines during this period was the Jacobs Wind Electric Company. Jacobs Wind Electric Company was founded by brothers Marcellus and Joe Jacobs. The Jacobs brothers started building wind turbines at their ranch near Wolf Point, Montana. Like the rest of the ranches in rural America, the Jacobs ranch was out of reach of the transmission network, so the Jacobs family got their energy from an old second-hand diesel generator. However, by 1922, the electric needs of the Jacobs family exceeded the output of their diesel generator. So the brothers rigged up a wind turbine by attaching a fan wheel from an American Windmill to the rear axle of a Ford Model-T. They mounted the fan wheel one end, and affixed a DC generator to the other end. They locked the differential with a pin and the first Jacobs wind turbine was born.[30]

It didn't take long for Marcellus and Joe to figure out that the American Windmill design wasn't very efficient for generating electricity. The high-solidity rotor didn't spin fast enough. Marcellus would later recall that it simply had too many blades that "got in each other's way."[31] After taking pilot lessons and learning to fly in 1926—like Stuart and Fales before him—Marcellus concluded that airplane propellers would be better suited for wind turbines than the flat blades of the American Windmill. The Jacobs brothers experimented with a two-bladed airfoil design, but they found that their two-bladed design had vibration problems that couldn't be remedied. Germany's Ulrich Hütter would later run into the same problem in the 1950s with his two-bladed wind turbine. Hütter would solve the problem by enabling the hub to "teeter" on the tower. In the 1920s, the Jacobs brothers solved the problem by simply adding another blade. They settled on a three-bladed wind turbine design in 1927. Three-bladed wind turbine designs would eventually become the technology standard. The blades of the Jacobs wind turbines were made from aircraft quality, vertical grain spruce. According to his own account, Marcellus used to go out and select the lumber personally and have it shipped back to Montana.

Of course, as soon as they solved one wind turbine design problem, the Jacobs brothers encountered another one. Their next challenge was to control the speed of their wind turbines in order to avoid over-speed in high wind conditions. They solved this problem by using a fly-ball governor, like the type invented by James Watt in 1788. They mounted weights on the hubs of the turbine blades, so the centrifugal force of higher speeds would twist all three blades identically. This automatically feathered the blades in high winds. Using this design, the Jacobs brothers built twenty to twenty-five three-bladed wind turbines with fly-ball governors between 1927 and 1937 and sold them to ranchers in the Wolf Point area. That would have been the end of it too, if the Jacobs wind turbines didn't work so darn well.

In 1931, they sold their ranch and set up a wind turbine manufacturing line in Montana. They later moved the operation to Minneapolis. Along the way, they were awarded twenty-five patents for wind turbine innovations. They redesigned a DC generator to allow its load to increase with the rotational speed of the wind turbine, they developed new, longer lasting generator brushes that minimized sparking as they moved from one coil to the next, they invented a "reverse current relay" that regulated voltage to ensure that the wind turbine didn't overcharge the batteries, and they assembled customized bearings cases—relying on car parts again—that lasted twenty years instead of two, which was the standard at the time. The fact is, the Jacobs brothers were extraordinary innovators.

And so, from their Minneapolis factory, the Jacobs Wind Electric Company manufactured 2.5- and 3-kW DC wind turbines between 1933 and 1956. At their peak, they employed 260 workers that could produce eight to ten small wind turbines a day working one shift. According to Marcellus's own account, they ultimately built and sold approximately US$50 million worth of small wind turbines in twenty-five years.[32] The wind turbines that the Jacobs brothers built and sold represented the state of the art in the 1930s. They were unique because they employed the design features that they invented themselves. These features made Jacobs' machines more durable—and more expensive— than other small wind turbines and earned their wind turbines a reputation as the "Cadillac" of wind turbines.[33]

Figure 2.1 A 2.5-kW Jacobs wind turbine. The small wind turbines built by Marcellus and Joe Jacobs represented the state of the art in the 1930s and 1940s. The Jacobs wind turbines contained more features and were more expensive than others, which earned them the reputation as the "Cadillac" of small wind turbines.

In a testament to the durability of their wind turbines, the Jacobs brothers provided a wind turbine for installation at a weather station jointly operated by the United States and the United Kingdom at the Eureka research base in the Arctic Circle, and for the Byrd Expedition at Little America in Antarctica in 1933. The wind turbine at Little America was found in operable condition twenty-two years later—even though the snow pack covered the tower to within several feet of the blades.[34]

The Jacobs wind turbines provide a glimpse into the cost of small wind power in the United States in the 1930s. Marcellus Jacobs would later state that the cost of his 2.5-kW wind turbine was US$1,025 or US$410/kW.[‡‡] That's about US$5,740/kW in today's dollars.[§§] According to Jacobs, his turbine would

---

‡‡ Jacobs (1973). This estimate excludes both installation and maintenance costs. In this case, the installation cost for the Jacobs turbine included the labor of two men for two days. Ongoing maintenance costs were typically limited to just battery replacement.

§§ See Office of Management and Budget [OMB] (2017, table 10.1). The ratio of the GDP price index between 1940 (the earliest year available) and 2017 is fourteen.

typically produce between 400 and 500 kWh per month in most areas in the "Western half of the United States."[35] This translates to an average annual capacity factor of 25 percent, which means the wind turbine would produce about 5,400 kWh per year.[***] If the initial investment cost is annualized and divided by annual electricity generation, then the average cost of electricity from the Jacobs wind turbine turns out to be 21 ¢/kWh in today's dollars.[†††]

Thus, the cost of wind power from the Jacobs systems in the early 1930s was close to the average cost of electricity from electric utilities at the turn of the twentieth century.[‡‡‡] However, by 1932, utilities had succeeded in driving down costs even further through technology innovations, economies-of-scale, and lower fuel costs. The steam turbines that GE built for the Fisk Station in 1903 were 500 kW and operated at 1,200 rpm. By the 1930s, GE had built the first triple-tandem compound turbine generator, which operated at 1,800 rpm, with steam pressure and temperature of 600 lb/in^2 and 750°F, respectively. Total plant output was 315 MW—over 100,000 times greater than the Jacobs' 3-kW model.[36] Together, these steam turbine innovations, along with increases in the scale of utility holding company operations, reduced the price of electricity to 5.6 ¢/kWh by 1932.[§§§]

Wind power was chasing a moving target. Despite continuous innovation, a wind power electricity production cost that would have been competitive in 1900, was too expensive by 1932. This was the beginning of a pattern that would repeat itself throughout the twentieth century: innovations in wind power would continue to be offset by downward movements in the price of electricity from conventional electricity generators due to combinations of declining fossil fuel prices and technology advances. It wouldn't be until the second decade of the twenty-first century before wind power could rightly claim the throne as the lowest cost source of electricity.

Still, 21 ¢/kWh was an attractive price for rural customers without access to electricity. By the early 1930s, wind power had settled into a nice niche market providing electricity to Americans in remote locations. This market appeared secure at the time because electric utilities weren't investing to extend the transmission network to remote areas because it cost more money than they could collect through electricity sales.

*** The capacity factor is the power generated by a wind turbine divided by its rated power output. Capacity factor is the most commonly used measure of a wind turbine's utilization rate.

††† If the initial capital cost is US$5,740/kW and the rated capacity is 2.5 kW, then the total initial investment in current dollars is US$14,350. If the wind farm cost is amortized at 5 percent over thirty years, the annual carrying charge is 8 percent. Thus, the annual cost of the system is US$1,148. If the annual capacity factor is 25 percent, then annual output is approximately 5,400 kWh. Thus, the average cost per kilowatt-hour is 21.3 ¢/kWh since US$1,148 divided by 5,400 kWh is 21.3 ¢/kWh.

‡‡‡ According to EIA (2000, 111), the average cost of electricity was 17 ¢/kWh at the beginning of the twentith century.

§§§ See EIA (2000, 112). Also, according to Hein (2003), between 1907 and 1927 the price of electricity in the United States declined by 55 percent.

2.5 Rural Electrification

In the United States, electric holding companies thrived from the time of their formation at the turn of the twentieth century through the 1920s. By the late 1920s, the sixteen largest holding companies controlled 75 percent of all generation.[37] These holding companies were exempt from regulation because they operated across state boundaries and were out of the purview of individual state public utility commissions. Some holding companies were known to abuse the electric utilities that they owned by charging high rates for engineering and management services. This led to higher electricity prices for consumers. The Federal Trade Commission launched a six-year investigation of these holding company abuses in 1928. After the stock market crash of 1929 reduced the value of holding company securities, there was a public wave of antagonism against electric holding companies.

Franklin Roosevelt campaigned on a reform platform in 1932 and promised to end electric industry abuses. He also vowed to create the jobs necessary to lead the country out of the Great Depression. After his election, he attempted to combine these policy goals by reforming the electric power industry starting with the passage of the Public Utility Holding Company Act of 1935 (PUHCA). PUHCA placed restrictions on electric holding companies and put them under the regulatory authority of the Federal Power Commission (FPC). The regulatory powers of the FPC were later transferred to the Federal Energy Regulatory Commission (FERC) in 1977. Although PUHCA placed electric holding companies under federal regulatory authority, it left the primary tenet of their original political bargain in place—electric utilities would continue to operate in a protected environment that provided them with a guaranteed rate of return in exchange for an obligation to serve. This arrangement would lead to a fifty-year period of stability and expansion for the US electric power industry.

Roosevelt's efforts were not limited to industry restructuring. To create jobs, the Roosevelt Administration also sought to expand the role of public power, which accounted for just 6 percent of total US generation at the start of Roosevelt's first term.[38] With the Administration's support, Congress passed the Rural Electrification Act of 1936, which created the Rural Electrification Administration (REA). The goal of the REA was to encourage the growth of rural electric cooperatives that would sell electric power generated at newly constructed public power plants to remote customers at their cost of production. The REA was very effective. Between 1933 and 1941, over half of all new power plants built in the United States were publicly owned; and by 1941 public power accounted for 12 percent of the total US generation.[39]

To expand the role of public power, rural customers would have to be brought within reach of the transmission system. To accomplish this, REA made transmission expansion loans to state and local governments and electricity cooperatives. Cooperatives also received lower property assessments, exemptions from federal and state income taxes, and exemption from both state and federal regulation. Because of these incentives, the transmission system was able to successfully expand into rural areas during this period. The percentage of US farms with electricity service increased eightfold between 1930 and 1950, rising from 10 to nearly 80 percent.[40]

As a precondition for transmission interconnection, farms with wind turbines were required to destroy them. If the federal government was going to such great lengths to expand the demand for public power, they didn't want any wind turbines hanging around cutting into the market. Although it would take decades to play out, this would gradually lead to the end of the era of small wind in America. It was an unfortunate turn for wind power. After playing a limited role in the electric power system through the first two decades of the twentieth century, small wind power systems had finally found a place in rural America starting in the 1920s. For a brief period, they thrived out of reach of the transmission system. However, the grid's inevitable expansion gradually eliminated the last viable market for small wind power in the United States.

The only other option left was for wind power to somehow transform itself into a large-scale technology that was compatible with the central station model and that AC transmission system. But was that even possible? The world would soon find out.

Notes

1 U.S. Census Bureau, 1960, 510.
2 Smithsonian Institution, 2014.
3 Hein, 2003.
4 Cubillo, 2011, 2.
5 Hughs, 1983.
6 Righter, 1996, 66.
7 Hütter, 1954, 3.
8 Ibid., 2.
9 Pedersen, 2010, 8.
10 Righter, 1996, 74.
11 Ibid.
12 Pedersen, 2010, 8.
13 Stuart and Fales, 1931, figure 6.
14 Ibid.
15 Hütter, 1954, 3.
16 Ibid.
17 Karstens, 2010.

18 Current Opinion Editors, 1925.
19 Hütter, 1954, 3.
20 Wikipedia, 2010.
21 Bilau, *Wind Power in Theory and Practice*, 1927; Bilau, *Windmill Construction One and Now*, 1933; Bilau, *Wind Utilization for Power Generation*, 1942.
22 Karstens, 2010.
23 Ibid.
24 Bilau, *The First Exact Map of Atlantis*, 1932.
25 Betz, 1966.
26 Ragheb and Ragheb, 2011, figure 6.
27 Burton et al., 2001,173.
28 Fateyev, 1948, 2.
29 Sagrillo, 1992.
30 Mother Earth News Editors, November/December 1973.
31 Jacobs, 1973.
32 Ibid.
33 Righter, 1996, 97.
34 Marier, 1985.
35 Jacobs, 1973.
36 Zink, 1996.
37 EIA, 2000, 5.
38 Ibid., 7.
39 U.S. Census Bureau, 1960, 510.
40 EIA, 2000, 8.

Bibliography

Betz, A. *Wind Energy and Its Use by Windmills*. Göttingen, Germany: Vandenhoeck & Ruprecht, 1926.
Betz, Albert. *Introduction to the Theory of Flow Machines*. Oxford, UK: Pergamon, 1966.
Bilau, Kurt. *Wind Power in Theory and Practice*. Berlin, Germany: P. Parey, 1927.
——. "The First Exact Map of Atlantis." *Monatsschriften* (R. Voigtländers Publishing House) 8 (1932): 38–43.
——. *Windmill Construction One and Now*. Leipzig, Germany: M. Schäfer, 1933.
——. *Wind Utilization for Power Generation*. Berlin, Germany: P. Parey, 1942.
Burton, Tony, David Sharpe, Nick Jenkins, and Ervin Bossanyi. *Wind Energy Handbook*. Chichester, UK: John Wiley, 2001.
Cubillo, Diego Ibeas. *Review of the History of the Electricity Supply in Spain from the Beginning up to Now*. Madrid, Spain: Universidad Carlos III de Madrid, 2011.
Current Opinion Editors. "Now for the Aerodynamic Windmill." *Current Opinion* 78 (April 1925): 468.

EIA. *The Changing Structure of the Electric Power Industry: An Update.* Washington, DC: Energy Information Administration (EIA), 2000.

Fateyev, Ye. M. *Wind Engines and Wind Installations.* Translated by Leo Kanner. Associates Sponsored by NASA. Moscow, Russia: State Publishing House of Agricultural Literature, 1948.

Hein, Jeff. *Shining Light on the Utility Industry's Earliest Foundings.* Lakewood, CO: Western Area Power Administration, 2003.

Hughs, Thomas P. *Netowrks of Power: Electrification in Western Society 1880-1930.* London, UK: Johns Hopkins University Press, 1983.

Hütter, Ulrich. "The Development of Wind Power Installations for Electrical Power Generation in Germany." *Zeitschrift BWK* 6, no. 7 (1954): 270–8.

Jacobs, Marcellus L. "Experience with Jacobs Wind-Driven Electric Generating Plant, 1931-1957." In *Wind Energy Conversion Systems*, edited by Joseph M. Savino, 155–8. Washington, DC: NSF/NASA, 1973.

Karstens, Uwe. "Kurt Bilau. Approach to a Visionary." *Association for the Conservation of Windmills and Water Mills in Schleswig-Holstein and Hamburg.* e.V. Band 2ndSelf-published, Ascheberg 2010., 2010.

Marier, Donald. "Marcellus Jacobs." *Alternative Source of Energy* 75 (1985): 6.

Mother Earth News Editors. "Wind Power History: Marcellus Jacobs Interview." *Mother Earth News*, November/December 1973. Accessed August 23, 2017. http://www.motherearthnews.com/renewable-energy/wind-power-history-zmaz73ndzraw.

Office of Management and Budget (OMB). "Historical Tables." *The White Hosue*, 2017. Accessed August 23, 2017. https://www.whitehouse.gov/sites/whitehouse.gov/files/omb/budget/fy2018/hist.pdf.

Pedersen, Jørgen Lindgaard. *Science, Engeering and People with a Mission: Danish Wind Energy in Context 1891–2010.* Schumpeter Conference Report, 1–22. Copenhagen, Denmark: Technical University of Denmark (DTU), 2010.

Ragheb, Magdi, and Adam Ragheb. "Wind Turbines Theory: The Betz Equation and Optimal Rotor Tip Speed Ratio." In *Fundamental and Advanced Topics in Wind Power*, edited by Rupp Carriveau. London, UK: InTech Europe, 2011.

Righter, Robert W. *Wind Energy in America: A History.* Norman, OK: University of Oklahoma Press, 1996.

Sagrillo, Mick. "How It All Began." *Home Power* 27 (1992): 14–7.

Shepard, Dennis G. *Historical Development of the Windmill.* Government Report. Cleveland, OH: Lewis Research Center, National Aeronautics and Space Administration (NASA), 1990.

Smithsonian Institution. *Powering a Generation of Change.* July 2014. Accessed August 23, 2017. http://americanhistory.si.edu/powering/.

Stuart, Harve R., and Elisha N. Fales. Stuart Turbine. United States Patent US1802094 A, April 21, 1931.

U.S. Census Bureau. *Historical Statistics of the United States, Colonial Times to 1957: A Statistical Abstract Supplement.* Government Report. Washington, DC: U.S. Census Bureau, 1960. Accessed April 30, 2018. https://www2.census.gov/library/publications/1960/compendia/hist_stats_colonial-1957/hist_stats_colonial-1957.pdf.

Wikipedia. *Kurt Bilau,* April 22, 2010. Accessed August 23, 2017. https://de.wikipedia.org/wiki/Kurt_Bilau.

Zink, John C. "Steam Turbines Power an Industry." *Power Engineering* (PenWell Corporation) 100 (1996): 24.

3

The Birth of Big Wind[a]

Generating electricity on a large scale from wind is structurally impossible and uneconomic. None of the giant plans have been carried out, and none could have been carried out.

—Kurt Bilau (1937)

3.1 The High Altitude Turbogenerator

In the United States, the national rural electrification effort that extended transmission lines into rural America starting in the 1930s set in motion the gradual decline of small wind turbines. The expansion of the transmission and distribution system enabled the federal government to provide farms and rural customers with low-cost power from coal plants and hydropower stations. This would not have been possible without the presence of abundant and low-cost coal and hydro resources, and the US government's massive subsidies. Indeed, these resources made it possible to build large central station power plants and deliver cheap electricity to the far-flung towns that dotted the American West.

Some European countries were similarly well-positioned when it came to abundant energy resources. Like the United States, Germany is also endowed with an abundance of coal that made the production of low-cost coal-fired

---

[a] Of course, the term "big" is subjective here, but in this chapter, it means wind turbines with a power rating of 100 kW or greater. The focus is on electrical capability—not rotor diameter or tower height—although these characteristics are indeed positively correlated with a wind turbine's rated capacity. Prior to the 1930s, no wind turbines with a rated power capacity of 100 kW had been constructed.

The Wind Power Story: A Century of Innovation that Reshaped the Global Energy Landscape, First Edition. Brandon N. Owens.
© 2019 by The Institute of Electrical and Electronics Engineers, Inc.
Published 2019 by John Wiley & Sons, Inc.

electricity possible. However, at the time, some German thinkers believed the presence of large coal resources also created an opportunity to consider alternative electricity generation options because, as the thinking went, coal could be exported instead of used to generate electricity. This would generate much needed revenue for the German government, which was deeply in debt in the 1930s. In 1938, German engineer Hans Witte summarized this sentiment by noting that: "According to the latest drilling results, we still have enormous coal reserves in the form of brown coal and stone coal which will provide an ample supply of energy over a long-time period. It would not be necessary to consider other types of energy because our coal reserves are ample for electricity production. However, over the last few years it has been established, particularly in the period after the war, that coal is one of our most valuable and purest commodities and therefore represents a substantial part of the wealth of our people." and that the country "should not be burning it up in fire boxes."[1]

Wind power was one of the sources of alternative energy that was under consideration. Recall that Germany had been introduced to wind power by Kurt Bilau, who applied his aerodynamic blades to windmills throughout Europe and built a small wind turbine at the Göttingham Aerodynamic Research Institute in 1925.[2] But could wind really be developed at a large enough scale—and at a low enough cost—to become a viable alternative to coal-fired generation? The Germans were intrigued.

German engineer Hermann Honnef for one believed that wind power could indeed become a viable replacement for coal-fired electricity, and he developed an ambitious plan that outlined the possibilities. Honnef published his plans in multiple books and articles and provided a roadmap for the development of large wind turbines in Germany.[3] He envisioned the development of a series of wind turbines 300 m tall with rated capacities of 20 MW each. Each wind turbine would have three to five rotors, and each rotor would contain two hubs with multiple several 160-m diameter blades. If possible, Honnef recommended that the wind turbines be erected offshore where the winds were stronger. It was a bold vision. Honnef wrote about megawatt-scale wind power development at a time when the largest turbine constructed to date was Bilau's 45-kW unit.* Fact was, Honnef was proposing to build a single wind turbine that was over 400 times larger than the one that Bilau had built.

In truth, Honnef's wind turbine designs were more science fiction than science. However, his plans did include several innovative features that foreshadowed the future of wind power technology. For example, he designed the first gearless—or direct drive—wind turbine that used a ring generator.

---

* Although he developed his designs prior to Adolph Hitler's rise to power in 1933, Honnef was apparently not shy about expressing his enthusiasm for the potential of wind power to the Nazi regime. According to Matthias Heymann (1995), Honnef sent a telegram to propaganda minister Goebbels pleading for a favorable word from Adolph Hitler on his proposal.

He also proposed a power conversion approach to produce AC output at a fixed frequency. In addition, he suggested erecting turbines at heights that would allow them to reap the benefits of higher wind speeds at greater altitudes.

The two rotating hubs on Honnef's wind turbines were designed to "contra-rotate," or rotate in opposite directions. The hubs served as the primary components of the turbine's ring generator. One hub was to be fitted with a copper coil and would serve as the generator's armature. The other would be fitted with magnets. When the wind blew, the hubs would contra-rotate, and the armature would move across the magnetic field to generate electricity. One hub would rotate at 10 rpm, and the other would rotate at 17 rpm. Together they would combine for a generator speed of 27 rpm.[4] By integrating the electric generator directly into the rotating hubs in this manner, Honnef was able to eliminate the gearbox entirely. However, the contra-rotating hubs in Honnef's design didn't rotate fast enough to generate high frequency electricity. Therefore, in his design, Honnef increased the number of magnetic poles in the generator. By doing this, he increased the frequency of the electric current because the direction of the current changed every time it passed a magnetic pole. In this manner, the large, slowly rotating ring generator in Honnef's design would be able to produce AC at the required frequency. It was ingenious.

Honnef's intent was to connect his wind turbines directly to the transmission system. Until this point in time, small wind turbines were typically connected to DC generators that produced an electric output that was not compatible with the AC transmission system. Recall, transmission networks required power at a fixed frequency of 50 Hz in Europe and 60 Hz in the United States. Simply substituting an AC generator for a DC generator would not be sufficient because the AC generator would produce power at varying frequencies depending upon the wind speed.

Providing AC at a fixed frequency is a challenging task for a wind turbine because it is the rotational speed of an electric generator that determines the power frequency. Contrast this to a steam turbine where the generator's rotational speed is controlled by adjusting the amount of steam that enters the turbine through a control valve. Unfortunately, the rotational speed of a generator connected to a wind turbine is determined solely by the speed of the oncoming wind. There is no control valve for Mother Nature.

Honnef designed a way around this challenge. He would allow his wind turbines to operate at variable speed and produce AC at an uncontrolled frequency, but before transmitting power to the transmission network, the power would be routed to a rotary converter that would change the uncontrolled AC to DC. The DC would then be converted back to AC at a fixed frequency using a second rotary converter running in "inverted" mode.

The need to transform AC to DC—and vice versa—arose as soon as AC became the established power standard for the transmission system in the late nineteenth century. Islands of Edison-style DC power systems had already been built and their DC output needed to be converted to AC to integrate with broader power networks. In addition, several other applications such as subways, streetcars, railways, some industrial processes such as aluminum production, and variable-voltage power supply systems for radio transmitters, all required DC. So there was already a need to convert AC power to DC in many instances.

The machine of choice for such conversion was the rotary converter. They were manufactured by both GE and Westinghouse and were an effective mechanical solution to convert AC to DC. One of the key features of the rotary converter is that it could be run in "inverted" mode and convert DC to AC as well. Both rotary converters and—another conversion solution—mercury-arc rectifiers were eventually replaced by solid-state silicon devices starting in the 1960s. Interestingly though, rotary converters were so durable that some of those used by New York City Metropolitan Transit Authority weren't shut down until the end of 1999.[5]

So, in his wind turbine design, Honnef had relied upon state-of-the-art power conversion devices to make wind power suitable for the AC grid. However, power conversion was inefficient and therefore costly, and it entailed losses that were compounded by the double conversion in Honnef's design. Clearly, the approach to power conversion in Honnef's design would have been too expensive at the time. In fact, power conversion would not become an economically viable option for wind turbines until the 1980s.

Perhaps the most defining characteristic of Honnef's wind turbine was the proposed tower height. At a staggering 300 m, his wind turbines would be as tall as the Eiffel Tower. Having previously constructed a 259-m radio tower in Berlin, Honnef had direct experience building tall towers.[6] This experience taught him that the wind blows faster at higher altitudes. He was also familiar with Albert Betzes' work, which showed that the energy output from a wind turbine increases in proportion to the cube of the wind speed. This means that a doubling of the wind speed would yield an eightfold increase in power output. From this, Honnef concluded that wind turbines should be built as high as possible.

In addition to his practical experience building radio towers, Honnef's knowledge of the relationship between wind speed and altitude came from his familiarity with the wind resource assessments of German scientists Assmann and Hellman. Professor Assmann was the director of the aeronautical observatory in Lindenberg, and had conducted wind resource assessments between 1907 and 1910. At the end of this period, he published a book entitled *The Winds in Germany* based on more than one million individual observations made at altitudes between 40 and 120 m over a thirty-year period.[7] During World War I, Professor Hellmann—of the Reichs Weather Service—built upon Assman's work by conducting an analysis of wind flows across

Germany. Hellmann was one of the first to consider wind conditions at high altitudes. He discovered that average wind speeds increased with altitude and provided a mathematical formula to describe the phenomenon.[8] He also found that wind turbulence decreased as a function of height. This is a point that Honnef focused on when he wrote that in order "to remove the turbines from the effect of the surface turbulence zone, the tower must be about 300 meters high."[9]

In practice, the choice of tower height is based on the economic trade-off between increased energy capture and increased tower construction cost. Wind turbine output increases not only because of the higher wind speeds at higher altitudes, but also because wind turbine blades can be made longer to capture more wind. However, it is more expensive to build high wind turbine towers because of the need for more materials and the need for specialized equipment to erect them. As of 2018, Enercon's E 126 wind turbine was the tallest onshore wind turbine with a hub height of 138 m. Most wind turbine towers today are built with a hub height in the range 80–100 m.[10]

A 2014 study used a mathematical model to examine the tradeoff between tower construction costs and energy output found, that the optimal tower height for wind turbines falls within the range 60–120 m, and that the optimal height varies depending upon the average wind speed.[11] So building tall towers makes sense, but—given the incremental construction costs and the need for specialized materials—it requires exceptional wind speeds at higher altitudes to make it work. Thus, Honnef was on the right track when it came to tower height, just as he was with his direct drive and power conversion concepts. However, like many great visionaries, he was a bit too far ahead of reality. Honnef's wind turbine design elements would finally be implemented when German wind turbine manufacturer Enercon released its gearless 500-kW E40 wind turbine in 1993—sixty years after Honnef proposed his wind turbine designs.

Not everyone in Germany was convinced that high altitude wind turbines were the wave of the future. Kurt Bilau—Germany's premier wind power expert at the time—was skeptical. In 1937, Bilau said, "My experiments on the utilization of wind power over the last fifteen years have convinced me of one thing: generating electricity on a large scale from wind is structurally impossible and uneconomic. None of the giant plans have been carried out, and none could have been carried out."[12] Others, such as German engineer G.W. Meyer also believed Honnef's designs were unrealistic. According to Meyer "the construction of giant plants—such as those planned by Honnef—with towers at least 250 meters tall, is still today a leap into the unknown." He suggested that "before this takes place, an experimental plant with a vane diameter of about thirty meters should first be built to first determine the most favorable vane profile."[13]

Across the Atlantic, *The New York Times* weighed in too: "Engineers who have built skyscrapers will readily concede that the construction of a steel, latticed tower 1,000 feet high presents no technical difficulties. But they are not so sure of the frame that tilts automatically in all directions. There are no precedents for such a structure. No one knows how it will behave in so fickle a medium as the wind, which may vary from a zephyr to a howling gale. Vibrations may be set up in the frame as a whole and jam the rather nicely balanced windwheel-frame."[14]

Honnef's designed captured the public imagination and the search for energy alternatives in Germany led to the founding of the Reich Working Group for Wind Energy (RAW), in which scientists and inventors collaborated on the development of wind power. With the support of RAW and the government Board of Trustees for Wind and Water Power, Honnef built a model of his wind turbine concept on a test field on the Mathiasberg hill north of Berlin in 1941.[15] The rotor blade and hub were later excavated from Honnef's test ground and are currently on display in the German Museum of Technology (DTMB) in Berlin.

RAW also worked on a large wind turbine developed by engineer Frank Kleinhenz. Kleinhenz's plans took shape with the help of a German manufacturing company, Maschinenfabrik Augsburg-Nürnberg (MAN). The proposed MAN-Kleinhenz wind turbine was also very large, with a rated capacity of 10 MW and a 130-m diameter rotor.[16] The MAN-Kleinhenz wind turbine was more technically refined than Honnef's proposal, and with a 250-m tower height, it was closer to Earth—both figuratively and literally speaking. However, actual construction of the MAN-Kleinhenz wind turbine was delayed by the onset of World War II as MAN turned its attention to building armaments. Instead of wind turbines, the MAN factory in Augsburg produced diesel engines for U-boats while the MAN factory in Nuremberg built Panzer tanks.[17]

3.2 The Soviets Advance

While the Germans discussed the merits of wind power—1,000 miles to the east—the Soviet Union was actively pursuing its own wind power research program that would lead to the construction of a 100-kW wind turbine. Because the Soviet economy was not integrated with the economies of Europe and the United States, Moscow's wind research efforts were not hindered by the Great Depression and its wind power research continued uninhibited throughout the 1930s.

The ability of the USSR to fund wind research during this period was not lost on the Russians. At the time, Russian scientist V.R. Sektorov of the Central Wind-Power Institute noted that "the history of capitalist technology does not reflect any significant attempts for using wind energy for permanent power stations. In the USSR, socialist economy makes it possible to build wind power."[18]

Sektorov was either unaware of the wind power progress that had already been made in the United States and Germany, or unwilling to acknowledge it. A favor that would later be returned by American wind turbine innovators in the 1940s. Nonetheless, Sektorov's observation was insightful. The socialist economy did indeed provide a natural buffer that allowed the Soviet Union to engage in high-risk research that had the potential to pay big dividends if successful regardless of the external economic environment. This buffer enabled the Soviet Union to advance wind power development in the 1930s and gain a brief theoretical and technological lead over Europe and the United States.

Fact was, the Soviet Union had been actively engaged in wind research since 1918. In the aftermath of the Russian Revolution, Lenin instructed the Academy of Sciences to include wind research in its reorganization plan. As a result, an investigation into wind power was started in 1920 at the National Aerodynamic Institute. In 1931, wind research was transferred to a special Central Wind-Energy Institute, which focused on research related to wind turbine aerodynamics and control. This work led to the development of a wind power roadmap that included the proposed construction of 5 MW-scale wind turbine serving as a complement to existing power systems in remote regions. This vision was articulated in 1932 in the Second Five-Year Plan for the National Economy of the Soviet Union.[19]

Russian scientists were familiar with Honnef's 20-MW wind turbine design, but they concluded that the cost, complexity, and risks associated with building high-altitude wind turbines was simply too great.[20] They reasoned that Honnef focused on building tall wind turbines primarily because high altitude winds were stronger and less turbulent; however, the Russians themselves had just conducted research on the integration of wind turbines into their transmission system and concluded that successful integration could occur without difficulty, so there was no need to build tall towers. In other words, it didn't matter to the Russians that wind turbines produced variable output at lower elevations because they had studied the problem and believed it would not be difficult to accommodate variable wind power within the power system by adjusting the output of other generators. At low volumes, accommodating wind power simply was not a problem.

The focus on wind integration into the transmission system was a core component of the Soviet Union's research program. The vision was to integrate wind turbines into agricultural regions with an existing transmission and generation network. Wind would provide supplemental power for agriculture use. In the words of researchers Shefner and Ivanov: "The clustering of lower- and medium-power stations not requiring high-power transmission lines is of particularly great importance in the electrification of agricultural areas."[21]

The Soviet Union's research on wind integration in the 1930s was just as forward-looking as Honnef's wind turbine designs. They discovered that the key to integrating wind power into the grid was to ensure that existing hydropower stations and steam generators were flexible enough to dispatch around the wind generation to ensure that the demand for electricity would be met. Shefner and Ivanov provide a window into their thinking on the topic: "For the sake of effective utilization of a windmill it is desirable to have a connection between the individual stations of the cluster flexible enough so that during periods of high winds the windmill will be loaded to full capacity and relieve the load on the other stations in the structure."[22] The Russian's were spot on—so much so that these discussions wouldn't be revisited seriously again until the 1980s and 1990s.

The Soviet Union also studied the advantages of geographic distribution of wind turbines. The basic idea was that by spreading out the placement of wind turbines their combined output is much less volatile than that of a single wind turbine because of differences in wind speeds over large geographic areas. By taking advantage of geographic dispersion, it is easier to integrate wind generation into the broader electricity network. Through empirical measurement of the wind speeds in three different locations, they discovered that the ratio between the maximum and minimum output of a wind turbine at a single location was 4.4, but with proper geographic spacing, this ratio could be reduced to 1.8.[23]

The development of a prototype wind turbine that would operate with existing generators was the first step to testing their theories about wind integration. They decided that the best approach to achieve their 5-MW vision was to start with a series of smaller 100-kW prototypes. If successful, they could build progressively larger machines with the goal of constructing a 5-MW unit. This was exactly what German researcher G.W. Meyer had proposed in response to Honnef's 20-MW wind turbine proposal. Start small and scale up. Thus, at a time when German scientists were still dreaming of erecting large wind turbines, the Soviet Union turned the dream into reality.

To start, the Soviets built a small test wind turbine in Kursk in 1930. They followed up by building a 100-kW wind turbine in the Ukrainian town of Balaklava, in Crimea on a cliff overlooking the Black Sea.[†] The 100-kW turbine was known as WIME D-30 and it was put into operation in May 1931. The Balaklava wind turbine was constructed under the direction of Russian scientist N.V. Krasovsky. It was a three-bladed horizontal-axis wind turbine and produced 100 kW of power at a wind speed of 10 m/s. It weighed 49 tons and was mounted on a circular track which had a small motor that was used to

---

† Balaklava is a former city on the Crimean Peninsula and part of the city of Sevastopol. It was a city until 1957 when it was formally incorporated into the municipal borders of Sevastopol.

rotate the turbine along the track to face the wind. Perched atop a 25-m tower, it was the largest wind turbine ever built at the time.

The circular track and accompanying motor used to rotate—or yaw—the wind turbine in response to changes in the direction of the wind was interesting. The wind turbine tower was affixed to the ground, but it also contained an additional tail that sat upon a cart which itself rested upon on rail laid around the tower in a circular radius of 20.5 m. The nacelle was affixed on top of the tower and the tail, in a manner that enabled it to rotate 360°. Powered by a 1.5-kW motor, the cart moved along the rail when the direction of the wind changed thereby causing the nacelle to rotate toward the wind.

In order to detect changes in the direction of the wind, a weather vane was installed atop the nacelle that was connected to the motor at the bottom. It was a unique approach, but it wasn't used in any subsequent wind turbine designs either in the Soviet Union or elsewhere. Wind turbine designers discovered that it was easier to yaw the wind turbine toward the oncoming wind using a rotating nacelle powered by a yaw drive mounted atop the tower.

Figure 3.1 The Soviet Union's 100-kW wind turbine at Balaklava in the Crimea. It was the first 100-kW wind turbine ever built and it was successfully interconnected to the local transmission network and co-dispatched with an existing thermal power plant. It was likely destroyed in 1942 when the Russians defended Sevastopol from the Germans.

3.3 Victory

Monitoring during the first two years of operation revealed that the average annual energy output was 279,000 kWh. This means that the annual capacity factor was nearly 32 percent.[24] The machine had an efficiency of 24.2 percent, which was obtained at a tip-speed ratio of 4.75. The turbine's rated wind speed was eleven meters per second. An *Electrical Times* journal article in 1935 reported that "the Russian engineers are so pleased with their actual working experience of this windmill, extending over two years, in parallel with the local Crimean network, that they are designing plants of greater capacity along similar lines."[25]

However, earlier laboratory tests indicated that the efficiency should have been at least 35 percent. Thus, despite some impressive numbers for a prototype, the wind turbine still did not perform as well as expected. According to Russian scientist V.R. Sektorov—who was involved with the development of the turbine—the difference between theoretical and actual efficiency could be attributed to "the narrow blade chosen in order to reduce the weight of the rotor and increase the uniformity of the movement, and also by some imperfections in making the blades, and by the badly crumpled sheathing made from five kilograms of roofing iron, due to the large number of alterations which the blade underwent during the installation period."[26] It appears that they crushed the blade when they tried to install it, which reduced the efficiency of the wind turbine.

Power from the Balaklava wind turbine was increased from 220 to 6,300 V using a step-up transformer and was fed into the local transmission network. Russian engineers faced the same fundamental design challenge as their German counterparts did—namely, how to build a wind turbine that provided constant frequency given the fact that generator's frequency was determined by the variable wind speed. In his designs, Honnef solved this problem on paper using power converters. He converted the uncontrolled AC into DC, and then back into fixed-frequency AC. The Russians rightly rejected this as an expensive and inefficient solution given available power conversion technologies. Instead, they developed a hybrid approach that employed a mechanical control system.

The mechanical control system used hinged surfaces, or ailerons, mounted on the wind turbine blades, just like those used on airplane wings. It was similar to the one German Kurt Bilau designed for his wind turbines. When the rotational speed of the wind turbine exceeded a predefined level, counterweights—mounted on struts—would extend the ailerons and slow down the rotational speed. By using this control system, the rotational speed was kept within a target range of 30 rpm.

However, given its rudimentary nature, the mechanical control system couldn't quite keep the rotational speed precisely at 30 rpm as the wind speed changed. This was a problem because synchronous generators—which were typically used in large power plants at the time—operated by maintaining a frequency that is synchronous with the frequency of the transmission system. They are very "stiff," meaning that when they are coupled directly to the grid they must operate at the exact grid frequency or else they will fall out of synchronism. This meant that the wind turbine would have to be kept at an exact rotational speed if it was going to use a synchronous generator like other grid-connected power plants. Either a more refined control system would be required, or some other more precise method of controlling rotor fluctuations would be needed.

To overcome this challenge, the Russians fitted the Balaklava wind turbine with an induction or asynchronous generator instead of a synchronous one. This has several advantages. Construction of induction generators is less complicated as it does not require brushes and a slip ring arrangement. This reduces the manufacturing cost of the generator. Furthermore, electricity in an induction generator is produced within a range of rotational speeds depending upon the load—rather than a single point—and this provides a natural buffer that is useful in absorbing changes in the wind turbine's rotational speed that are caused by fluctuations in the wind.

The Russians used a rheostat to adjust the resistance in the rotor and create more slip. Hence, to connect the Balaklava wind turbine to the grid, the Russians used a generator that was not as stiff as a synchronous generator to account for the variations in the wind turbine's rotational speed. However, one drawback of induction generators is that they draw reactive power from the grid, dragging down voltage levels and choking a transmission line's current-carrying capacity. This problem wouldn't be addressed until the 1980s when wind turbines were equipped with capacitors that compensated for their reactive power consumption.

The Russians were ahead of the game here, as induction generators would eventually become the preferred generator type of large wind turbines toward the end of the twentieth century. However, unlike the Balaklava machine, which rotated at a constant rate at maximum output, most wind turbines today are variable-speed machines that use power converters to produce fixed-frequency output. The most common type of induction generator used in variable speed-wind turbines today is the doubly-fed induction generator, which is a useful configuration because it minimizes the amount of power that needs to be converted before the power is transmitted to the grid.

The Balaklava wind turbine operated successfully for ten years. The wind turbine was operated in coordination with a 20-MW, peat-burning power plant at Sevastopol twenty miles away.[27] Increases in the output of the Balaklava wind turbine were met with corresponding decreases in the output of the steam unit in Sevastopol. It was the first instance of successful co-dispatch between wind and another electric generator.

Building upon their success with the Balaklava wind turbine, the Soviet Union began construction of additional experimental units to test other design features. As of 1937, there was "one experimental plant under construction at Rostov with a twenty-nine-kilowatt power rating and another in Peressadovka in Ukraine with a seventy-five-kilowatt power rating."[28]

At the request of the Soviet Main Energy Authority, Russian researchers also completed a preliminary design for a "super powered wind turbine." The three-bladed wind turbine was intended to have a rotor diameter of 100 m and a tower height of 65 m. The constant-speed machine would operate at 12 rpm and would be connected to two 2.5-MW induction generators operating at 600 rpm. The machine was designed to have an efficiency of 32 percent and would a rated wind speed of 16.5 m/s. It was intended to be installed at Markhotsk Pass in the Novorossiysk region of Krasnodar in Russia, where the average wind speed was measured to be 9.5 m/s.[29]

However, the 5-MW machine was never constructed. The Soviet Union's wind turbine development plans were deprioritized toward the end of the 1930s. The Soviet economy was focused on supporting the military effort. By March 1939, the Red Army was on the march, and the Soviet Union's wind power research was abandoned.

There is some uncertainty surrounding the fate of the 100-kW Balaklava wind turbine. In 1949, Russian researcher Ye. M. Fateyev stated that machine was destroyed by the Germans during World War II.[30] Fateyev's claim is not entirely improbable, since Balaklava was the southernmost point in the Soviet-German line during the war. Later, in 1952, V. R. Sekterov—who was generally a more reliable source for historical details on the Balaklava machine—simply noted that the machine was "destroyed in 1941."[31] In 2014, another writer, Viktor Elistratov, stated that the Balaklava machine was "...working in a power system simultaneously with a steam power plant until 1942 when it was blown up during the occupation of the Crimea."[32] Elistratov noted that the wind turbine served as an observation post in defense of Sevastopol before it was destroyed.

The details surrounding the demise of the 100-kW Balaklava wind turbine remain one of the great mysteries in the history of wind power. However, one thing is clear: the Russians were the first out of the gate when it came to building large-scale wind turbines connected to the surrounding transmission system. It would now take a Herculean effort to surpass the Russians' wind research accomplishments. As it turns out, just such an effort was already underway in America.

Notes

1 Witte, 1938, 1373.
2 *Current Opinion*, 1925.
3 For example, see Honnef (1939).

4 Putnam, 1948, 108.

5 Blalock, 2013.

6 *Current Opinion*, 1925.

7 Witte, 1938, 1374.

8 Witte, 1938, 1374 and Honnef, 1939, 501.

9 Honnef, 1939, 504.

10 Ryan and Bolinger, 2017.

11 Lee, Cho, and Lee, 2014.

12 Meyer, 1941/1942, 111.

13 Ibid.

14 *Power from the Wind*, 1933.

15 Beurskens, 2014, 19.

16 Hau, 2013, 31.

17 MAN, 2017.

18 Sektorov, *The Present State of Planning and Erection of Large Experimental Wind Power Station*, 1933, 9.

19 Sektorov, *The Present State of Planning and Erection of Large Experimental Wind Power Station*, 1933; Stein, *Utilization of Wind Power in Agriculture in the USSR* 1941; Stein, *Wind Power Plants in Russia and the United States*, 1941 provide brief reviews of the Soviet wind research program in the 1920s and 1930s.

20 Shefner and Ivanov, 1941, 22.

21 Ibid.

22 Ibid.

23 Sektorov, *The Present State of Planning and Erection of Large Experimental Wind Power Station*, 1933, 10.

24 Ibid., 6.

25 As reported in Parsons (1953).

26 Sektorov, *The Present State of Planning and Erection of Large Experimental Wind Power Station*, 1933, 12.

27 Putnam, 1948, 105.

28 Sauer, 1937, 947.

29 Ibid.

30 Fateyev, 1948, 359.

31 Sektorov, *Using the Energy of the Wind for Electrification*, 1953, 2.

32 Elistratov, 2014, 446.

Bibliography

Beurskens, Jos. "The History of Wind Energy." In *Understanding Wind Power Technology: Theory, Deployment and Optimization*, edited by Alois Schaffarczyk. Translated by Gunther Roth. West Sussex, UK: John Wiley, 2014.

Bilau, Kurt. "Generation of Electricity Through Giant Wind Power Plants." *Elektrizitatswirtschaft (Frankfurt am Main)* 18 (1937).

Blalock, Thomas J. "The Rotary Era, Part 2: Early AC-to-DC Power Conversion." *IEEE Power and Energy Magazine* 11 (2013): 96–105. http://magazine.ieee-pes. org/novemberdecember-2013/history-10/.

Current Opinion. "Now for the Aerodynamic Windmill." *Current Opinion*, 78 (April 1925): 468.

Elistratov, Viktor. "The Development of the Wind Power Industry in Russia." In *Wind Power for the World: International Reviews and Developments*, edited by Preben Maegaard, Anna Krenz, and Wolfgan Palz, 443–53. Singapore: Pan Stanford Publishing, 2014.

epoznan.pl. 2012. "Nowy Tomyśl: the highest windmills in the world created!" December 3.

Fateyev, Ye. M. *Wind Engines and Wind Installations*. Translated by Leo Kanner Associates for NASA. Moscow, Russia: State Publishing House of Agricultural Literature, 1948.

Hau, Erich. *Wind Turbines: Fundamentals, Technologies, Appications, Economics*. Berlin, Heidelberg: Springer-Verlag, 2013.

Heymann, Matthias. *Die Geschichte der Windenergienutzung 1890–1990 (The History of Wind Energy Utilization 1890-1990)*. Frankfurt: Campus Verlag, 1995.

Honnef, Hermann. "High Altitude Wind Power Plants." *Elektrotechnik und Maschinenbau* 57, no. 41–42 (1939): 501–6.

Lee, Jaehwan, Woojin Cho, and Kwan-Soo Lee. "Optimization of the Hub Height of a Wind Turbine." *Journal of Industrial and Intelligent Information* 2, no. 4 (2014): 275–9.

MAN. *History*. August 22, 2017. Accessed August 29, 2017. https://www. corporate.man.eu/en/company/history/man-timeline/MAN-Timeline.html.

Meyer, G.W. "Progress in the Utilization of Wind Power." *Elektrizitdtsverwertung* 16, no. 6/7 (1941/1942): 109–13.

Parsons, H. E. "Wind Power: History and Present Status." *Engineering Journal* (1953): 19–21.

Popular Science. "New Schemes for Harnessing the Winds." (August 1939): 100–1.

Putnam, Palmer Cosslett. *Power from the Wind*. New York: Van Nostrand Reinhold Company, 1948.

Ryan, Wiser, and Mark Bolinger. *2016 Wind Technologies Market Report*, 82. Government, Office of Energy Efficiency and Renewable Energy. Washington, DC: U.S. Department of Energy, 2017.

Sauer, Th. "Wind Power Plants in Russia." *Zeitschrift* 81, no. 32 (1937): 947–8.

Sektorov, V. R. "The Present State of Planning and Erection of Large Experimental Wind Power Station." *Elektrichestvo* 2 (1933): 9–13.

———. "Using the Energy of the Wind for Electrification." *Elektrichestvo* 3 (1953): 11–6.

Shefner, K. I., and A. A. Ivanov. "Lines of Development of Rural Wind Power Plants." *Elektrichestvo* no. 5 (1941): 21–2.

Stein, Dimitry. "Utilization of Wind Power in Agriculture in the USSR." *Elektrizitdtswirtschaft* 40, no. 4 (1941): 54–6.

———. "Wind Power Plants in Russia and the United States." *Elektrizitatswirtsch* 40, no. 16 (1941): 17–8.

The Montreal Gazette. "Power from the Wind." (March 13, 1933): 11.

Witte, Hans. "Economy and Practical Applications of Large Wind-Driven Power Plants." *Elektrotechnische Zeitschrift* 50, no. 51 (1938): 1373–6.

4

Wind Power's Giant Leap

It seemed to me that the most economical design would be a larger wind turbine than any yet built...

—Palmer Putnam (1948)

4.1 The Winds of Cape Cod

Scientists and engineers in the United States stood idle as the Germans and Russians moved forward with large wind power in the 1930s. No large wind power proposals like those of Germany's Herman Honnef were considered in America, and the US government took no interest in funding prototypes like the Russian 100-kW wind turbine in the Crimea. Furthermore, as the United States plunged deeper into the Great Depression in the 1930s—and energy prices fell—conducting research into higher cost energy alternative technologies was not on the public or private research agenda. Wind turbines were a nonstarter in America in the 1930s.

In fact, if it wasn't for one man—Palmer Cosslett Putnam—America would have stayed on the sidelines of wind power development as the Europeans and the Russians moved ahead. Due almost entirely to the efforts of Putnam—by 1941—America burst onto the wind power scene with a wind turbine that was bigger and bolder than anything that had ever come before. It was a giant leap in the evolution of wind power that would not be duplicated on a comparable scale until a quarter century later, when—in 1979—GE built the 2-MW MOD-1 with funding from the US National Aeronautics and Space Administration (NASA).

Putnam was an engineer who initially became fascinated with the power of wind during his stint as a World War I pilot. When he settled in Cape Cod later in life, his interest was rekindled. It seems that he was not entirely happy with his home on Cape Cod because it was too windy. At the same time, he was

The Wind Power Story: A Century of Innovation that Reshaped the Global Energy Landscape,
First Edition. Brandon N. Owens.
© 2019 by The Institute of Electrical and Electronics Engineers, Inc.
Published 2019 by John Wiley & Sons, Inc.

faced with electricity bills that he felt were too high. Could he come up with something that would solve both problems simultaneously?[1]

After carefully considering the problem, he concluded that wind turbines were the answer to producing low cost electricity. Although large wind turbines were too big for his own personal residence, he realized that they could provide a solution for the world at large. Thus, he began a six-year quest to build the world's largest wind turbine. He ultimately succeeded in the fall of 1941 when the 1.25-MW "Smith-Putnam" wind turbine started feeding electric power into the Central Vermont Public Service Corporation (CVPSC) electrical system. This feat led Time magazine to predict that New England will someday "rival Holland as a land of windmills."[2] Because of Putnam's accomplishment—in 1941—it felt as if the world had embarked upon a new era in electricity production.

The Smith-Putnam was no ordinary wind turbine—in part—because its chief architect was no ordinary engineer. Putnam's personal history reads like an action hero out of a 1930s film serial. Born in 1900, he was the son of the American soldier, publisher, and author, George Haven Putnam. He was the grandson of publisher George Palmer Putnam. His mother—Emily—was a noted scholar, writer, and historian. His cousin—George Palmer—was also a publisher and explorer, and was married to American flyer Amelia Earhart.

Palmer Putnam served as a member of the British Royal Air Force in World War I. After the war, he went to college and graduated from the prestigious Massachusetts Institute of Technology (MIT) in 1924 with a master's degree in Geology. Finished with his studies, he set out to explore the volcanoes of Central America. He served the Belgian Government as a geologist in the Belgian Congo.[3] After this adventure, he succeeded his father as the President of G.P. Putnam and Sons Publishing Company. Putnam settled in Cape Cod where he became an active sailor and helped found the "Stone Horse Yacht Club" of Harwich on Cape Cod. He resigned from G.P. Putnam and Sons in 1932 and began pondering his next move. He had already squeezed more living into a thirty-year-old body that most people did in their whole lives.

He asked his neighbor Elisha Fales for insight on wind power. Recall—a decade earlier—Fales and his colleague Harve Stuart successfully applied the aerodynamic lessons of World War I to wind turbines by adding twisted airfoil-shaped blades to wind turbines. Fales encouraged Putnam to investigate recent German and Russian wind power efforts. Putnam studied Honnef's plans and eventually concluded that he had "exaggerated the importance of height."[4] By studying the tradeoff between increasing costs and performance with height, Putnam determined that there was no justification for a tower more than fourteen meters high. He also noted that the large air gaps in Honnef's proposed wind turbine could not be properly maintained. Putnam concluded that Honnef's elaborate machines would likely fail under strong winds.

Despite this knowledge of the failings of Honnef's design, it turns out that the 1.25-MW design wind turbine that Putnam would ultimately put into operation in 1941 would be ill-equipped to handle strong winds itself—Thank goodness neither of them actually erected a wind turbine at the heights that Honnef had proposed. It would have been blown to bits. In fact, it wasn't until the 1990s that engineers were able to accurately quantify and adapt to the immense loads faced by large wind turbines. Until this breakthrough occurred, MW-sized wind turbines like the Smith-Putnam machine could not withstand the mighty force of Mother Nature.

When Putnam assessed the Russian 100-kW wind turbine in Crimea, he failed to fully appreciate what the Russians had accomplished. He thought that it was "bold and practical" but noted that it was also limited by the conditions "imposed upon the designers by the state of the industry in Russia, where heavy forgings, large gears, and precision instruments were unavailable."[5] The Russian wind turbine machine was indeed a crude prototype. However, it was also the first and only 100-kW wind turbine to feed electric power into an AC transmissions system. Further, the Russian's use of an induction generator—rather than a synchronous one—was an innovation, not a limitation. Induction generators would eventually become the preferred generator type for large wind turbines. Furthermore, Russian researchers understood both the potential of wind power as well as the unique challenges and made a lasting contribution to the development of large wind turbines. These accomplishments were lost on Putnam and the rest of the world at the time.

After three years of study, Putnam developed a design for a wind turbine. It was a two-bladed, constant-speed, horizontal-axis wind turbine attached to an AC generator that fed synchronous power into the transmission system. The blades would be built on steel spars and covered with a stainless-steel skin. Putnam's idea was to build the wind turbine and connect it to a local power system that relied heavily on hydroelectric power. In this manner, hydropower could provide baseload energy to account for wind power's variability. He reckoned that from the point of view of the hydro system, "the energy in the wind then became merely increased stream flow."[6]

Putnam presented his design to Vannevar Bush, who was then dean of engineering at MIT before he became President Roosevelt's science advisor in 1942. Bush referred Putnam to Tom Knight, a vice president at GE. The two struck it off immediately and Knight agreed to work with Putnam to refine his design. They engaged additional engineers to determine the right size of the wind turbine, to produce a layout, and to examine the potential economics of the wind power. Thanks to Knight, GE was on-board with Putnam's plan.

After the preliminary engineering and economic plans were completed, Putnam still had a dilemma. Who would pay to turn his dream into reality? Knight mentioned Putnam's wind turbine idea to GE's hydroelectric specialist—Alan Goodwin—who offered some insight. Goodwin had been promoting hydropower development in the Northeastern US in collaboration with hydro-power turbine manufacturers such as S. Morgan Smith since the mid-1920s. Fact was, the late 1930s were a tough time for hydropower manufacturers like S. Morgan Smith. The federal development of large dams created a shortage of new projects. These conditions forced hydropower turbine manufacturers to close shop or explore new ways of using their engineering and manufacturing capabilities. The S. Morgan Smith Company—a manufacturer of controllable-pitch hydraulic turbines—was struggling to survive and they needed a way to unlock new markets. They were willing to consider all ideas, including Putnam's wind turbine concept.[7]

Beauchamp Smith—President of the S. Morgan Smith Company—would later recall: "We were concerned about the dwindling market for hydropower because most of the commercially feasible sites had already been developed. Our organization was already deeply involved in the promotion of pumped storage. It seemed to us that wind power in combination with pumped storage would be a natural partnership; and, if it could be proven technically and eco-nomically sound, it would give us both a new product and an expanded market for our existing lines of hydraulic turbines and pump turbines, which in 1939 supplied the lifeblood for our company."[8]

Goodwin had worked closely with Howard Mayo of the S. Morgan Smith Company and he was aware of S. Morgan Smith's need for a new business opportunity. Mayo arranged a meeting between GE's Goodwin and S. Morgan Smith Company executives. Within a week, two S. Morgan Smith vice presi-dents—Beauchamp and Burwell Smith—came to see Knight in his GE office in Boston. The Smith-Putnam wind turbine project was born in Knight's office in October 1939 when Beauchamp and Burwell Smith agreed to fund the construction of Palmer Putnam's wind turbine. GE agreed to develop and furnish the electrical equipment at cost.[9] Putnam's vision was about to become a reality.

Yet, as soon as this hurdle was overcome, another important question needed to be answered. Who would buy the electricity produced by the wind turbine? The ever-resourceful Knight provided Putnam with the solution to this prob-lem as well. Knight sent Goodwin to see Walter Wyman, President of the New England Public Service Corporation (NEPSC). Wyman was the one man in the region who had the authority to push such a large project. Goodwin met with Wyman and successfully convinced him to authorize the purchase of electrical output from a large wind turbine at a yet-to-be-determined location within the NEPSC service territory. Wyman subsequently arranged for CVPSC to be the guinea pig and provide the site, transmission tie-in facilities, and ultimately own and operate the wind turbine.

Putnam's basic wind turbine design was converted into a functioning wind turbine in a two-year time span after the meeting in Knight's office in October 1939. The Smith-Putnam wind turbine became operational in August 1941.

The journey was hurried by international events. By the end of 1939, Europe was at war, and the US government had begun to mobilize for its possible entry. In 1940, the United States instituted the first ever peacetime draft and, in December, President Franklin D. Roosevelt proclaimed that America would be the "Arsenal of Democracy" by selling arms to Britain and Canada. If the Smith-Putnam wind turbine was going to get built—it needed to happen quickly, or else there would be no spare industrial capacity available to manufacture the wind turbine parts.

Putnam assembled a team of scientists and engineers to design, build, and erect the wind turbine at an accelerated pace. In Putnam's words "Beauchamp Smith had insisted on gathering together the leading men in each field; without a doubt he succeeded. Each in turn came under the peculiar spell of the project, which roused the enthusiasms of all of us."[10] Assembling a group of world-class experts, who were enthusiastically engaged in the project was one of Putnam's greatest achievements. It is clear in hindsight that the reason the Smith-Putnam project was transformed from a vision in 1934 to reality by 1941 was not just because Palmer Putnam was a talented engineer—he also had the uncanny ability to bring people together and turn them into champions of his cause.

The Smith-Putnam team's first task was to review the preliminary design and develop cost estimates to settle upon the final design. In early 1940, the S. Morgan Smith Company decided upon a "downwind", two-bladed 1.25-MW configuration with a rotor diameter of 53 m and a hub height of 38 m. Downwind machines have the rotor placed on the lee side of the tower, whereas upwind turbines have the rotor positioned facing the oncoming wind in front of the tower. This size was selected because the design team had determined that it was the smallest possible size that would have enough inertia to provide smooth electrical regulation when it was connected to the grid. Like the Russian's 100-kW turbine in the Crimea, the Smith-Putnam turbine would be a constant-speed machine. The Smith-Putnam turbine would maintain constant rotational speed by adjusting the angle of the blades in response to changes in wind speed. The wind turbine would be connected through a gearbox and a hydraulic coupling to a GE synchronous generator operating at 600 rpm that was connected directly to the transmission system through a high voltage transmission line.[11]

S. Morgan Smith Company itself did not have the spare capacity to fabricate the major components. Beauchamp Smith would later explain that "World War II was already being fought in Europe and our manufacturing facilities in America were loaded with orders both for our regular products and for various military munitions. So, when the time came to begin manufacturing of the experimental unit, we had no capacity available in our own facilities, and all components had to be farmed out for manufacture by other companies in order to meet the delivery requirements."[12]

Putnam selected Wellman Engineering Company of Cleveland for the job. Given Charles F. Brush's pioneering wind turbine work in 1888, Cleveland was a fitting location for the fabrication of the world's largest wind turbine in 1940. The on-rush of war created urgency, and the team worked quickly to provide Wellman with the final design specifications so that the wind turbine frame could be ordered by May 1940. The deadline was met, but the order had to be placed before the aerodynamic loads were fully estimated and the stress analysis was complete.

When the stress analysis was completed later in 1940, it indicated that the blade and the blade connection point on the frame were designed too small to withstand the expected force of the wind for a prolonged period of time. Later still—in 1944—a more thorough stress analysis indicated that the actual loadings experienced by the Smith-Putnam turbine in practice were even higher than those estimated in the 1940 stress analysis.

So the Smith-Putnam team knew in 1940 that larger blades and stronger blade connections would need to be used to avoid the risk of failure. However, because the fabrication of the wind turbine frame was already underway, thinner blades would have to be used. They moved forward, hoping that this mistake would not prove to be catastrophic.

The rest of the components were successfully designed and fabricated. In March 1941, the turbine components were shipped from Cleveland to the chosen site for the turbine on a mountaintop near Rutland, Vermont. After five months of construction—punctuated by a truck overturning on its way up the mountainside—the turbine was ready for operation by the end of August 1941.

4.2 Grandpa's Knob

Any modern-day wind farmer will tell you that a series of questions must be answered to determine whether a site is suitable for the construction and operation of a wind turbine. For example, does the site have access to transmission lines? Is the land flat enough to allow a turbine to be constructed on the site? Is the site accessible for construction equipment? Is the site in a national park or environmentally sensitive area that may not be appropriate for development? Who owns the land? Can it be purchased or leased to build the wind farm? Is there a town nearby so that the wind turbine operations staff has someplace to call home? These questions really haven't changed since Putnam and his team set out to find the right location for their wind turbine in 1939.

Once these questions are answered, the final—and most critical—factor is whether the level of wind resources at the site is strong enough to enable an acceptable level of production from the wind turbine. Is it windy enough? Answering this question—with a reasonable level of confidence—was extremely difficult for the Smith-Putnam team for a couple of reasons. First, no one had mapped the wind resource yet. Today, wind prospectors have access to industry-standard software package and government-sponsored wind resource maps for siting wind turbines and wind farms. Second, there was no established process or equipment for a wind resource assessment campaign. Today, wind resource monitoring, data analysis, and modeling are commonplace. The hardware and software can be purchased off the shelf. The Smith-Putnam team had to invent their own wind resource assessment process on the fly.

One thing about the wind resource assessment process that hasn't changed since 1939: to make a final site selection, at some point in the resource assessment process it is necessary to erect a wind speed measurement device at the candidate site and take measurements for at least twelve consecutive months. Although potential sites in the same vicinity typically have similar wind patterns, site-specific topology will ultimately determine the characteristics of the wind at a specific location. Putnam's team knew this, but—again, due to the onrush of war—they were obligated to select a final site for the Smith-Putnam wind turbine in the spring of 1940 without the benefit of any wind resource assessment data whatsoever. The lack of wind measurement data would turn out to be a big flaw in the Smith-Putnam wind turbine development process.

The wind turbine needed to be in the CVPSC service territory. The Smith-Putnam team consulted with meteorologists who believed that the higher and sharper the ridge, the more the wind would be accelerated. In addition, Putnam and his team walked through the area observing the shape of trees and hedges to determine if they had been impacted by the wind over many years. The ultimate result of this effort was the "Griggs-Putnam Index of Deformity." The Griggs-Putnam index grades tree deformity on a scale of one to seven. The index is still used as a local climatic wind indicator.[13]

Through this process, the team identified fifty potential mountain summits in the Green Mountains of Vermont that they believed were promising sites. One particular summit—called "Grandpa's Knob" because the landowner simply referred to the summit as "Grandpa's"—was within range of the transmission system and close to the town of Rutland.

However, no direct wind measurement had been performed at any of the sites, so the team had to rely upon wind speed data from a pilot balloon station in Burlington seventy miles to the north of Rutland. Based on this data—in June 1940—the team selected Grandpa's Knob as the future site for the Smith-Putnam turbine. Only after the final selection was complete was an anemometer erected to begin actual wind resource measurements.

Pilot balloons are small unmanned balloons that are set aloft to measure air velocity. To make a site selection, the Smith-Putnam team gained access to data from a series of pilot balloons in Burlington, Vermont, the most populous city in the state. Once the data was in hand, they adjusted it to account for differences in elevation between Burlington and Grandpa's Knob. They estimated that the average annual wind speed at Grandpa's Knob would be 7 m/s. Today, this would be recognized as a wind site with good wind resources.

Pilot balloon measurements are taken at a variety of altitudes as the balloon ascends, so the Burlington data provided a good record of changes in wind speed at different elevation levels. However, the Smith-Putnam team was confused. The Burlington data indicated that—instead of increasing with altitude—the average wind speed stop increasing at 600 m and didn't begin to increase again until around 1,200 m. This was another puzzle that needed to be solved.

The Burlington data was a concern for the team. However, they concluded that the flattening of wind speeds in the 600–1,200 m range was due to a systematic error in the pilot balloon data and didn't reflect actual wind patterns. The attractiveness of Grandpa's Knob as a wind turbine site hinged upon the team's assumption about the validity of the Burlington data. If they were correct, and the data contained a systematic error that underrepresented the true wind speeds at a 600-m elevation, then Grandpa's Knob was indeed an ideal location for the Smith-Putnam wind turbine. If they were wrong, and the wind speeds didn't increase between 600 and 1,200 m as indicated by the pilot balloon data, then the output of the Smith-Putnam wind turbine would be much less than expected. The difference would be about 2.25 m/s in average annual wind speed. Because output is a function of the cube of the wind speed, this difference would cut the expected output of the turbine in half.

It turns out that they were wrong and the Burlington data was correct. The prevailing winds flowing from east to west at Grandpa's knob are deflected by the Green Mountains, which run north to south across Vermont. The average height of the Green Mountains is 1,200 m, so winds below this altitude are diverted north and south around the mountains. A pronounced drop in the average wind speed starts around 600 m, which is the exact height of Grandpa's Knob.

Truth be told, there was no practical way for the Smith-Putnam team to determine if the Burlington data was correct or not given the time constraints. They had to take a best guess and made a reasonable assumption that since it was known that wind speed increases with height, the Burlington data was flawed. Unfortunately, wind speed is a critical determinant in the output of a wind turbine and the ultimate cost of electricity. Like the mistake made underestimating the loads, the wind resource estimate error would have big implications for the ultimate performance of the wind turbine.

4.3 A Dream Realized

Construction of the Smith-Putnam turbine was completed in August 1941. It wasn't esthetically pleasing, nor was it offensive to the eyes. It had clearly been built with an eye toward function rather than form. It was ready for

Figure 4.1 The 1.25-MW Smith-Putnam wind turbine was designed by Palmer Putnam and a world-class team of experts. It began operation in October 1941. After several years of intermittent operation, the turbine experienced a catastrophic blade failure in March 1945 and was put out of operation permanently.

operation, but the team wanted to be sure there would be no mishaps. They spent two months checking the controls, adjusting the machine, and carrying out test runs without feeding power into the transmission system. Finally, starting at 6:56 p.m. on October 19, 1941, power from the turbine was feed into CVPSC's transmission network.

The wind turbine operated without difficulty and provided 700 kW of constant frequency power into the transmission system. This was the first time that this was accomplished at this scale. Like the Russians before him, Putnam had proven that is was possible to successfully build and operate a large-scale wind turbine, and that it could be integrated into the AC electric transmission network. It was a noteworthy accomplishment.

However, things quickly went haywire. For starters, most of the oil seals leaked oil out of the hydraulic system; as a result, the packing glands had to be rebuilt, the oil head on the main shaft had to be entirely rebuilt, and oil seals needed to be added to the shafts in the gearbox. In addition, the generator bearing overheated and had to be returned to GE in Schenectady and redesigned. Mysterious creaking noises were traced to lose rivets in the assembly. The hydraulic coupling overheated because it had never been operated continuously at a rating corresponding to full load operation of the wind turbine. An ad hoc cooling system was developed to correct the problem. Later, the vertical shaft of the yaw mechanism sheared off several times under the stress of large yawing movements.

In May 1942, after less than one year of operation, the steel blades started to crack. The cracks were repaired by arc welding, but the Smith-Putnam team knew this was a temporary solution. Recall, the 1940 stress analysis indicated that larger blades and stronger blade connections would be needed to minimize the risk of failure. The cracks were simply the first visible confirmation that the blades were undersized and would eventually fail.

In the 1940s there was neither the awareness of the complex loads that a wind turbine was subject to, nor the proper knowledge or tools to understand the implications in terms of appropriate wind turbine and blade design. At the time, there were no national or international standards for designing wind turbines. It wasn't until the 1980s before standards for horizontal-axis wind turbines would appear. The first publication was a set of regulations for certification produced by Germanischer Lloyd in 1986. These rules were further refined and published by Lloyd in 1993. National standards were published in the Netherlands in 1988 and Denmark in 1992. The International Electrotechnical Commission published the first international standard in 1994.[14]

These standards were based on modeling and analysis of wind turbine loads and included consideration of aerodynamic loads, gravitation loads, inertia loads, and operational loads rising from the actions of the control system. The blades themselves are subject to deterministic loads and probabilistic loads due to turbulence, gravity, and internal loadings. Verification of the adequacy of a blade design required knowledge of the fatigue loading cycles expected over the lifetime of the machine. Without these standards, the Smith-Putnam team was flying blind.

But the blades weren't the only problem. In February 1943, the main bearing failed. A routine inspection revealed that it had cracked and operation was immediately halted. The failure of the main bearing marked the beginning of the end of the Smith-Putnam turbine. In the wartime environment, it took two years to secure a new bearing. Further, installing the new bearing required an expensive three-month disassembly job. The Smith-Putnam turbine was out of operation for twenty-five months.

It was placed back in service March 3, 1945. The blades failed three weeks later. On March 26, one of the eight-ton blades ripped off the hub and was tossed 229 m and landed on its tip. It's a miracle no one was injured. The Smith-Putnam turbine would never operate again. During its operations between 1941 and 1945, the turbine logged one thousand hours.

The Smith-Putnam team believed that enough operational experience had been gained to develop a redesigned wind turbine that could avoid the problems associated with the original Smith-Putnam turbine. From this perspective, the original goals of the Smith-Putnam turbine project had been achieved. Wind power could be successfully produced at a large scale and fed directly into the transmission network.

The challenge now was to develop a redesigned model that could be manufactured at scale to produce electricity at a lower cost than traditional power plants. Was that possible? CVPSC wanted to know the answer. CVPSC told the Smith-Putnam team that they were willing to buy the output from additional wind turbines if it could be produced at a cost that was equivalent to other sources of electricity. The team sharpened their pencils and set about to estimate the costs of a redesigned wind turbine.

4.4 A Lesson in Economics

In a 1945 study, the Smith-Putnam team provided an estimate of the cost of manufacturing twenty 1.5-MW wind turbines. The team assumed that the wind turbines would be redesigned to reduce costs and improve reliability based on the knowledge gained through the operation of the Smith-Putnam turbine from 1941 through 1945. They also assumed that the wind turbine would be located at a site with favorable wind resources, unlike Grandpa's knob. Even so, their cost and performance estimates indicated that the cost of electricity production from the wind turbines would be US$65/MWh in today's dollars.[*][†] However, CVPSC was only willing to pay up to US$45/MWh. The gap between what the utility was willing to pay and what the Smith-Putnam teams estimated that wind turbines would cost was too large. CVPSC decided to pass on any future wind turbine projects. This effectively terminated the Smith-Putnam research program.

Palmer Putnam was not satisfied. He still believed in the original conclusion that he reached in 1934: wind power could be produced more cheaply than

---

[*] According to Office of Management and Budget [OMB] (2017, table 10.1), the ratio of the GDP price index between 1945 and 2017 is eleven. Putnam calculated that the total capital cost would be US$204.75/kW in 1945 dollars. This would be comparable to US$2,274/kW in current dollars. The Smith-Putnam team assumed that the wind turbines would be in a very wind location and used 40 percent as the average capacity factor. That's a very generous assumption given the state of the technology at the time. For a 1.5 MW wind turbine, a 40 percent capacity factor translates to 3,504 MWh. If we apply a 10 percent rate to the capital cost to convert it to an annual equivalent, the cost of electricity from the wind turbines turns out to be about US$65/MWh in 2017 dollars.
[†] The price and cost of electricity is often expressed either in cents per kilowatt-hour (¢/kWh) or dollars per megawatthour ($/MWh). To convert from ¢/kWh to $/MWh, simply multiply by ten. Likewise, to convert from $/MWh to ¢/kWh, simply divide by ten.

other options. He set about to find ways to reduce the estimated cost of wind power below the level estimated in the 1945 cost study. In his 1948 book, *Power from the Wind*, he documented five avenues to reduce the cost of wind power: (1) competitive bidding of components; (2) design refinements; (3) major design modifications; (4) radical design departures; and (5) economies-of-scale from mass production.[15] Taken together, Putnam reckoned that the cost of wind could be reduced to US$35/MWh, well below the amount the CVPSC was willing to pay for the power.

Putnam's ideas for wind turbine cost reduction were realistic and many of them were realized decades later. However, all of them required two things that were not available in 1948: time and money. Significant additional investment would have been required by S. Morgan Smith Company—or some other benefactor—to redesign and test successive wind turbines. S. Morgan Smith Company entered this venture to develop a new business opportunity, not to fund a decade long research initiative with uncertain prospects for success. Funding this type of energy research is a job best suited for governments and perhaps large utilities, not small private manufacturing companies.

There are several useful lessons from the Smith-Putnam wind turbine experience. First, it provided a confirmation that wind power could be successfully integrated into the AC transmission network at the MW-scale. Second, the project underscored the importance of site selection. Site-specific topologies are so unique that it is impossible to accurately predict wind resource in a specific location without a history of anemometry data to draw upon. As a result of the Smith-Putnam team's misspecification of the wind speed at Grandpa's Knob, the Smith-Putnam wind turbine would only produce 30 percent of the energy output originally expected from the machine. Third, in Putnam's words, the development of wind power is a lot more "subtle and difficult, apparently, than many people have realized."[16] Advances would be required before anyone could build wind turbines to handle the dynamic loads created by the wind. Forth, large wind turbines could not yet be manufactured at a low enough cost to compete with coal-fired electricity and hydropower. Using the most optimistic assumptions at the time, Putnam estimated that his redesigned wind turbines could produce power at US$65/MWh, but the utilities were not willing to pay more than US$45/MWh for output.

However, the true miracle of the Smith-Putnam turbine was that it was even built in the first place. The project was financed entirely with private investment dollars in an era marked by low energy prices and a general indifference to alternative energy options. Putnam succeeded through ingenuity, persistence and—most importantly—by leveraging his vast network of social connections. It was his leadership that made it possible.

Would the technical lessons of the Smith-Putnam turbine be enough to convince decision makers in the US government to launch a full-fledged wind power research program in the wake of World War II? It turns out that part of the US government was already watching the Smith-Putnam project very closely and taking notes. An engineer at the FPC had been studying the effort since 1941, and he had some big ideas about the next generation of large wind turbines. But he'd need the help of an Arizona Congressman to turn his vision into reality. The next part of the wind power story takes place in the halls of Congress.

Notes

1 Putnam, *Power from the Wind*, 1948, I.
2 *Harnessing the Wind*, 1941.
3 Yergin, 2012, 600.
4 Putnam, *Power from the Wind*, 1948, 108.
5 Ibid., 105.
6 Ibid., 188.
7 Putnam, *Wind Power—Yesterday, Today, and Tomorrow*, 1982.
8 Smith, 1973.
9 *Harnessing the Wind*, 1941, 6.
10 Putnam, *Power from the Wind*, 1948, 9.
11 Ibid., 116.
12 Smith, 1973.
13 John and Hewson, 1979.
14 Burton et al., 2001, 209.
15 Putnam, *Power from the Wind*, 1948.
16 Putnam, *Wind Power—Yesterday, Today, and Tomorrow*, 1982.

Bibliography

Burton, Tony, David Sharpe, Nick Jenkins, and Ervin Bossanyi. *Wind Energy Handbook*. West Sussex, UK: John Wiley, 2001.
John, Wade E., and E. Wendell Hewson. "Trees as a Local Climatic Wind Indicator." *Journal of Applied Meteorology* 18 (1979): 1182–7.
Office of Management and Budget (OMB). 2017. "Historical Tables." *The White Hosue*. Accessed August 23, 2017. https://www.whitehouse.gov/sites/whitehouse.gov/files/omb/budget/fy2018/hist.pdf.
Putnam, Palmer Cosslett. *Power from the Wind*. New York: Van Nostrand Reinhold Company, 1948.
———. "Wind Power—Yesterday, Today, and Tomorrow." In *Large Horizontal-Axis Wind Turbines*, edited by Robert W. Thresher. NASA Conference Publication 2230, 7–22. Cleveland, OH: National Aeronautics and Space Administrations, Lewis Research Center, 1982.

Smith, Beauchamp E. "Smith-Putnam Wind Turbine Experiment." In *Wind Energy Conversion Systems*, edited by Joseph M. Savino, 5–7. Washington, DC: NSF/NASA, 1973.

Time. "Harnessing the Wind," *Time*, XXXVIII(10), September 8, 1941.

Yergin, Daniel. *The Quest: Energy, Security, and the Remaking of the Modern World*. New York: Penguin Books, 2012.

5

Wind Power in the Wake of War

> *It is time that our nation, traditionally the leader in technical advancement, utilize more of the tremendous, inexhaustible energy contained in the winds which blow across our prairies and our mountains.*
> —William E. Warne, Assistant Secretary of the US Department
> of the Interior (U.S. House of Representatives 1951)

5.1   The Wind Power Aerogenerator

World War II seemed to have destroyed just about everything in its path, including budding wind power technology developments in Russia and the United States. In America—in the run-up to war—the Smith-Putnam wind turbine component manufacturing was rushed resulting in a critical error. In addition, parts were hard to come by as US industry geared up for wartime production. During the war, shortages led to a two-year delay after the turbine's main bearing broke. It was hoped that after the fog of war lifted conditions to support alternative energy research might improve. In Russia, the Soviet Union's flagship—100-kW Balaklava wind turbine—was likely destroyed during the Crimea Campaign.

Unfortunately, as it turned out, the postwar environment was no better for wind power in America. The culprit this time was the broader energy landscape. In the United States, the price controls, fuel shortages, and embargos that accompanied the conflict soon gave way to an era of fuel abundance that initially stabilized and then gradually lowered energy prices. Low energy prices created an era of energy abundance that eroded the justification for continued wind power research.

Take coal for example, the primary cost input into coal-fired electricity. The average price of coal in the United States in 1949 was US$4.90/short ton.

The Wind Power Story: A Century of Innovation that Reshaped the Global Energy Landscape, First Edition. Brandon N. Owens.
© 2019 by The Institute of Electrical and Electronics Engineers, Inc.
Published 2019 by John Wiley & Sons, Inc.

Coal prices followed a downward trajectory through the 1950s and 1960s and hit a low point of US$4.45/ton in 1965. In real terms, this represented a price decline of one-third.[1] Likewise, domestic oil prices remained low throughout the 1950s and 1960s because of increasing domestic oil production. Average domestic oil prices in the United States in 1949 were US$2.54/bbl. By 1965, the nominal price sat at US$2.86/bbl. The nominal price increase of 28 ¢/bbl translated into a 20 percent reduction in real oil prices.[2]

To the extent that any alternatives were being considered at all, the focus was squarely on nuclear power. Starting in 1946, there was a concerted international effort—led by the United States—to harness the power of nuclear fission for peaceful purposes. In the United States, the Atomic Energy Commission (AEC) was created to foster and control the development of atomic science and technology.[3] Nuclear power became the focal point of energy research during this period. AEC authorized the construction of Experimental Breeder Reactor I at a site near Arco, Idaho, which first generated nuclear power in December 1951.[4] The Soviet Union also launched efforts to develop nuclear power. In June 1954, the 5-MW Obninsk Nuclear Power Plant became the world's first nuclear power plant to feed electricity into a transmission system. The United Kingdom followed suit by opening the Calder Hall nuclear generating station in October 1956.

If nuclear power was the future of electricity generation in the postwar era, the mighty steam turbine was the workhorse of the present and it was sorely needed to keep up with rising electricity demand. Electricity consumption grew rapidly in the postwar era due to rapid economic growth and continued electrification. In the United States, a period of economic prosperity emerged after the end of World War II in 1945 through the early 1970s. Rising electricity consumption in homes was the primary driver of the national electricity demand surge.

Electricity consumption became embedded in American culture in the 1950s. Although electric appliances were introduced in the first decades of the twentieth century, it wasn't until the 1950s when they became standard accessories in "modern" kitchens. The electricity requirements of large appliances increased due to the addition of new features like refrigerator ice makers; and new small appliances appeared in droves—hand-held mixers, electric skillets, toasters, toaster ovens, ice crushers, and milkshake makers. Automated clothes washers became common, and air conditioning brought cool temperatures on hot nights. In 1953 alone, one million air conditioning units were installed in the United States.[5]

Electricity demand jumped 14 percent between 1946 and 1947 alone, but there was not enough equipment or labor to meet this demand. As the immediate postwar constraints were relieved, electricity consumption settled into a growth rate of 8.8 percent. Electricity use in 1950 was 255 billion kWh. By 1970, it had climbed to 1,392 billion kWh. At this rate of growth, electricity consumption more than doubled every decade. If that rate of growth had persisted, electricity consumption in the United States in 2010 would have been 41,479 billion kWh—instead of actual 2010 consumption of 3,886 billion kWh.[6]

This period is known as the "Golden Years" for US electric utilities. To meet growing electricity demand, utilities built large, coal-fired steam turbine power plants as quickly as possible. Thermal electric capacity in the United States increased from 49 GW in 1949 to 265 GW in 1970.[7] GE and Westinghouse offered electricity utilities steam turbines in increasingly larger sizes. The largest steam turbines in 1928 were around 100 MW. By 1960, 575-MW steam turbines were available; and in 1965, the first 1-GW power plant was installed by the Consolidated Edison Company of New York.[8]

Coal-fired steam turbines didn't just grow, they became more efficient as well. Back in 1882, Thomas Edison's Pearl Street Station converted 2.5 percent of its raw input energy into electricity.* By 1920, state-of-the-art steam turbines had a thermal efficiency of 20 percent. By 1960, the thermal efficiency of new steam turbines in the United States was 40 percent.[9] As a result, American homes, businesses, and factories were eventfully flooded with copious amounts of increasingly cheap electric power.

However, all was not lost for wind power just yet. An engineer at the United States Federal Power Commission had studied the Smith-Putnam wind turbine and had some ideas on how to launch a broader federal wind power research program. FPC was an independent commission created in 1920 by the Federal Power Act to provide licensing for hydropower projects on federal lands. The FPC was also responsible for regulating interstate electric power trade. Beginning in 1938, the agency also become responsible for regulating interstate gas trade because of the 1938 Natural Gas Act. Later—in 1977—as part of the Department of Energy Organization Act, the FPC was renamed the Federal Energy Regulatory Commission (FERC), which remains the regulatory body for interstate energy sales and pricing in the United States today.

Back in the 1940s, the FPC was focused on encouraging electric utilities to expand in rural areas and finding ways to keep electricity costs low. A new technology like the Smith-Putnam wind turbine—which produced power using an abundant and free domestic energy resource—was bound to catch FPC's attention. Officially, Section 311 of the Federal Power Act provided the FPC with authority to "conduct investigations regarding the generation, transmission, distribution and sale of electric energy, however produced, throughout the United States and its possessions whether or not otherwise subject to the jurisdiction of the Commission." As soon as word reached the FPC that the Smith-Putnam turbine was feeding power into the CVPSC network, an investigation into the subject of wind power was launched.

---

* The percent of input energy that is converted into electricity is known as a power plant's "thermal efficiency."

What started as an initial investigation in 1941, turned into a ten-year odyssey by FPC Senior Engineer Percy Holbrook Thomas with the purpose of advancing wind power in the United States. Like Putnam Palmer, Percy Thomas was an MIT-trained engineer. He had worked his way up FPC's engineering ranks based on his skill and passion for electric power technologies. He was so well respected within the FPC that FPC's lead attorney would later introduce Thomas to a Congressional committee as a "respected engineer, beloved by all of his associates."[10] Thomas was not some dreamer toiling in obscurity in a back room of the FPC building while others focused on more pressing matters of the day. He was their top guy and he was onto something big. Thomas summoned the key members of the Smith-Putnam team to his New York office for regular debriefing sessions.[11]

Thomas's decade-long investigation ultimately produced four books: *Electric Power from Wind: A Survey* (1945), *The Wind Power Aerogenerator Twin-Wheel Type* (1946), *Aerodynamics of the Wind Turbine, Adapted for use of Power Engineers* (1949), and *Fitting Wind Power to the Utility Network* (1954).

In *The Wind Power Aerogenerator*, Thomas laid out his design for a next generation wind turbine. He blended the most favorable elements of the Smith-Putnam turbine from the 1940s and German Herman Honnef's proposal from the 1930s. Like Palmer Putnam and the Russians before him, Percy Thomas envisioned wind power operating within the context of the larger electricity network. In fact, Thomas's vision for wind power's eventual role within the electricity system was quite similar to the eventual role that wind power would actually play in the power networks of the twenty-first century. He "envisioned the prospect of supplying ultimately very cheap power, up to perhaps 25 percent of the total utility energy and 20 percent of the aggregate capacity of the interconnected whole. This 20 percent firm capacity is secured partly from the diversity in the fluctuations of the wind velocities between well-separated turbine sites, and partly by utilizing wind energy to form free hydro generating capacity on the peak, in hydro storage systems."[12]

Thomas's basic assertion was that wind power—which started out as a small-scale technology—could be readily adapted to "operate in synchronism with a transmission circuit in the usual way, and delivering power to the system in such a measure as it is developed by the wind, will be entirely acceptable as an auxiliary source of utility power."[13]

Thomas envisioned two different models: a 6.5-MW version and a 7.5-MW version. Each version would have twin rotors on a single tower. Each rotor would contain two or three blades depending upon the design, and the blades would be 24 m long. The tower height would be 145 m. The maximum rated wind speed would be 12–15 m/s. At maximum speed, the wind turbine would operate at 42–47 rpm. Sixty-cycle AC would be delivered to the transmission network at 1,200 V. The entire structure would be made of steel, except for the airfoil blades which would be made of "light metal alloys" to save weight and improve performance.[14] Thomas proposed the use of thin spruce planks as blade skins instead of sheet metal because he believed that wooden planks could be more easily shaped than sheet metal.

However, Thomas quickly ran into the fundamental problem of the "fluctuating frequencies." As described previously, this is the challenge associated with converting the kinetic energy in wind into electrical energy at the fixed frequency required by the AC transmission network. Mother Nature didn't deliver wind at the same speed all of the time, and changes in the speed of the wind altered the rotational speed of the wind turbine, which resulted in fluctuations in the frequency of the electrical output. The Russians addressed this problem in 1931 by using aileron-type wings on each of their blades that were driven by a weight system controlled by the centrifugal force of the rotating blades. Ten years later, in America, the Smith-Putnam team used a mechanical system that adjusted the angle of the entire blade.

In Honnef's high altitude wind turbine design, the wind turbine rotated at a variable speed and produced AC electricity with a variable frequency. Electrical conversion equipment would then be used to convert the variable frequency AC into DC and then back into fixed-frequency AC in two steps. Thomas proposed a variable speed wind turbine that relied on power conversion to produce fixed frequency output. However, to avoid a two-stage power conversion, Thomas's design employed a DC generator. DC electricity could then be converted to AC in one step. This would simplify the process and reduce the inefficiencies associated with power conversion. According to Thomas's calculations, the savings from moving from constant to variable speed outweighed the cost and inefficiency associated with electric power conversion.

Thomas also estimated the cost of manufacturing his wind turbine. His calculations were based primarily on the manufacturing experiences of the Smith-Putnam team and Palmer Putnam's subsequent post-mortem cost assessment as published in his book *Power from the Wind*. Putnam's cost assessment was not published until 1948, but was available to Percy Thomas through his conversations and meetings with Beauchamp Smith. Thomas estimated that the tenth wind turbine manufactured would cost US$68 to US$75/kW. Thus, the total cost of constructing ten 7.5-MW wind turbines was estimated to be in the range of US$5 million. Thomas's estimate was well below Palmer Putnam's already optimistic figure of US$204.75/kW for a manufacturing run of ten improved Smith-Putnam wind turbines. Palmer Putnam's numbers indicated that wind power was still too expensive, particularly in light of the falling cost of production from coal-fired steam turbines. But at Thomas's estimated cost of US$68–$75/kW, wind power would be able to compete directly with hydropower and coal-fired steam turbines.

The FPC supported Thomas's work, and—in 1949 and 1950—the agency applied for, and received, patents for the wind turbine designs. The challenge was how to obtain funding for the research in order to design, build, and test the wind turbines. The S. Morgan Smith Company had elected not to embark upon any additional wind power research after the Smith-Putnam turbine believing it was too costly and there was no commercial market for wind power, so funding was unlikely to come from the private sector.

The only real option was to work with the US government to set up and fund a wind power research program. The United States Department of Interior, which is responsible for the management and conservation of most federal land and natural resources, was the most natural home for wind power research. A bill would have to be introduced in Congress to authorize funding. A Congressional supporter was needed to introduce the bill. But who would carry the torch for Thomas and the FPC?

5.2   The Search for Tomorrow

As luck would have it, Arizona Congressman John Murdock was an ideal candidate to support wind power research. Murdock was the Congressional representative for Arizona and the Chairman of the Committee on Interior and Insular Affairs in the Eighty-second Congress (1951–1953). Arizona was in the midst of a severe drought in 1951 that had diminished surface-water availability and the ground-water basis in Central Arizona had been overdeveloped throughout the 1940s. Murdock was on the lookout for methods to conserve the region's depleted hydropower resources and he believed that wind power could potentially play a valuable role. He wanted to investigate "...the possibility of firming the power output of hydroelectric systems by the use of wind power during periods of low stream flow."[15]

Working with the FPC, Murdock introduced House Resolution (HR) 4286 in 1951 to authorize the "...investigation, research and development work by the secretary of the interior and the construction and operation of facilities, not including more than one demonstration plant, to determine and demonstrate the economic feasibility of producing electric power and energy by means of a wind-driven generator operated in conjunction with an electric power system." Murdock's bill, HR 4286, sought to allocate US$5 million for wind power research.

In September 1951—the same month that the long-running American soap opera *Search for Tomorrow* first aired on American television—Murdock was leading the search for tomorrow's energy source in the halls of Congress. With the support of the DOI and FPC, Chairman Murdock convened a hearing on September 19, 1951 to discuss HR 4286. Murdock had invited written testimony from the Assistant Secretary William Warne of DOI, the Chairman of the FPC, and Percy Thomas. All three came out in strong support of wind power research, and Thomas's 1949 book *The Wind Power Aerogenerator*

Twin-Wheel Type was added to the Congressional record. The committee apparently needed convincing that wind power was a serious subject, so Secretary Warne affirmed the importance of the topic: "I suppose some may think it a little fantastic that we come here and seriously discuss harnessing the winds, but it is quite the reverse for the reason that windmills are as old a power mover as we have had in this country."[16]

Thomas spent the morning describing the plans for the Wind Power Aerogenerator as well as explaining more broadly exactly how wind power worked. The Committee's reaction was mixed. On the one hand, Thomas's ideas were praised, and Thomas was given accolades for his "honesty, humility, and scientific brain, and the patriotic pursuits of the scientific idea."[17] On the other hand, Committee members questioned the need for American wind power research given that research was already being conducted in Europe by Germany and Denmark. Representative Wesley D'ewart asked that if "this experimental work had already been done in other countries, in addition to Grandpa's Knob, and it is only a question of accumulating the data to get the very research material we need in this country. It seems to be that doing that would save quite a bit, perhaps US$5 million."[18] In other words, why spend American money when the country could simply piggyback on European wind power research efforts of the 1950s?

During his testimony, Representative Wayne Aspinall of Colorado predicted that the bill would not pass because of the high cost of wind power research. He was correct. The bill never made it out of the Committee. Representative Murdock lost his reelection bid in 1952 and didn't have an opportunity to reintroduce the bill at a subsequent date. The prospects for federal wind power research in the United State started and ended with HR 4286. Thus, despite the efforts of Percy Thomas and John Murdock, wind power research was dead in America.

Interestingly though, as it turns out, the US government did fund some wind power research in the 1950s—just not in America. The 200-kW Gedser wind turbine built in in Denmark by Johannes Juul in the 1950s was funded through a grant through the Ministry of Public Works. The source of funds for the grant was the US Marshall Plan. Had they known that American funds would sponsor European wind research, this certainly would have ruffled the feathers of the Committee members on the morning of September 19, 1951.

5.3 The British Experiments

As this drama is unfolded in the United States, across the Atlantic, in Great Britain, both the government and electric utilities expressed interest in developing wind power. Wind power research efforts in the postwar era in Great Brittan were driven by the British Electrical and Allied Industries Research Association, more commonly known as the Electrical Research Association

(ERA). ERA was a government-funded research entity founded in 1920. Its purpose was to provide a facility for co-operative electrical research. ERA was diverted during World War II to focus on the development of radar and mine detection equipment, but after the war, a wind power committee was formed within the ERA to conduct wind power research.[19]

ERA subsequently provided funds to the British Electrical Authority (BEA) to develop two wind turbines starting in 1950.[†] Wind power legend E.W. Golding led ERA's wind power research program. Born in Northwich, Cheshire in 1902, Golding was educated at the Manchester College of Technology and worked for two years in the research department of the Metropolitan-Vickers Electrical Company. He was a lecturer in electrical engineering at Nottingham University between 1926 and 1945, and joined the ERA in 1946. A prolific writer, Golding wrote the first authoritative text book on wind energy, *The Generation of Electricity by Wind Power* in 1955.[20] Golding's book was recognized as the most definitive account of wind power research undertaken up until that date. In the Introduction of his book, he explained that Britain's interest in wind power in the postwar years was rooted in the "difficult economic and political conditions of the postwar years tending to make countries depend upon their own resources for the generation of power rather than upon imported fuels."[21] He also noted the contemporary "realization that coal and oil resources are being used up at an increasingly high rate and that they can be put to better use than burning them as fuels."[22]

Fact is, wind power research generated more interest across all of Europe than it did in America in the postwar period, in part because fuel prices stabilized and then dropped more rapidly in the United States than they did in war-torn Europe. The subject of wind power was discussed at international meetings such as the World Power Conference, and several national conferences in Europe; and, in 1950, the Organization for European Economic Cooperation (OEEC)—the predecessor to the Organization for Economic Co-operation and Development (OECD)—established a wind power group to promote cooperative wind power research between European countries.[23] Europe was at the vanguard of wind power research at this time.

The first ERA-funded wind turbine was constructed on Costa Head, a prominent headland on the northern coast of the main island in the Orkney island chain in Scotland. With ERA support, BEA worked with the local utility—the North Scotland Hydroelectric Board—to engage the John Brown Company to design and construct the 100-kW wind turbine, which became known as the Costa Head experimental wind turbine. The John Brown Company was a shipbuilding firm from Glasgow that had built the RMS Queen Mary in 1934.[24]

---

† BEA had been established as the central electricity authority in Great Brittan in 1948 and took over operations of the country's electricity generation, transmission, and distribution assets. (Great Britain, Parliament, House of Commons 1948).

Figure 5.1 Scotland's 100-kW Costa Head wind turbine was installed in Orkney, Scotland in 1953. The wind turbine was situated on a windy cliff overlooking the ocean in an area with strong and turbulent winds. It worked only intermittently before failing within two years.

The Costa Head wind turbine was built in 1953 and it attracted the attention of the highest levels of the UK government. Winston Churchill's Minister of Fuel and Power—Geoffrey Lloyd—publicly praised the project.[25] It used a synchronous AC generator and fed power into the local transmission network. It was a constant speed wind turbine with three blades and a rotor diameter of 15 m. Its rotational speed was regulated using a hydraulic control system that adjusted the pitch of the blades. In several ways, it was similar in design to present-day wind turbines, except for the short blades that resembled airplane propellers. When the construction was complete, it looked like someone had placed the nose cone of a British Spitfire atop a 24-m lattice tower.[26]

The wind turbine was situated on a windy cliff overlooking the ocean in an area with strong and turbulent winds. Given its location next to the salty sea, it quickly ran into "operational problems."[27] It operated only intermittently before failing entirely in 1955. According to Golding—who summarized the project in 1960— "Its design—undertaken with but little information to draw upon as a guide to the building of a machine of this type and size—was found to be unsuitable and the plant was eventually dismantled."[28] The results were disappointing, and the UK government redirected its attention to another—more promising— wind turbine that had also been under development since 1950.

The Andreau-Enfield wind turbine in St. Albans was a bold departure from all of the wind turbines that had been constructed to date. It was designed by French automotive inventor and engineer Jean Édouard Andreau. Andreau was born in Pontacq, France in 1890. He received an engineering degree at the Saint Cyr Military School. He graduated in 1913, just in time to be conscripted into World War I. After the war, his interest in aerodynamics led him to focus on automotive engineering. In 1924, he designed the world's most efficient variable-stroke engine and attracted the attention of André Citroën who hired Andreau to help design his new line of cars. Andreau's most well-known work was the design of the Peugot 402 in 1936. The Peugot 402 was recognized for its dominant tail fin that gave it a futuristic appearance.[29]

In the 1940s, Andreau applied his talents to designing wind turbines. In 1946, he applied for a patent for a wind turbine that was "better adapted to meet the requirements of practice than those used up to the present time." He believed he had designed a wind turbine that was superior in terms of "simplicity of construction and efficiency." The idea was actually very simple. The wind turbine had a hollow tower and blades with open tips. When the blades rotated, the centrifugal flow of air through the blades pulled air up from the base of the tower. Andreau placed a high-speed turbine affixed to an electric generator at the base of the tower. The vertical air flow through the tower powered the turbine and drove the electric generator. Andreau had designed the world's first pneumatic wind turbine.

In a conventional wind turbine, a gearbox is used to convert the slower rotational speed of the hub to the higher rotational speed required to drive an electric generator. In Andreau's design—instead of a gearbox—he used air flow to drive the turbine directly. If the air flowed through the tower efficiently, then the overall efficiency of the Andreau wind turbine would only be limited by the combined efficiency of the turbine and generator. On paper, it appeared that Andreau had designed the world's most efficient wind turbine. Furthermore, given its simplicity, it would likely have a lower construction cost than conventional wind turbine designs.

Andreau worked with France's national electric authority—Electricité de France (EDF)—to build a small prototype in 1947 near Orléans, France.[30] The prototype operated for two years and produced promising results that caught the attention of the British. In 1950, BEA hired Enfield Cables to design and construct a 100-kW wind turbine based on Andreau's design. De Havilland Propellers designed the hollow blades for Enfield. The ERA-funded 100-kW Andreau turbine was erected in St. Albans in 1952.

The Enfield-Andreau wind turbine was a two-bladed, constant speed machine that used a hydraulic motor to control blade pitch to maintain constant rotational speed. However, the wind turbine was initially located in the middle of a forest where wind speeds were weak and uneven. It was later relocated to a more favorable site in Algeria in 1957 for continued testing. The results of the Andreau-Enfield wind turbine tests in 1957 were not positive. The design was characterized by an overall low efficiency of 22 percent due to air flow inefficiency in the inlets, ducts, and outlets.[31]

As it turned out—in practice—air flow inefficiency outweighed the potential gains of the direct drive transmission system. The Andreau-Enfield wind turbine was disassembled in the fall of 1957. It had a brief second life after being acquired by Electricité et Gaz d'Algérie and re-erected at Grand Vent, Algeria. Algerian engineers improved the design and operation of the wind turbine, but the modifications weren't enough to generate any new interest in the Andreau design.[32] A project that started with a promising new design intended to improve wind turbine efficiency and lower costs, had succumbed to the practical realities of constructing an operating a wind turbine in real world conditions.

By 1960, the inability of ERA's two experimental wind turbines to produce positive results after a decade of funding precipitated the UK's withdrawal from wind power research. Despite the early promise of these experiments, it had become clear that the pathway to eventual success would require a series of successive wind turbine experiments that provided incremental improvement; and by 1960, the UK government had exhausted its interest in wind power research.

5.4 The Wind Power Schism

Between 1900 and 1930, wind turbines were small systems with rated capacities less than 100 kW. Small wind turbines were manufactured in Denmark using la Cour's designs. In the United States, small wind turbine designs were improved by several manufacturers through the 1930s, most notably the Jacobs brothers. Starting the 1930s, larger wind turbines emerged. First, with the 100-kW Russian wind turbine in the Crimea in 1931, and then later, with the

1.25-MW Smith-Putnam wind turbine in the United States in 1941. Based on these wind turbines—and his own additional research—Percy Thomas confirmed that large wind turbines interconnected with the AC transmission network were the future direction of wind power. In the twenty years between 1930 and 1950, the rated capacity of wind turbines grew from actual sizes less than 100-kW to proposed sizes of up to 7.5 MW.

Percy Thomas's testimony during the 1951 Congressional hearing was an opportunity to reinforce this growth trajectory and confirm that the future of wind power lied in building increasingly larger machines. However, the manufacturers of small wind turbines were not yet ready to concede their demise. Their chief spokesperson—Marcellus Jacobs—took a keen interest in continuing to advance the cause of small wind even at the expense of big wind power research.

The divide between proponents of small wind and big wind would last through the end of the twentieth century. This wind power schism would become one of the most noticeable and least pleasant aspects of the wind power development in the second half of the twentieth century. Interestingly, as it would later turn out, both sides were right. The future of wind power did indeed lay in the production of MW-sized wind turbines; however, the pathway to MW-sized wind turbines was through the gradual and successive upsizing of small wind turbines throughout the 1980s and 1990s, rather than the spontaneous construction of MW-sized wind turbines as proposed by Thomas.

Against this backdrop, in October 1951, Marcellus Jacobs submitted a letter to Chairman Murdock to state his case. Jacobs argued that large wind turbines like Percy Thomas's were not necessary to help meet the country's electric power needs. All that was needed—according to Jacobs—were hundreds of 5-kW wind turbines mounted directly on utility transmission lines spread out through rural America. These small wind turbines would cost only US$1,000 each—or US$200/kW—so that the country could obtain 5 MW of proven and dependable distributed wind power capacity by investing US$1 million. Clearly then, there was no need to spend US$5 million experiment with unproven large wind turbines like Thomas's.

Jacobs had a point. Small wind power systems had been commercially available for decades and were able to provide the nation with reliable electricity at a known cost. If Congress was looking for solutions to complement hydropower in the Southwest, small wind turbines would have fit the bill. However, Jacob's argument in favor of small wind occurred at the expense of Thomas's big wind turbine research and may have been a factor in the HR 4286's failure to emerge from committee. Whether or not Jacobs's letter had an impact on the committee's deliberations is unknown. It was certainly timed for maximum effect and it left a bad taste in everyone's mouth.

In the end, Percy Thomas's wind turbine plans from the 1940s and 1950s were never pursued. Further wind power progress would require government support, but this support was not forthcoming in the United States. Proponents of wind power could only hope that European countries would continue to take up the baton and advance wind power. As illustrated by the UK's efforts, wind power research activities were certainly heating up across Europe. But would these efforts lead to a sustained push to develop wind power? Researchers in Denmark, France, and Germany were about to find out.

Notes

1 Energy Information Administration, 2012, 215.
2 Ibid., 161.
3 Buck, 1983, 1.
4 Ibid., 8.
5 Powell-Smith, n.d.
6 Energy Information Administration, 2012, 221.
7 Ibid., 258.
8 Smithsonian Institution, 2002.
9 Yeh and Rubin, 2007.
10 U.S. House of Representatives, 1951, 25.
11 Lines, 1973, 15.
12 Thomas, 1946, 1.
13 Ibid.
14 Ibid., 5.
15 U.S. House of Representatives, 1951, 16.
16 Ibid., 12.
17 Ibid., 34.
18 Ibid., 24.
19 Edif Group, 2017.
20 *E. W. Golding Obituary*, 1965.
21 Golding, 1955, 2.
22 Ibid., 3.
23 Ibid., 291.
24 John Brown, 2017.
25 *Electricity by Windmill Generation*, 1953.
26 Nelson, 2009, 7.
27 Ibid.
28 Golding, 1960.
29 Digital Mechanism and Gear Library, n.d.
30 Rapin and Noël, 2010, 238.
31 Bonnefille, 1974, 35.
32 Delafond, 1961.

Bibliography

Bonnefille, R. *Wind Projects of the French Electrical Authority.* Report F40/74, No. 4, Atmospheric Exchange and Pollution. Paris, France: French Electrical Authority, 1974.

Buck, Alice L. *A History of the Atomic Energy Commission.* Government Report, DOE/ES-003/1, 28. Washington, DC: Office of the Executive Secretary, History Division, U.S. Department of Energy, 1983.

Delafond, F. "Solar Energy, Wind Power and Geothermal Energy: Problems Concerning Automatic Connection of An Aerogenerator to a Network." *United Nations Conference on New Sources of Energy*, Rome, United Nations. 390–394.

Digital Mechanism and Gear Library. n.d. "Andreau, Jean-Edouard (1890–1953)." Accessed September 8, 2017. http://www.dmg-lib.org.

"E. W. Golding Obituary." *The Engineer* 6, no. 11 (1965): 1003.

Edif Group. 2017. "Heritage." Accessed August 31, 2017. http://www.edifgroup. com/about/heritage.

Electrical Journal. 1953. "Electricity by Windmill Generation." July 13.

Energy Information Administration. *Annual Energy Review 2011.* DOE/ EIA-0381(2011), Energy Information Administration. Washington, DC: U.S. Department of Energy, 2012. DOE/IEA, 370. https://www.eia.gov/totalenergy/ data/annual/pdf/aer.pdf.

Golding, E.W. *The Generation of Electricity by Wind Power.* London, UK: E. & F. N. Spon, 1955.

———. "Paris Symposium Paper No. 11: Energy from Wind and Local Fuels." United Nations Educational, Scientific, and Cultural Organization: General Symposium on Arid Zone Problems, Paris, 1960.

Great Britain. Parliament. House of Commons. *Electricity Act, 1947.* London, UK: Proquest LLC, 1948.

John Brown. 2017. "Brief History of the John Brown Company." Accessed August 31, 2017. http://www.johnbrown.eu/history.asp.

Lines, Charles W. "Percy Thomas Wind Generator Designs." In *Wind Energy Conversion Systems*, edited by Joseph M. Savino, 5–7. Washington, DC: NSF/ NASA, 1973.

Nelson, Vaughn. 2009. *Wind Energy: Renewable Energy and the Environment.* Boca Raton, FL: CRC Press.

Powell-Smith, Michelle. n.d. "1950s Popular Appliances." Accessed September 13, 2017. http://homeguides.sfgate.com/1950s-popular-appliances-85038.html.

Rapin, Marc, and Jean-Marc Noël. *Énergie Éolienne.* Paris, France: Dunod, 2010.

Smithsonian Institution. 2002. "Post-War "Golden Years"." *Powering The Past: A Look Back.* September. Accessed September 13, 2017. http://americanhistory. si.edu/powering/past/history2.htm.

Thomas, Percy H. *The Wind Power Aerogenerator*. Washington, DC: Federal Power Comission Office of the Chief Engineer, 1946.

U.S. House of Representatives. 1951. "Production of Power by Means of Wind-Driven Generator." *H.R. 4286*. Washinton, DC, September 19.

Yeh, Sonia, and Edward S Rubin. "A Centurial History of Technological Change and Learning Curves for Pulverized Coal-Fired Utility Boilers." *Energy* (Elsevier Ltd.) 32 (2007): 1996–2005.

6

Wind Power's Invisible Solution

Our wind conditions here in Denmark and our experience with building windmills and operating wind-driven electricity generating plants can win us a prominent position in this field and create great opportunities for Danish labor.

—Johannes Juul (1947)

6.1   The Death of Windmills

Poul la Cour would have been aghast! The Danish electric power system did not evolve in the manner that he envisioned. Professor la Cour's government-sponsored wind research at Askov starting in 1891 gave Denmark a jump on wind energy use, and positioned the country as a global leader in wind turbine manufacturing. La Cour's objective was not only to accelerate the use of wind power in Denmark, but also to help ensure that the country didn't travel down a path that led to foreign fossil fuel dependence. La Cour recognized that the lack of domestic fossil fuel resources exposed Denmark to the risk "of paying increasingly high rates for the purchase of fuel from foreign countries...."[1] However, the allure of cheap fuel from abroad was simply too great, and Denmark gradually increased its dependence on foreign fossil fuels as the twentieth century progressed.

Nonetheless, la Cour did plant the seeds that would eventually turn Denmark into a global wind power leader by the end of the twentieth century. In addition to his wind turbine research, la Cour taught wind power engineering courses through the Danish Wind Electricity Society (DVES), which he founded. A young Danish engineer took la Cour's course in 1904. Fifty years later, that young student—Johannes Juul—would become Denmark's leading proponent of wind power and would oversee the development of a pioneering 200 kW wind turbine in the southern town of Gedser.

The Wind Power Story: A Century of Innovation that Reshaped the Global Energy Landscape, First Edition. Brandon N. Owens.
© 2019 by The Institute of Electrical and Electronics Engineers, Inc.
Published 2019 by John Wiley & Sons, Inc.

Along the way, Juul invented a new solution to the problem of wind power's fluctuating frequency. Recall, AC transmission systems require electricity at a fixed frequency (60 Hz in America, 50 Hz in Europe). However, wind power generates electricity at varying frequencies depending upon the rotational speed of the wind rotor, which itself depends upon the speed of the oncoming wind. The Russians and Americans solved this problem using relatively crude mechanical solutions that helped ensure that the wind rotor maintained a constant rotational speed. Johannes Juul took another approach by developing a solution that relied upon blade aerodynamics and was invisible to the eye. The method of regulation that Juul invented became known as "passive stall control." It is a subtle, invisible, control process that keeps the power output from a wind turbine constant without any active regulation mechanism.

Through the end of the World War I, la Cour's research efforts led to advances in wind power in Denmark. A company named Lykkegaard began manufacturing wind turbines based on la Cour's designs. In 1908 (the year of la Cour's death), seventy-two 10- to 20-kW wind turbines—all manufactured by Lykkegaard—were installed across rural Denmark. Ten years later, one hundred and twenty small wind turbines were in operation.[2] During this time, another Danish wind turbine manufacturer—Agricco—was founded by two Danish engineers Poul Vinding and Johannes Larsen. Taking a cue from American researchers Elisha Fales and Harve Stuart—who first applied an airfoil blade to a wind turbine in 1917—the Agricco wind turbines also used airfoil blades. In 1919, Vinding and Larsen succeeded in creating a 30 kW test turbine that produced AC electricity using an induction generator and delivered power directly to the grid.[3] Larsen also proposed the use of a converter to send DC output from a wind turbine to the AC transmission network.[4]

After the Great War, coal and oil prices fell from their wartime peaks. Diesel engines were used increasingly in rural Denmark to provide electricity to farms and coal-fired steam turbines were relied upon to power Copenhagen. Oil remained low during the interwar years, falling in price by 40 percent from 100 to 60 Danish krone per ton (kr/t). Coal prices held steady near 20 kr/t for the entire interwar period.[5] Coal prices were kept low because the waterways from coal producing regions in Poland, Germany, and England were short, which kept the shipping costs from these coal mines to Danish harbors low. In fact, coal prices in Denmark were half of those in countries like Czechoslovakia and Sweden, which were closer to coal producing regions but had to send coal over rail instead of by water.[6]

Low fossil fuel prices translated to low electricity costs and wind power struggled to compete on an economic basis. Wind power's economic situation at the time was later summed up by Russian wind power pioneer Dmitri Stein, who noted that "wind pays off in all cases, as long as diesel oil is not on hand."[7] But wind power faced additional challenges in Denmark at the time beyond pure economics. In particular, wind turbines were not equipped to supply power to new farm machinery. Using wind power with new grinding machinery caused the operation to be "appreciably impaired not only by the delivery of low and irregular power but particularly by fluctuations in rpm."[8]

The Danish power system evolved in a manner that was different from the United States and the rest of Europe where the central station model dominated. Because Denmark remained largely agricultural and had a distributed population, the electrical system never settled on a single approach to power production and delivery. Both central station steam turbines and distributed diesel engines were installed throughout the country. The transmission network also remained a patchwork of both DC and AC transmission lines.[*] By 1937, there were approximately 800 power plants in Denmark and only thirteen of these produced more than 5 million kW per year.[†] The rest were smaller systems scattered throughout the country. These figures paint a picture of a hybridized generation and transmission system that was unique to Denmark.[9]

Rural electrification spread in Denmark starting in the 1920s, just as it had in the United States and other European countries. Danish farmers were increasingly faced with the option of ditching their small wind DC turbines and plugging directly into the AC grid—and this is exactly what many of them did. Small wind turbines were rejected in favor of diesel engines, or—in some cases—both of these distributed generation options were discarded in favor of a connection with the transmission system.[10] The twenty-year period between 1920 and 1940 represented the nadir for wind power in Denmark and became known as the "Death of Windmills."[11] By 1936, coal and oil together accounted for 95 percent of electric power generation. Wind amounted to less than 1 percent of total electricity production. The challenges faced by wind power—low fuel prices, intermittency, and the expansion of the transmission system—were simply too much to overcome.

Denmark's era of cheap fuel ended abruptly with the Nazi occupation starting in April 1940. The occupation also created a shortage of oil and coal. As an occupied country, all available fuel was commandeered to support the German war effort. Further, fuel imports were halted as Denmark lost its largest trading partner—the United Kingdom—which initiated a blockade against Germany and its occupied territories. Although the Danish government had the foresight to stockpile coal and oil on the eve of the war, these stockpiles were not enough to avert widespread fuel shortages. Diesel quickly became unavailable within Denmark. Some quantities of coal were still available, but it was scarce enough that it settled at a price of three times greater than the pre-war level.[12]

---

[*] As of 1943, 79 percent of generators produced AC, 21 percent produced DC. Sixty percent of the power was distributed as AC, 40 percent was DC. See (No author 1943, 107).

[†] For comparison purposes, a single 1.5-MW wind turbine can produce over 5 million kWh of electricity annually.

The country responded to the crisis in several different ways. Energy conservation was first. Danish electricity consumption in 1937 was 1.1 million kWh. In 1940, total consumption dropped by 21 percent to 900,000 kWh. By 1945, electricity consumption was 1 million kWh, still below the 1937 level.[13] Some of these reductions were achieved through greater interconnections within the transmission system and better coordination of existing power plants to improve system efficiencies. The next response was to develop alternative fuels to replace coal in steam turbines and diesel in diesel engines. Peat was the natural replacement for coal. Peat—which is partially decayed vegetation—is Denmark's oldest source of energy and was used over 2,500 years ago as a fuel source during the Iron Age. In the twentieth century, peat was used to mix with coal for combustion in steam turbines. Before World War II, Danish production of peat was less than 500,000 tons per year, but during the war production soared to 6.1 million tons per year.[14]

Peat was also used to create "producer gas," which became a substitute fuel for diesel engines during the war. Producer gas is created with Peat through a gasification process known as "pyrolysis," where organic material is converted into a combustible gas under the influence of heat. Producer gas was used for heating as early as 1840 in Europe, and, by 1884, it had been adapted to fuel engines in England. In Denmark during World War II, diesel engines were run on producer gas. The cost of using producer gas was a decrease in efficiency and a greater difficulty of operation.[15] Ninety-five percent of all of Denmark's mobile farm machinery, tractors, trucks, stationary engines, and fishing and ferry boats were also powered by producer gas during World War II.[16]

However, these efforts were not enough to alleviate Denmark's energy crisis. Small wind power systems provided an additional piece of the puzzle. Lykkegaard—which was still manufacturing wind turbines after two decades—built a series of new wind turbines for use in homes and businesses that normally got their power from diesel engines. The new Lykkegaard wind turbines were similar to the 10- to 20-kW models that they'd been already been manufacturing, except—at 30 kW—they add a greater electrical capacity.

The 30-kW Lykkegaard wind turbines were classic Dutch machines with four wooden blades. They had an 8-m rotor diameter and sat atop a 20-m tower. The drive train was connected to a DC generator that charged a bank of batteries. These were variable speed machines with no need for a control mechanism to maintain a constant rotational speed, but the blades contained flaps that could be extended through a spring mechanism to serve as a brake in high winds. They had a rated wind speed of 11 m/s and a tip-speed ratio of 2.5. These were slowly rotating machines in contrast to the larger Russian and Smith-Putnam turbines that were being built around the same time. Lykkegaard's new wind turbine was an instant success and helped solve Denmark's energy crisis during the war.

The success of Lykkegaard's 30-kW wind turbine and the urgent need for power in Denmark during the war attracted another manufacturer to the Danish wind market. Founded in 1882 by Frederik Læssøe Smidth—who operated out of his mother's apartment in Copenhagen—FLSmidth (FLS) was a global engineering firm with deep expertise in cement and related industries with offices in Copenhagen, Paris, New York, and Berlin. The company was temporarily cut off from international markets during World War I. Having learned from the experience, FLS had the foresight to relocate its executive operations to New York and London in 1938, just prior to the Nazi occupation of Denmark.

The bulk of FLS employees left in Copenhagen rolled up their sleeves and helped solve Denmark's energy crisis by digging up peat.[17] The company had some wind power-related experience; they supplied components to aircraft and—along with their business partner Kramme and Zeuthen—studied Agricco wind turbines as part of their work.[18] Company executives hoped that they could parlay this experience into a new wind turbine manufacturing enterprise.

They focused on building wind turbines greater than 30 kW because they didn't want to compete directly with Lykkegaard, and they were hoping to turn a profit by exploiting an unserved market segment. In August 1940, a prototype wind turbine was tested at FLS's cement factory in Aalborg. The wind turbine was designed by K.G. Zeuthen, an engineer and co-founder of a small airplane factory. The test turbine had a 17.5-m diameter rotor and was connected to a 60-kW generator through a gearbox manufactured by FLS for the cement industry.[19] It sat atop a 24-m lattice steel tower.

The FLS aerogenerator was introduced in late 1940. Whereas, Lykkegaard relied on four wooden blades, the FLS turbines initially used two airfoil-shaped blades just like the Agricco wind turbines that were built after World War I. The FLS aerogenerator had a 17.5-diameter rotor and was connected to a 60-kW generator. The FLS turbine was connected to a DC generator that produced output that charged batteries, flaps in the blades served as brakes in high winds, and a small auxiliary wind turbine was used to yaw the machine.

As expected, the airfoil blades used by FLS resulted in a much greater rotational speed and power output compared to the flat wooden blades used by Lykkegaard. However, FLS's 60-kW wind turbine experienced unexpected increases in the aerodynamic loads in the tower, shaft, and nacelle. To correct this problem, FLS built a second three-bladed 70-kW model. This model had a tip-speed ratio of nine, almost four times greater than the Lykkegaard turbines. The only technical problem experienced by 70-kW turbines was excessive wear and tear of the concrete towers. As a result, some of the concrete towers had to be reinforced.[20]

There were two reasons for the tower problems. First, it turned out that the frequency of the blades matched the natural frequency of the tower, which caused tower vibrations and cracks in the concrete. Second, concrete was simply not the right material for wind turbine towers. Concrete towers provide unlimited flexibility in terms of size and they don't corrode or have stability problems. However, the material is subject to fatigue and cracking. Tubular steel towers would eventually become the standard for wind turbines because steel can absorb a lot more stress before getting bent out of shape. Steel also has the highest degree of manufacturing automation and prefabrication making it cheaper to produce. In the 1980s, Danish wind turbine manufacturer Nordtank—originally a manufacturer of road tankers for oil companies—used its knowledge of rolling and welding steel tank sections to introduce wind turbines with tubular steel towers. Nordtank's towers would be adopted by other wind turbine manufacturers and would eventually displace the steel lattice towers that had become commonplace at that time.

Due to sales of both the Lykkegaard and the FLS wind turbines, wind power installations in Denmark grew from sixteen in July 1940 to sixty-five by 1941. Monthly wind power production rose from 20 to 220,000 kWh during this time.[21] Denmark's response to the energy shortage during the Nazi occupation moved the country back toward that path that la Cour had envisioned forty years earlier. In the process, Danish wind turbine manufacturers leveraged the latest innovations in wind turbine design like the airfoil blade to increase the performance of their designs.

6.2 The Return of Johannes Juul

In the midst of World War II—as he tramped through the mosses of South Sjaelland in the stiff March wind in search of peaty soil as a replacement for coal in the Masnedø Island coal plant—Chief Engineer Johannes Juul knew there was a better way to harvest fuel for electricity generation. He'd been through the same exercise during the Great War, and he believed that if Denmark remained dependent upon foreign coal supplies after this war, then the next time an energy shortage occurred there would be no more moss, and Danish electricity demand would be even higher. In Juul's mind, Denmark's electricity system was lurching from one crisis to another because of its dependence upon fossil fuels. It had to stop.

Juul knew a thing or two about electricity. In 1904—at the age of seventeen—Juul was the youngest participant at la Cour's wind power engineering courses from the Danish Wind Electricity Society (DVES) held at Askov Folk High School. After graduating from la Cour's wind engineering school in 1905, he studied electrical engineering at the Elsinore Technical College. After graduating from Elsinore, he started his own business manufacturing high-voltage electricity equipment. His main customer was South Jutland Electricity Laboratory (SEAS). SEAS was so impressed with his electrical knowledge that they later offered him a full-time position. Juul agreed to work for SEAS under the condition that he could pursue his own research. Juul invented a low-voltage stove that went into production in 1934. In 1940, he joined the Danish Engineering Association and was provided with the title of engineer. By the time the Danish energy crisis struck during World War II, Juul had worked his way up to Chief Engineer at SEAS.

After World War II ended in 1945, Juul watched as the number of wind turbines installed in Denmark began to decline. After peaking at sixty-five in 1941, by 1947, only fifty-seven wind turbines were installed and Denmark was once again being supplied with electricity from coal-fired steam turbines and several hundred smaller diesel engines. These plants—of course—operated on imported foreign coal and oil. Thus, the transitory movement toward wind power during World War II had reversed.

To short-circuit Denmark's return to energy dependence, Juul made an argument for rapid wind power deployment in a 1947 article in the Danish technical journal *Elektroteknikeren*. The Chief Engineer made multiple arguments in favor of wind power: "Windmills in the necessary numbers will ensure Denmark's electricity supply especially in the event of critical times, such as we have lived through twice in one generation. Apart from this, there will be many other good reasons for setting to work on the task. Our wind conditions here in Denmark and our experience with building windmills and operating wind-driven electricity generating plants can win us a prominent position in this field and create great opportunities for Danish labor."[22]

The fact is, Denmark did have strong wind power manufacturing capabilities, a compatible power network, and good wind resources. Truth be told, these reasons alone were enough to provide a case for wind power support. However, Juul went even further by claiming that wind power was the least expensive way to meet Denmark's growing electricity needs. According to Juul "the exploitation of wind power to the greatest possible extent will mean cheaper electricity."[23] By Juul's reckoning, the cost of electricity from a coal-fired steam turbine in 1947 was 0.06 Danish kroner per kilowatt-hour (kr/kWh), or 6 Danish ore per kilowatt-hour (ø/kWh).[‡] He compared this to the cost to wind power, which he estimated to be 23 ø/kWh. Wind had a higher capital cost of 800 kr/kW, the same operations cost of 1 ø/kWh, but no fuel cost and—according to Juul—wind turbines could operate up to 3,000 hours per year. This was a bit high, given 1950s wind technology. Juul concluded that wind power was 3.7 ø/kWh cheaper than coal-fired generation. Based on this analysis, Juul suggested that Denmark immediately begin a program of wind power experimentation and he identified the west coast of Jutland as the most suitable location because of the frequent and high wind speed in that region.

[‡] Since 1 kr = 100 ø.

With the shock of the energy crisis in recent memory, Juul's appeal for wind power drew immediate attention. Not all of it was positive. Objections to the plan were raised in the pages of *Elektroteknikeren*. One respondent pointed out that Juul had underestimated the cost of wind power—which should actually be 12 ø/kWh—making it more costly than electricity generated by coal. He also pointed out that the total investment in wind turbines to achieve an annual production rate of 5 million kWh would be just as great as that required for 360 million kWh of production from coal steam power plants. Another raised an objection to Juul's statement that wind power would not require the full reserves of thermal power plants to account for periods when there is no wind. This argument—not entirely accurate—would ultimately play a major part in the cessation of wind power research in Denmark. If you have to build a coal plant to back up every wind turbine, what's the point of wind power anyway?

It would later be widely acknowledged that wind power doesn't need to be backed up by conventional power plants on a one-to-one basis. This is because wind power provides a fraction of electricity during periods of high electricity demand, thereby avoiding the need for power plants that are constructed specifically to generate electricity during these high demand periods. Further, it turns out that if wind turbines are aggregated and geographically disbursed, the entire collection of wind turbines will provide an even greater level of capacity value than a single wind turbine because some fraction will be operating at any given time. The capacity value currently assigned to wind power by electricity system today planners ranges from 5 to 40 percent.[24] Thus, Juul was optimistic in his response on the subject stating that "where wind power plants are to be found in sufficient numbers, it will be possible to relieve heat power plants of sixty to seventy percent of their production."[25]

Others expressed misgivings about building wind turbines on the west coast of Jutland because it was feared that the wind was too gusty and the air too salty. These concerns were off base. Gusty, the winds of Jutland are, but they are also consistent and powerful. As later evidenced by both wind resource assessments and the construction of wind turbines in the area, the Jutland coast would prove to be a favorable location for both on- and off-shore wind farms.[§] Juul had identified the right location to test wind turbine in Denmark. Jutland would later become the location of Østerild, Denmark's national test center for large turbines.

---

§ In 2009, the offshore wind farm Horns Revs 2 2012 was completed off the western coast of Jutland. Upon completion it was the largest offshore wind farm in the world. In 2015, Jutland became the home of Denmark's largest wind farm, the 70.4 MW Klim wind farm.

6.3   Learning to Stall

Juul concluded that the technical and economic uncertainties provided an even stronger justification for Denmark to embark upon wind power research by building upon the work of la Cour, Lykkegaard, and FLS. Time was of the essence if another energy crisis was to be avoided. It didn't take much convincing within SEAS for the utility to agree to allow him to develop an experimental wind power program. SEAS authorized Juul to construct an experimental wind turbine.

The initial goal of the SEAS program was to develop a small wind turbine in an area with favorable wind speeds. The SEAS team wanted to connect the wind turbine to the local AC transmission network in order to determine exactly how much electricity could be produced, and they wanted to know how to regulate the wind turbine in coordination with the transmission system. In addition, they wanted to test Juul's new blade shapes, which were designed to automatically control the rotational speed of the wind turbine without the need to pitch, or rotate, the blades out of the wind when wind speeds increased beyond the desired maximum.

In February 1950, SEAS built the Vester Egesborg wind turbine on Sjaelland Island under the direction of Juul. It was a two-bladed, upwind wind turbine powered by a 15-kW induction electric generator. Upwind machines have the rotor facing the wind. The basic advantage of upwind designs is that one avoids the wind shade behind the tower. Upwind configurations are the most common design type in the twenty-first century. Downwind configurations were gradually abandoned because they have the disadvantage of creating added stress on the blades as they pass through the shadow of the tower.

The Vester Egesborg wind turbine had an 8-m diameter and sat atop a 12-m concrete tower. It reached full output at a rotational speed of 38 rpm when the wind speed was 7 m/s. At speeds greater than 7 m/s, the blades were shaped in such a way that excessive turbulence slowed the rotational speed back to 38 rpm, not only resulting in a loss of efficiency but also ensuring that the rotational speed remained constant as required by the induction generator and the electrical network. The induction generator has enough slip to absorb variations in rotational speed. The wind turbine also included another innovation—brake flaps on the blade tips that rotated perpendicular to the direction of the oncoming wind. Juul patented these flaps in 1952. The flaps could also be used to slow the rotational speed of the wind turbine and protect against over-speed in high wind conditions.

The grid-connected 15-kW wind turbine provided tangible proof of Juul's design approach. At the time, it would have been easy to overlook the importance of this accomplishment because wind power grid connection had already been achieved as early as 1921 in Denmark, again by the Russians in 1931 in the Crimea, and for a third time by the American Smith-Putnam 1.25-MW machine in 1941. Also—at the same time that Juul was experimenting—the Germans, British, and French were in the process of testing grid-connected wind turbines. The SEAS experimental wind turbine merely confirmed what everyone already knew by this time—wind power could be successfully inter-connected to the AC transmission system with available technologies.

But this self-regulating business was something new altogether. The concept of aerodynamic stall itself was well known, but not in the context of wind power. A stall is a reduction in the lift coefficient generated by airfoil as the angle-of-attack increases. This happens when the angle that produces the max-imum lift is exceeded. Aviators have long been familiar with the perils of stall. Early aviator Otto Lilienthal died while flying a glider in 1896 because of a stall. Wilbur Wright encountered stall while flying his glider in 1901. In fact, awareness of stall motivated the Wright Brothers to design their aircraft to recover from stall more easily. Fast forward to 1950, and Juul used stall to his advantage in the design of wind turbine blades and avoided the need for complex and crude mechanical blade pitch regulation methods like those used by the Russians and the Americans in their large wind turbines.

Beyond blade design and position, the other important part of passive stall is the use of an induction generator. The induction generator naturally limits the wind turbine's rotational speed by providing resistance if the speed falls above or below the synchronous speed. Furthermore, induction generators provide a degree of slip that allows for slight variations in the wind turbine's rotational speed. That's why the Russians used an induction generator in 1931 on the Balaklava wind turbine in the Crimea.

The disadvantage of stall control are loud noises and large blade stresses. These disadvantages are less important for small wind turbines, but for tur-bines greater than around 1 MW, the disadvantages tend to outweigh the advantages of stall control. These disadvantages become evident even with the first experimental machine in 1950, which experienced a blade break on August 2, 1950.

The three-bladed, upwind, passive stall wind turbine design coupled with an induction generator became known as the "Danish Concept" in deference to Juul's combination of these technologies starting with the experimental wind turbine in 1950. Machines built using the Danish Concept dominated the wind turbine market through the 1980s. However, as wind turbine sizes increased to the megawatt scale and variable speed operation emerged as a viable option starting in the mid-1990s, passive stall became less common. Today, large wind turbines either use pure pitch control systems, or they employ "active stall" systems that blend elements of pitch and stall control.

There was a lot of interest in SEAS's wind power experiments. In 1950, the Danske Elvaerkers Forening (Danish Electricity Producers' Organization [DEF]) formed a wind energy committee consisting of representatives from across the electric power industry. The members of the committee included the director of civil engineering at SEAS, the director of civil engineering for the City of Frederikshavn, Mr. Lykkegaard—the wind turbine manufacturer—a representative from FLS, a civil engineer from the Labor and Public Works Ministry, and, of course, Mr. Juul himself. It was a "Who's Who" of Danish wind power. The committee met with the Ministry of Foreign Affairs and decided to establish contact with other corresponding organizations abroad and with international organizations. The goal of the committee was to facilitate the development of a large wind turbine that could be manufactured for both domestic and international use.

The creation of the DEF wind committee occurred while there was growing interest in wind power across Europe. In 1950, the Organization for European Economic Cooperation (OEEC)—which was created in 1948 to run the US-financed Marshall Plan—established a wind power group to promote research and cooperation across Europe. OEEC would become the Organization for Economic Co-operation and Development (OECD) in 1960 after the United States and Canada joined. The group held several meetings in Paris and London to discuss wind power, including a ten-country meeting in London on November 9, 1950 to discuss the economic and technical aspects of wind power and to coordinate research. The United Nations Education, Scientific and Cultural Organization (UNESCO) also displayed an interested in wind power and sponsored some preliminary work. In 1954, UNESCO held a symposium on wind and solar energy in New Delhi. The World Power Conference (WPC) devoted its sectional meeting in Brazil to the subject of wind power in July 1954.

In 1952, the DEF wind committee became aware that SEAS assumed ownership of one of the FLS wind turbines on the island of Bogø. With the approval of the committee, SEAS converted the FLS DC wind turbine to AC, and outfitted it with a three-bladed hub and rotor based on the same design that Juul used on the experimental Vester Egesborg wind turbine. The upgraded 45-kW wind turbine had 13-m blades and sat atop a 22-m concrete tower. The Bogø wind turbine produced an average annual output of 80,000 kW per year, which indicates that the machine operated at an average annual capacity factor of around 20 percent. This was double the output achieved from the previous incarnation of the same wind turbine. The Bogø wind tubine would go on to operate for another ten years with only minor maintenance.[26]

With the Bogø wind turbine up and running, the committee turned its attention to the development of a larger machine, one that would prove Juul's concept at a greater scale. The committee settled upon 200 kW as "the largest size that would be reasonable under the circumstances could be built." Furthermore, "it was decided to have a wind motor with a twenty-four-meter propeller diameter mounted on a twenty-four-meter tower."[27] There was interest in the project, and Juul's experimental wind turbines had reduced the chances of failure, but who was going to pay for the new 200-kW wind turbine?

This was beyond the financial means of SEAS, and besides, if successful, the wind turbine would only benefit the country of Denmark.

At this time in Europe, there was a source of funds that were intended to promote the "...revival of a working economy in the world so as to permit the emergence of political and social conditions in which free institutions can exist." Nearly US$400 million of the US$12.7 billion that the United States supplied to Europe between 1948 and 1951 as part of the Marshall Plan had made its way to Denmark. Marshall funds were available for technical and scientific research through the Danish Ministry of Public Works. The committed applied for—and received—a grant of 300,000 kr for the development of the 200-kW wind turbine. Later, in 1957, the committee received another grant for 225,000 kr for the acquisition of more equipment that caused the cost of the plant to exceed the grant. The total amount of Marshall funds received for the project was 525,000 kr, or approximately US$76,000, or about US$700,000 in today's dollars. The ultimate cost of the project was 552,000 kr, and DEF elected to pay the difference. American efforts "... to assist in the return of normal economic health to the world..." funded the development of one of the most advanced wind turbines constructed in the post-World War II era.[28]

The grant was made in May 1954, and the funds were designated for the support of experiments related to the utilization of wind power. As part of the grant conditions, the Danish Academy for the Technical Sciences (ATV) requested to be represented on the DEF wind power committee. ATV appointed four scientists to the committee to help advice the committee. To facilitate the design and construction of the new wind turbine, a smaller five-person working group—which included Juul—was established. The working group reviewed the design for Juul's existing experimental wind turbines and confirmed the decision to develop a 38-m diameter, 200-kW three-bladed wind turbine with passive stall control and movable blade flaps. They elected the three-bladed design used on the Bogø wind turbine, rather than the smaller two-blade design Vester Egesborg machine to avoid the negative impacts of the strong loads associated with the two-bladed design as experienced at Vester Egesborg.

To determine the right location for the wind turbine, anemometers were installed in three towns: Gedser, Tune, and Torsminde. In 1956, after collecting one year's worth of wind speed measurement, the working group decided to site the wind turbine in Gedser, a town at the southern tip of the Danish island of Falster. Gedser is the southernmost town in Denmark and it is home to some of the most favorable winds in Denmark. Gedser would later become the home of the offshore wind farms Rødsand I and II, built in 2003 and 2010, respectively. Construction of the Gedser wind turbine began in 1956 and proceeded through the summer of 1959. Construction progressed as planned and no major difficulties were reported by the working group.

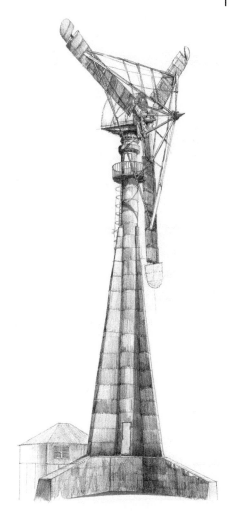

Figure 6.1 The 200-kW Gedser wind turbine started operating in 1959. With its three blades and its passive control system, it was a groundbreaking design that provided the basis for the "Danish Concept," which would become the standard design through the 1980s.

6.4   Gedser Economics

The Gedser wind turbine was put into operation in June 1959. The turbine used Juul's passive stall design to regulate rotational speed and feed constant frequency power into the SEAS electrical system. The blades, the electric installation, the transformer station, and its coupling to the electrical network were manufactured by SEAS. The other wind turbine elements and the nacelle were manufactured by the Aarhus Machine Factory.

By all accounts, the wind turbine operated as expected. For the calendar year 1960, the turbine generated 353,600 kW; and for the calendar year 1961, it generated 339,020 kW. The levels were adequate, but below the expected

400,000 kWh of annual output because of unexpected three-week outages during each year. A comparison between the Gedser and Bogø wind turbines revealed that Gedser had 5.1 times more electrical output than Bogø per year.[29] This was to be expected given the more favorable wind conditions at Gedser, and the larger blade diameter.

By 1962, it had become clear that the Gedser wind turbine had operated as expected. It was up to the committee to evaluate the project and determine if the economics were promising enough to proceed with additional wind power research. Given the wind turbines' technical success, expectations were that research would continue. The committee issued its final report in May 1962. In the report, final construction costs were estimated to have been 320,000 kr. The committee also estimated that the cost could be reduced to 240,000 kr if it were to be mass produced as per the original objective of the committee. Based on actual production figures, the committee estimated that the expected annual electricity production of the wind turbine would be 410,000 kW. That means that the capacity factor was 20 percent, about the same as the earlier Bogø wind turbine. Adding an estimated operations cost of 1 ø/kWh, and amortizing the capital costs over twenty-five years at 6 percent, yielded an estimated production cost of 7 ø/kWh. While certainly not the 2.3 ø/kWh that Juul estimated back in 1947, it was still a reasonable cost for a new source of electricity. Further, with a capacity factor of 20 percent—and first-of-a-kind construction costs—there was plenty of room to improve upon this if research continued.

The problem was that by 1962 the cost of coal had declined so much that the cost of electricity production from coal-fired steam turbines was now too low for wind power to compete. Coal prices had fallen from 100 kr/ton during the war to about 20 kr/ton by 1960.[**] As a result, the cost of electricity from new coal plants dropped from 8 to 4 ø/kWh.[††] Coal prices would have had to increase backup to 75 kr per ton again to make wind power economic. Juul's dire predictions of high coal prices didn't materialize.

The committee decided to terminate the wind research program and recommended against any additional research. This conclusion—of course—was not acceptable to Juul. There is no historical record that details the committee deliberations that led to the creation of the final report. But one can imagine that unpleasantries were exchanged when the committee concluded that "wind power electricity produced by a plant of the Gedser type, under the present

---

[**] (Juul 1947). By 1960 coal prices had fallen back to the level experienced during the inter-war years.

[††] Assumptions: 800 kr/kW capital cost amortized over twenty-five years at 6 percent, 4,000 kcal/kg coal energy content, 2,132 kcal/kW reflecting a 40 percent conversion efficiency and a 1 ø/kWh operations costs. At a fuel cost of 100 kr/t this amounts to a production cost of 8 ø/kW, at a fuel cost of 25 kr/t this amounts to a production cost of 4 ø/kWh.

price structure, is unable to compete with steam power electricity."[30] Juul was unable—and unwilling—to agree with the committee conclusions. He knew that progress had been made and believed that with additional research wind power would be within the range of competitiveness. It would only take the next energy crisis to make the economics work.

The Gedser wind turbine would operate another five years before being decommissioned in 1967. Johannes Juul died in November, 1969 in Haslev, Denmark. Like Poul la Cour before him, Juul had done as much as a single person could do to bring wind power to fruition in Denmark. He advanced the technology to the point where economic competitiveness at the 200-kW scale could be achieved through mass production.

It would take a shift in the global energy landscape for decision makers to understand the value of what Juul had created. Within five years after Juul's death, the American space agency NASA offered to pay to put the Gedser wind turbine back together again. It would be the second check that Uncle Sam would write to put the same wind turbine into operation. The Gedser wind turbine was resurrected in the mid-1970s to provide a new set of lessons for the next generation of wind researchers. The nacelle and rotor of the turbine are now on display at the Electricity Museum in Bjerringbro, Denmark.

In the annuals of wind power history, the Gedser wind turbine and Mr. Johannes Juul loom very large.

Notes

1 Champly, 1933, iii.
2 Johansson, 1974, 1.
3 Pedersen, 2010, 8. Also see No author (1943, 109).
4 No author, 1943, 106.
5 Juul, 1947, 128, figure 1.
6 Ibid., 138.
7 Stein, 1942, 392.
8 Ibid., 358.
9 No author, 1943, 106.
10 Juul, 1947, 139.
11 Stein, 1942, 358.
12 Juul, 1947, 138.
13 Lagendijk, 2008, 55, table 2.3, and 119, table 4.2.
14 Stenkjaer, 2010.
15 No author, 1943, 106.
16 Stenkjaer, 2010.
17 FLSmidth, 2015.
18 Pedersen, 2010, 9.

19 Christensen, 2013, 52.
20 Pedersen, 2010, 9.
21 No author, 1943, 106–7.
22 Juul, 1947, 148.
23 Ibid., 144.
24 See Milligan and Porter (2008) for a review of wind power capacity values in the United States and Europe.
25 Kromann and Juul, 1949, 714.
26 Christensen, 2013, 57.
27 Dahske Elvaerkers Association, 1962, 6.
28 OECD, 2017.
29 Dahske Elvaerkers Association, 1962, 10.
30 Ibid., 20.

Bibliography

Champly, R. *Wind Motors: Theory, Construction, Assembly and Use in Drawing Water and Generating Electricity.* Translated by Leo Kanner Associates Sponsored by NASA. Paris, France: Dunod Publishers, 1933.

Christensen, Benny. "History of Danish Wind Power." In *The Rise of Modern Wind Energy: Wind Power for the World*, edited by Preben Maegaard, Anna Krenz, and Wolfgang Palz, 642. Singapore: Pan Stanford Publishing, 2013.

Dahske Elvaerkers Association. *Report of the Wind Power Committee*, 107. Committee Report. Copenhagen, Denmark: Dahske Elvaerkers Association, 1962.

FLSmidth. *History of FLSmidth.* April 12, 2015. Accessed October 3, 2017. http://www.flsmidth.com/en-US/About+FLSmidth/Our+History/History+of+FLSmidth.

Golding, E. W. *The Generation of Electricity by Wind Power.* London, UK: E.& F. N. Spon, 1955.

Johansson, M. *Exploiting Wind Power for the Production of Electricity*, 54. Technical Report, Danske Elverkers Forening Investigation Fund, Lundtoftevej. Washington, DC: National Aeronautics and Space Administration, 1974.

Juul, Johannes. "The Application of Wind Power to Rational Generation of Electricity." *Electro-Technician* 43 (August 1947): 137–48.

Kromann, C, and J Juul. "Investigation of the Possibilities of Using Wind Power." *Elektroteknikeren* 45 (December 1949): 711–4.

Lagendijk, V.C. *Electrifying Europe: The Power of Europe in the Construction of Electricity Networks.* Eindhoven, the Netherlands : Technische Universiteit Eindhoven, 2008.

Milligan, Michael, and Kevin Porter. *Determining the Capacity Value of Wind: An Updated Survey of Methods and Implementation*, 27. Technical Report. Golden, CO: National Renewable Energy Laboratory, 2008.

No Author. "Utilization of Wind Energy in Denmark." *La Technologie Moderne* 35, no. 13–14 (July 1943): 106–9.

OECD. *The "Marshall Plan" speech at Harvard University*, 5 June 1947. 2017. Accessed October 3, 2017. http://www.oecd.org/general/themarshallplanspeec hatharvarduniversity5june1947.htm.

Pedersen, Jørgen Lindgaard. *Science, Engeering and People with a Mission: Danish Wind Energy in Context 1891–2010*, 1–22. Schumpeter Conference Report. Copenhagen, Denmark: Technical University of Denmark, 2010.

Stein, Dimitry. "Importance and Progress of Wind Power Utilization in Denmark." *Elektrizititswirtschaft* 41, no. 17 (September 1942): 390–2.

Stenkjaer, Nicolaj. March 2010. "Peak." *Nordic Folkecenter for Renewable Energy*. Accessed October 3, 2017. http://www.folkecenter.net/gb/rd/biogas/biomass-energy-crops/peat/.

7

The French Connection

There is no doubt that considerable amounts of energy can be obtained from wind power.
— Pierre Ailleret, Vice-General-Director of EDF (1946)

7.1 The Duke's Invention

Charles-Marie-Michel de Goyon—the third Duke of Feltre—wouldn't be denied. Like contemporaries Charles F. Brush in America and Poul la Cour in Denmark, he was certain that a windmill could be coupled with an electric generator to produce electricity. One of France's most distinguished diplomats and politicians, the Duke had gained notoriety for his escape as a prisoner of war during the Franco-German War of 1870. Now ensconced near Normandy, the retired politician clearly understood the importance of being inventive to solve tough challenges.

De Goyon was most certainly aware of the power plant's delivering electricity to customers in America. The world's first hydropower plant—Schoellkopf Power Station at Niagara Falls—opened in 1881, and, Thomas Edison's Pearl Street Station in New York City, opened in 1882. Electricity offered French homes and farms in Normandy the possibility of electric light and all the accompanying benefits. Normandy's nineteenth century tourist beach resorts would also benefit from an influx of electric power.

So—in 1887—de Goyon attached a Halliday windmill built by the US Wind Engine and Pump Company of Batavia, Illinois, to two electric generators and used the resultant power to charge batteries that he would distribute to French homes. The Duke's wind power experiment took place and at La Hève Cape near Le Havre, the largest harbor in Normandy. The "La Hève System" was later manufactured and sold by several companies in the United States and France until World War I, including US Fritchle, Lewis, Chêne, and Beaume.[1]

The Wind Power Story: A Century of Innovation that Reshaped the Global Energy Landscape,
First Edition. Brandon N. Owens.
© 2019 by The Institute of Electrical and Electronics Engineers, Inc.
Published 2019 by John Wiley & Sons, Inc.

However, the Duke's innovation suffered the same fate as all other global wind power projects after World War I. Low diesel prices after the wartime surge made diesel generators the most economical solution in areas with transmission and distribution lines. In the interwar period, Louis Constantin of the French Government's Aeronautics Ministry took up the baton of wind power in France. Like Stuart and Fales in the United States, Constantin applied his knowledge of aviation to wind turbines by attaching an airplane propeller to a test wind turbine. Constantin experimented with small test wind turbines between 1923 and 1926.[2]

Concurrent with Constantin's efforts, the chief engineer of the Compagnie Électro-Mécanique—George Darrieus—conducted research to better understand wind power starting in 1925. Darrieus's work was documented in 1933, when the Paris publishing house Dunod released a comprehensive wind power history and textbook.[3] The book provided a description of existing wind turbines and discussed the advantages of different wind turbine configurations.

Darrieus envisioned the development of a vertical-axis wind turbine (VAWT) and received a French patent for a VAWT design in 1929, which now bears his name. Unlike the more common horizontal-axis wind turbines (HAWTs)—which have a horizontally rotating shaft—VAWTs use two or more curved blades mounted at the top and the bottom of a vertically rotating shaft. The wind blows over the contours of the blades and creates an aerodynamic lift, which pushes the blades around the vertical shaft. The main advantage of VAWTs is the placement of the drivetrain and control system at the base of the tower. This improves accessibility and reduces maintenance costs. Another advantage is that there is no need for a yaw mechanism to turn the rotor into the wind because VAWTs operate independent of the direction of the wind.

Another VAWT design was also proposed by Sigurd Johannes Savonius—a Finnish architect and inventor—around the same time that Darrieus conducted his work. Savonius invented what became known as the Savonius wind turbine and received a Finnish patent for it in 1926. To develop his wind turbine, Savonius worked with German engineer Anton Flettner, who had built a ship that was propelled by two large cylindrical rotors. Savonius first described the details of his research in his 1926 book *The Wind-Rotor in Theory and Practice*.[4] Darrieus's VAWT design uses lift forces to create rotation, whereas Savonius's VAWT uses drag forces to create rotation. Although VAWTs—in general—are less efficient at converting wind energy to electricity than their horizontal-axis brethren, Darrieus's VAWT design has been found to be more efficient than Savonius's design. Much later, both Canada and the United Kingdom launched VAWT research programs starting in the 1970s. Canada focused on the curved-blade Darrieus VAWT, and ultimately installed a 230-kW VAWT on an island in the Gulf of St. Lawrence. The United Kingdom developed the Musgrove VAWT, which was tested at Carmarthen Bay, South Wales as part of a national wind power test program.

Back in the 1920s, while Darrieus dreamed of building vertically oriented wind turbines, he actually went on to build the horizontally oriented ones that dominate the twenty-first century landscape. Starting in 1927—under his guidance—the Compagnie Électro-Mécanique (EMC) built three wind turbine prototypes with capacities ranging from 1.5 to 12 kW.[5] EMC finished construction of the wind turbines in 1931. Later—in 1935—Darrieus attempted to build a 50-kW wind turbine in North Africa, but the company was unable to complete any of the installations because of the high cost of construction.[6]

France's involvement in World War II put an end to Darrieus's efforts. However, during the war, a small group began building wind turbines in France to make up for electricity shortages. The group—originally led by Lucien Pabion—developed small 700–800 W machines in Tunis in 1941. Pabion was later captured and sent to a concentration camp in Germany. Pierre Gane picked up the cause and re-started production in 1943. Gane survived the war and started a wind manufacturing company called ENAG that eventually built hundreds of small wind turbines in the range of 1–9 kW. ENAG's wind turbines were used in small remote sites until the mid-1980s.[7]

World War II left France and the rest of Europe economically devastated. In addition to the destruction of its critical infrastructure—such as the railroad network—France also faced ongoing coal shortages caused by a reduction in coal production in key producing regions. At this time coal still dominated the energy systems of all major European economies. France's coal shortages occurred against the backdrop of postwar reconstruction, which heighted the need for enregy alternatives.[8]

Fossil fuel shortages both during and after the war meant that wind power—and the desire for energy independence—never strayed far from the French consciousness. In general, coal had its drawbacks that made it a potentially unreliable energy source. It required skilled labor to produce and miners can be displaced as they were during the war. In addition—because it was a solid fuel—supplies could easily be commandeered or disrupted.[9] France's experiences during the war planted the seeds for a movement away from coal in the postwar years.

The Vichy Regime established the Committee for Organizing Electric Energy to address the coal and oil shortages. French engineer, researcher, and teacher Pierre Ailleret was selected to lead the committee. Ailleret was an engineer who had been educated at the prestigious French universities, National School for Roads and Bridges (Polytechnique, Ponts et Chaussées), and the French Graduate School of Engineering (Supélec). Ailleret established a technical committee for studying wind energy and set about studying the potential of wind power. The results of this research were published in a 1946 article in which Ailleret concluded that that "there is no doubt that considerable amounts of energy can be obtained from wind power: the value of this power can be as high as a thousand kilowatt-hours per year per square meter of surface exposed to wind."[10]

Broader responsibility for diversifying France's energy sources after the war rested with France's state-owned utility, Electricité de France (EDF). EDF was formed in 1946 after the French government decided to nationalize the production and distribution of electricity. This was part of a wave of nationalizations of key industries in France and elsewhere in Europe following the end of World War II. The main reason for the government's decision to consolidate the electricity industry into a single nationalized utility was its determination to speed up industrialization and urbanization after World War II.[11]

EDF developed a plan to diversify away from coal and achieve greater energy independence. It involved two distinct strategies. The first strategy was to meet growing electricity demand by developing new indigenous energy resources. EDF moved quickly to develop indigenous hydroelectric power. Seven new hydroelectric installations were built in 1949, ten more in 1950, eight in 1951, and another eight in 1952. By 1957, a further fifteen hydroelectric facilities were brought into service. EDF's move into hydroelectricity was successful. By 1960, hydropower accounted for 71.5 percent of EDF's total electricity production.[12]

The second strategy was to explore the development of energy alternatives. EDF's attention quickly turned to the prospects for wind power. A 1949 technical review published in the French technical journal *La Technique Moderne* summarized the prevailing mood: "Installations producing power on the order of one thousand kilowatts have been projected or constructed in other countries. The current fuel shortage has only increased the already existing interest in such installations in France, and various official or private initiatives now offer the prospect that the use of wind energy may soon enter a more active phase."[13]

Citing the projects in the Crimea and Grandpa's Knob, the French believed that the fundamental technical questions surrounding wind power had been resolved. Wind power was already a viable electricity generation technology that could be integrated into utility transmission networks. The primary issue in the French view was one of economics. Could wind power be produced at a cost that was competitive with conventional generating options?

The answer came from the pages of *La Technique Moderne*. In a 1949 article, French engineer G. Lacroix advanced the idea that "From a technical standpoint, therefore, there is nothing to prevent the immediate construction of high power wind engines, at least to a maximum of 1,000 kW. The only problems which could be encountered are economic in nature."[14] Lacroix recognized that "the construction of wind powered electrical plants will be possible only if the cost per kilowatt-hour produced is comparable to that for hydroelectric or thermal plants."[15]

Lacroix's examination of the cost of the Smith-Putnam turbine led him to believe that wind power was not an economically competitive power generation option. His proposed solution was to conduct more wind power research to drive the costs down. Lacroix believed that additional research was needed to make wind power more reliable and less costly so that it could compete with conventional options. In Lacroix's words "the problem is no longer to invent wind engines, but to build and refine them—and also, of course, to finance them."[16]

If Grandpa's Knob was too costly, and Percy Thomas was unable to secure funding for additional wind power research in the United States in the postwar era, then France would have to build its own wind turbine. The charge would be led by Ailleret, who was appointed director of Studies and Research at EDF in 1946. Ailleret was later appointed vice-general-director of EDF, a role that he would hold until 1967.[17] One of Ailleret's first acts in his new role was to create a Wind Energy Department within EDF. The department had two goals: first, develop an estimate of France's wind power resources; second, initiate the development of wind turbine prototypes.

To accomplish the first task, Ailleret and his team set about to develop a new way to measure wind energy. Progress had been made in wind measurement instruments since the days when Putnam Palmer walked the Green Mountains in Vermont looking at tree deformations to find sites with high wind speeds. Several different instruments were available for wind speed measurement by this time: the Sheppard Anemometer based on the classic Dines pressure-tube anemometer but with small conical cups instead of tubes; cup generator anemometers that could measure instantaneous wind speeds measured in miles per hour but were slow to respond to changing wind speeds; and windmill-type anemometers, which were basically small wind turbines connected to DC generators.

But Ailleret and his team were looking for a simpler solution. They wanted an instrument that was easy to use and maintain. They were interested in identifying sites with the greatest wind potential—they weren't as concerned about the hourly or even daily pattern of the wind, so they didn't need an instrument that recorded the instantaneous wind speed, they just needed something that could tell them which sites has the greatest total wind power potential over a specified period of time.

They eventually settled upon a device developed by Montrouge Meter Company that would later became known as the "Ailleret Anemometer." The Ailleret Anemometer is a cup anemometer connected to a small generator and an energy meter. Through the meter, the anemometer can keep a running count of the amount of electricity generated at a particular site. By placing these anemometers as prospective wind sites, the Ailleret team could determine the most favorable sites by comparing the total metered electricity over a period. The drawback to this approach was that there was no record of exactly when the wind blows and its speed.

Starting in October 1946, 350 of these devices were produced and deployed in France and overseas.[18] They were placed on existing structures such as lighthouses at heights above the ground varying from 6 to 300 m. One was even placed atop the Eiffel Tower. This instrument was used as the basis of comparison for all the rest.[19] The highest reading was recorded at the top of a hill northwest of the city of Perpignan in Southern France. EDF's wind resource assessment program was then used to construct the first comprehensive map of France's wind resources. The assessment found that France's best wind resources were located near the coast.

To accomplish EDF's second wind power development goal—building prototypes—they established research alliances. Ultimately, EDF funded the development and testing of several large horizontal-axis wind turbines between 1948 and 1966. The work was carried was carried out by EDF in collaboration with its research partners: the Bureau des études scientifiques et techniques (BEST) and Neyrpic Grenoble. Neyrpic was a pioneer in hydroelectric power that was founded in the 1860s. The company was later sold to Alstom in 1967. BEST was an upstart engineering consulting company founded by Lucien Romani in 1946. EDF's wind power partnership with BEST would prove to be the most fruitful. In fact, BEST's Romani—along with Ailleret—would become the two driving forces behind France's wind power development efforts in the 1950s and 1960s.

7.2 The Seeker

Lucien Romani was born in 1909. As a child—during the Great War—he was sent by his parents to the countryside to study at a village school where he earned his primary studies certificate. This was both the beginning and the end of his formal education. Between the two world wars he spent his time making a living as an architect, an amateur chess player, and studying whichever topics interested him. He was interested in Astronomy and reached out to French astronomer Henry Minor. Minor and Romani hit it off and Minor eventually recommended Romani for a job working for Professor Joseph Peres at the Laboratory of Electrical Analogies of the Sorbonne. This led to Romani's wartime occupation as a mathematician working in the basement of the Ecole Normale Supérieure under the guidance of the French mathematician Henri Villat.

Romani excelled at studies related to wind power and mastered aerodynamics. In 1946, he founded BEST, which was an aerodynamics engineering consulting firm. EDF funded BEST to develop experimental wind turbines in the mold of the Smith-Putnam turbine in the United States. To accomplish this task, Romani reached out to the Institut Aerotechnique (IAT). Founded in 1910 with a grant from the Université de Paris, IAT was an aeronautical research institute that housed some of the largest wind tunnels in the world at the time. In the mid-1940s, IAT was primarily engaged in research related to military aircraft.[20] Romani arranged to use the IAT facilities to test the performance of small wind turbines.

Between 1950 and 1955, the BEST research team tested several small wind turbine prototypes. The largest was an 8-kW generator that was constructed by Aérowatt Company. These prototypes were a prelude to Romani's larger vision of constructing a full-scale wind turbine that could feed power into the transmission network. It was the same vision held by Palmer Putnam before him. Putnam was able to develop a megawatt-scale wind turbine and connect it to the local transmission system. Putnam's wind turbine was ultimately derailed by miscalculations made during the fabrication phase of the Smith-Putnam wind turbine project. Romani had a few tricks up his sleeve that would help his wind turbines avoid the same fate as the Smith-Putnam turbine.

The testing of the 40-kW wind turbine provided enough favorable results to encourage EDF and BEST to pursue the development of a much larger 800-kW wind turbine. It would have four times the capacity of Denmark's Gedser wind turbine, which was under development at the same time. Construction began in 1956 and the wind turbine was completed in March 1958. The 800-kW wind turbine was located near Nogent Le Roi, a small country village 120 km southwest of Paris, in a wide flat plain. The site was selected not for its high wind speeds—but for the steady and non-turbulent quality of winds in the area. Nogent-le-Roi also provided a good interconnection point for France's transmission network.

The wind turbine had three 30.2-m long fixed-pitch blades constructed out of light alloy. It sat 32 m above the ground and weighed 160 tons. The wind turbine connected to an 800 kW, 3-kV synchronous AC generator. The electrical output was stepped up to 15 kV using a transformer and the power was sent to France's 50 Hz electrical network through a 12 km transmission line.

The Romani wind turbine was a demonstration of France's ability to use state-of-the art materials and techniques to build and operate a large wind turbine. However, the Romani wind turbine did have one interesting feature—or rather it lacked a specific capability—which made it unique. Recall, one of the fundamental challenge of generating wind power and feeding into the transmission system is to provide electrical output at a fixed frequency. This requires constant rotational speed of the wind turbine rotor; however, given the variable nature of the wind, there must be some mechanism to keep the blade rotation constant in fluctuating winds.

In 1931, the Russians used flaps on the blades of the 100-kW Crimean wind turbine to maintain constant rotational speed, albeit rather crudely. A decade later in America, Putnam Palmer and his team used a more sophisticated variable blade pitch system to accomplish the same task with a bit more precision. Both solutions were mechanical in nature. In Denmark, in the 1950s Johannes Juul pioneered another method known as "passive stall" that would rely on the shape of the wind turbine blades to achieve the same effect. Today, power electronics are used to accomplish the same task, either by using converters that transform variable frequency AC into DC and then back into fixed-frequency

Figure 7.1 The 800-kW Nogent-le-Roi wind turbine. The wind turbine was located near Nogent Le Roi, a small village 120 km southwest of Paris. It began operation in March 1958, and it represented France's first significant contribution to the development of wind power.

AC that is suitable for the grid. This was also the method that was proposed by Germany's Herman Honnef in the 1930s and America's Percy Thomas in the early 1950s.

However, if the synchronous generator is connected to the transmission system and other generators are active on the system, then the grid can have enough inertia to force the wind turbine to rotate at the frequency of the electrical system. Thus, simply by connecting a variable wind turbine with a synchronous generator to a strong electrical grid—assuming it is small enough relative to the size of the electrical system—the generator will transform itself into a constant speed machine. The BEST team recognized this and built the Romani wind turbine as a variable speed wind turbine that would maintain constant rotational speed through its synchronous connection with the electrical grid. This was innovation by omission of an active mechanical control system that regulated the speed of the wind turbine. However, it only works when the wind turbine is connected to a large and strong transmission network with a lot of inertia from other rotating machines. Otherwise, wind turbine faults can put the entire transmission system at risk.

There are, of course, problems with this approach. Sometimes, the strength of the wind is such that overspeed cannot be avoided and the clash between the wind turbine speed and the fixed frequency requirements of the electrical system would strain and ultimately damage the generator. The wind turbine operated sporadically from March 15, 1958 through September 1963. During this time, the machine experienced numerous mechanical problems. The wind turbine frequency experienced overspeed conditions, including an evening in October 1959 when the generator output reached 1,025 kW. This was not by design, and did not bode well for the generator, which had to be reassembled between November 1958 and April 1959.[21]

In April 1962, BEST decided to replace the rigid blades of the wind turbine with a new flexible set that was constructed of light alloy to increase the rotational speed from 47.3 to 71 rpm and reduce the speed at which the turbine started generating power. The blades were successfully replaced; however, blade flutter soon became a problem. Blade flutter is always a concern in dynamic systems moving through a fluid such as wind turbines. In a wind turbine, flutter is dynamic instability of the blade caused by positive feedback between the blade and the wind. Wind turbine blades have natural frequencies of vibration and flutter occurs when the blades rotate at that natural frequency. Therefore, wind turbine blades must be designed to avoid flutter. Today, flutter can be detected in wind turbine blade designs using computer simulations.

The blade broke off from the main hub and flew off into a nearby field on April 12, 1962. The same thing happened to the Smith-Putnam wind turbine. Indeed, this whole dangerous business of blades breaking apart and flying off wind turbines would become a reoccurring event throughout the development of wind power across the twentieth century—one of the hazards of being at the frontier of wind power development.

Although the damage was primarily limited to the blade, this incident effectively ended the Romani wind turbine project because EDF pulled the plug after the blade failure. In the end, the wind turbine operated 5,428 hours and generated 200,540 kWh of electricity.[22] Total generation was not as high as expected given the duration of the experiment from 1958 to 1962 due to ongoing mechanical problems, and also because the wind turbine's high cut-in speed of 7.2 m/s, which only occurred 5 percent of the time at the Nogent-le-Roi site.[23] The irony is that the wind turbine was operating successfully before the blade replacement. However, in an effort to reduce the cut-in speed and achieve higher generation levels, the team installed flexible blades that ultimately doomed the project.

Romani dissolved BEST in 1966 after the death of his brother. Aérowatt Company—who built the first 40-kW test wind turbine for BEST—acquired the wind power expertise, staff, and patents and continued to carry the wind power development torch for France.[24] Aérowatt begin manufacturing small wind turbines with capacities less than 100 kW that contained variable-pitch blades that maintained constant rotational speed using springs with a centrifugal governor.[25] Aérowatt maintains its focus on manufacturing small wind turbines today. The company was merged with JMB Énergie in 2013 to create Quadran. Quadran has since merged with the Direct Energie group to become the largest alternative energy supplier in France.

7.3 The Neyrpic Wind Turbines

During this same period, EDF's partnership with Neyrpic ultimately led to the development of a 1-MW wind turbine. Neyrpic had been experimenting with alternative energy technologies since the war ended and developed a 132-kW wind turbine that underwent testing in Toulouse and was installed in Saint-Remy-des-Landes near the English Channel in 1958. The Neyrpic wind turbine had a blade length of 21.2 m, a rotational speed of 56 rpm at a rated wind speed of 12.5 m/s. It was a variable-speed three-bladed wind turbine. The blades were constructed of welded aluminum with a plastic coating. As with the Romani wind turbine, mechanical challenges arose quickly. Full testing began in earnest early 1959. By June 1959, one of the blades had broken off the machine. The machine was restarted and ultimately ran through 1966, but electricity generation was sparse given frequent stoppages to repair mechanical problems.

Still, the 131 kW wind turbine provided enough encouragement to the Neyrpic team to enable them to move forward with a much larger wind turbine. With the approval of EDF, Neypric built a 1-MW wind turbine in the same location as the smaller prototype. The wind turbine was a three-bladed downwind machine with variable pitch blades that maintained constant rotation starting at the rated wind speed of 17 m/s. The 1-MW, 2-kV induction generator was mounted 30 m above the ground.[26] It weighed 95 tons.

The 1-MW wind turbine was connected to the electrical network in June 1963 and operated for one month before an electrical failure took it offline. Tests were resumed in October 1963 and the wind turbine ran until June 1964, at which time the bearings in the step-up gear broke. The breakage occurred after just 2,000 hours of operation rather than the 200,000 hours predicted. Because of this failure, the machine was left out of operation and eventually disassembled in 1966.[27] Ultimately the 800 -kW Romani wind turbine proved to be a more successful prototype than the 1-MW Neyrpic machine.

When considering EDF's full wind power development program of the 1950s and 1960s, it is also worth noting that EDF paid homage to Chief Engineer Darrieus's VAWT designs by building a 7-kW experimental VAWT in the Fressinades at Mondragon along the Donzère-Mondragon canal. It was in operation from August through October 1954 when it was destroyed by a failure in its braking system. This experimental wind turbine was found to be inefficient and unreliable.[28]

In fairness to the BEST and Neyrpic efforts, the reasons that wind turbines were abandoned in the 1960s had more to do with changes a foot in the global energy landscape than the mechanical failures of the turbines themselves. The failures just provided a convenient reason to end the projects at that time. By 1960, the world was awash in oil and the global glut of oil kept prices below US$3/bbl throughout the 1960s. Oil-fired generation had thus become a preferred option for electricity production during this time. In addition, France's nuclear ambitions were ramping up in the 1960s as well and EDF opened France's first nuclear power station in 1962.

Thus, France's flirtation with wind power in the 1950s and 1960s ended after EDF's support for the BEST and Neyrpic projects stopped. France's connection with wind power has changed over the intervening fifty years. Denmark, Germany, and Spain led the European wind power wave in the 1990s, while France took a backseat. France later entered the wind turbine manufacturing business in 2007 when Spanish wind turbine manufacturer Ecotècnia was purchased by French engineering company Alstom. Alstom's purchase by GE in 2014 led to the relocation of GE Renewable Energy's headquarters to Paris.

More recently, strong policy supports have pushed the country's installed base of wind power to nearly fourteen gigawatts, good for fourth place in Europe.[29] By 2023, the French Wind Energy Association expects the country's installed wind power capacity to approach thirty gigawatt. France is just beginning to regain its status as one of the forerunners of wind power innovation like it was in the 1950s. But it's hard not to wonder what might have been if France had maintained its interest in wind power development and continued to build upon the knowledge gained from the BEST and Neyrpic projects starting in the 1960s.

Notes

1 Jediczka, 2014, 377.
2 Rapin and Noël, 2010, 12–3.
3 Champly, 1933.
4 Savonius, 1926.
5 Jediczka, 2014, 355.
6 Lacroix, *Wind Power Part II*, 1949b, 109.
7 Jediczka, 2014, 377.
8 Milward, 1984, 135.
9 Risser, 2012, 245.
10 Ailleret, 1946, 103.
11 EDF, 2017.
12 Ibid.

13 Lacroix, *Wind Energy Part I*, 1949a, 77.
14 Lacroix, *Wind Power Part II*, 1949b, 6.
15 Ibid., 10.
16 Ibid., 20.
17 Merlin, 1996.
18 Bonnefille, 1974, 1.
19 Golding, 1955, 71.
20 Institute Aerotechnique (IAT), 2017.
21 Bonnefille, 1974, 16.
22 Ibid., 23.
23 Noel, 1973, 194.
24 Ibid., 187.
25 Ibid., 191.
26 Bonnefille, 1974, 34.
27 Ibid., 35.
28 Ibid., 40.
29 GWEC, 2018.

Bibliography

Ailleret, Pierre. "Wind Energy: Its Value and the Choice of Site for Exploitation." *Revue gendrale de l'lectricite* 55 (1946): 103–8.

Bonnefille, R. *Wind Projects of the French Electrical Authority*. Report F40/74, No. 4, Atmospheric Exchange and Pollution. Paris: French Electrical Authority.

Champly, R. 1933. *Wind Motors: Theory, Construction, Assembly and Use in Drawing Water and Generating Electricity*. Translated by Leo Kanner Associates Sponsored by NASA. Paris: Dunod Publishers.

EDF. 2017. *EDF History*. Accessed November 9, 2017. https://www.edf.fr/en/the-edf-group/who-we-are/history.

Golding, E. W. 1955. *The Generation of Electricity by Wind Power*. London: E.& F. N. Spon Ltd.

GWEC. 2018. *Global Wind Statistics 2017*. Brussels, Belgium: Global Wind Energy Council.

Institute Aerotechnique (IAT). 2017. *The history of IAT*. Accessed November 9, 2017. http://www.iat.cnam.fr/institut/histoire/histoire.htm.

Jediczka, Marc. 2014. *History, State-of-the-Art, and Future of Wind Energy in France*. In *Wind Power for the World: International Reviews and Development*, edited by Preben Maegaard, Anna Krenz, and Wolfgang Pals, 3, 373–91. Singapore: Pan Stanford Publishing.

Lacroix, G. "Wind Energy Part I." *La Technique Moderne* 41, no. 5–6 (1949a): 77–83.

———. "Wind Power Part II." *La Technique Moderne* 41, no. 7–8 (1949b): 105–10.

Merlin, A. 1996. "In Memory of Pierre Ailleret." *IEEE Power Engineering Review* 16 (December 1996): 31.

Milward, Alan S. *The Reconstruction of Western Europe, 1945-51*. London: Methuen, 1984.

Noel, John M. "French Wind Generator Systems." In *Wind Energy Covnersion Systems*, edited by Joseph M. Savino, 5–7. Washington, DC: NSF/NASA, 1973.

Organization of the Petroleum Exporting Countries (OPEC). 2014. "Oil and Gas Data." *Annual Statistical Bulletin 2014*. Accessed November 9, 2017.

Rapin, Marc, and Jean-Marc Noël. *Énergie Éolienne*. Paris, France: Dunod, 2010.

Risser, Nicole Dombrowski. *France Under Fire: German Invasion, Civilian Flight and Family Survival During World War II*. Cambridge, UK: Cambridge Unversity Press, 2012.

Savonius, Sigurd J. *The Wind-Rotor in Theory and Practice*. Helsinford, Finland: A/B Nordblad & Pettersson O/Y, 1926.

8

Germany's Timeless Beauty

These (wind turbines) must for that reason in a deeper sense be of a timeless beauty, so that they do not in three or four decades hence burden a later generation with the heavy task of removing angular skeletons....
—Ulrich Hütter (1942)

8.1 Wind Power in Wartime

Unlike the wartime interruptions in the United States and Russia, for the most part wind power development continued unabated in Germany during the initial years of the war. German wind power development efforts advanced for both small and large wind turbines. This left Germany better positioned to continue to explore wind power in the postwar period.

In Germany—in the 1930s—Herman Honnef developed plans to build high-capacity, gearless wind turbines atop towers up to 274 m tall. These plans generated detractors, who believed that coal prices were too low for any alternative energy sources to make sense; meanwhile, Honnef's supporters marveled at Honnef's revolutionary designs and the potential for wind power to solve Germany's energy challenges.

To validate his designs, Honnef oversaw tests on small-scale systems that were conducted in the wind tunnel of the Aerodynamic Research Institute at Göttingen and the Aerodynamic Research Institute in Berlin-Adlershof in the late 1930s. Based on these tests, a group of German scientists reported in 1939 that "this is an extraordinarily worthwhile project which not only shows the Honnef design to be valuable under specific technical feasibility conditions, but moreover has shown that Honnef, in his decades of preliminary work, had provided the basis for solving the wind power question in general, which must be sustained and extended at all costs."[1]

The Wind Power Story: A Century of Innovation that Reshaped the Global Energy Landscape, First Edition. Brandon N. Owens.
© 2019 by The Institute of Electrical and Electronics Engineers, Inc.
Published 2019 by John Wiley & Sons, Inc.

In 1940, Honnef built a testing ground on Mathias Hill in Botzow-Veblen near Berlin. Between 1940 and 1945, Honnef tested thirteen different small-scale designs with capacities up to 20 kW, tower heights up to 37 m, and blade diameters up to 9 m.[2] During this period, he discovered that the counter-rotating wind turbine ran as expected without incident, even during the harsh winter of 1941–1942.

Honnef's successful tests eventually caught the attention of Allgemeine Elektricitäts-Gesellschaft Aktiengesellschaft (AEG), the German electric company. AEG originated in 1882, when its founder Emil Rathenau acquired licenses to use some of Thomas Edison's lamp patents. AEG grew up along with twentieth century electrification and—by the 1920s—had become a global supplier of electricity equipment. Based on the successful test results at Mathias Hill, AEG agreed to build a 1-MW counter-rotating wind turbine with a three-phase synchronous AC generator and a blade diameter of 50 m. Construction of the turbine was complete in 1944. The turbine just needed to be delivered to and erected at a suitable site. With this project—finally—Honnef was poised to the change in the history of wind power.

The wind turbine was delivered to Rottland, a village in the town of Waldbröl about 50 km east of Cologne. AEG worked through 1943 to prepare the Rottland site for the Honnef wind turbine. Anemometers had been taking wind measurements at the site for several years and the site also had an overhead transmission line supplied by RWE, the German electric utility. Construction began in late 1944, but the project was never completed. According to P. Juchem—writing in 1955—after the buildings and the tower were constructed "further construction had to be suspended and the generator discounted and removed due to air-raids of increasing intensity."[3] Another source indicates that the site was purposely burned in 1945 as American soldiers were passing through the Bröltal and the Nutscheid.[4]

Thus, the Honnef wind turbine "could not be put into operation because of air raids and was lost because of Germany's collapse."[5] Honnef's test site in Botzow-Veblen near Berlin was also destroyed in 1945 by the Bombing of Berlin. Honnef's test turbines and the larger Honnef turbine were never recovered. According to Ulrich Hütter's 1954 account "because of confusion at the end of the, parts of his even larger projects were lost."[6] The war had finally caught up with Germany's wind power development efforts and Honnef's dream of building high-altitude, contra-rotating, gearless wind turbines was over.

But Honnef's research wasn't the only wind power research activity happening in Germany during the war. In fact, starting in 1941—using wind expert Kurt Bilau as an advisor—the firm Ventimotor GmbH built and tested several small wind turbine prototypes. Ventimotor was formed in late 1940 by Walther Schieber and Fritz Sauckel to develop wind turbines that could be used in the war effort. Drawing upon Bilau's expertise in designing highly efficient wind turbines, Ventimotor was able to build and test small wind turbines with efficiencies greater than anything built until that time. By 1945, Ventimotor had an operational test site that featured a 15-kW wind turbine with an 18-m blade diameter and a 30-m tower height. Ventimotor engineers had designs on developing a 500-kW wind turbine and installing it in Thüringer forest, but the project never came to pass.[7]

The automotive company Porsche KG in Stuttgart-Zuffenhausen also developed and tested wind turbines during this period. At the time, Porsche offered motor vehicle development work and consulting, but did not build any cars under its own name. Porsche built and tested both two-bladed and four-bladed wind turbines over a period of four years from 1940 to 1945. Based on the lessons learned from testing these turbines, the company designed a wind turbine for mass production. The 10-kW production unit had a hub diameter of 9.2 m, a tower height of 19.6 m, and a tip-speed ratio of seven.[8] The blades were built from curved sheet metal and used a special welding process to connect the sheets. However, this wind turbine design never made it into production. After the war, Porsche re-focused its attention on cars and released its first production model in 1948.

In addition, the firm Hein, Lehmann & Co. in Berlin built a 5-kW wind turbine that was designed to sit atop lattice frame towers ranging in height from 10 to 30 m. The wind turbine was designed by G. König, who also developed a centrifugal control system that was powered by the electric generator and displaced the turbine blades along their axis. By doing this, König's design limited the fluctuation in the rotation rate to within 3 percent. This was enough to enable the use of a synchronous AC generator interconnected with the electrical network. However, as with the Porsche design, the wind turbine development efforts of Hein, Lehmann & Co. were not pursued after the war.

8.2 Lighting Neuwerk Island

During the postwar reconstruction period, food, fuel, and material shortages were experienced across Europe, but were particularly acute in Germany. From an energy perspective, Western Germany had lost its eastern coal resources, and output in the war-torn west was only 50 percent of pre-war levels.[9] In 1948, the recovery and reconstruction of Europe was catalyzed by the US Marshall Plan, which provided US$13 billion in economic support to help rebuild Europe. However, energy shortages persisted even with this influx of resources. The price that the American oil majors charged European customers for oil doubled between 1945 and 1948, going from US$1.05 to US$.2.22/bbl.[10]

High oil prices created a challenge for those who relied upon diesel generators to meet their electricity needs. It also opened a window of opportunity for the development of alternative power sources. One wind power innovator—Dimitri R. Stein—used the opportunity to build a successful wind turbine on the tidal island of Neuwerk, located in the shallow coastal waters of the North Sea, northwest of Cuxhaven in Germany.

With just a few farms in 1945, the island had modest power requirements. Electricity was needed primarily to power the island's lighthouse. The lighthouse tower had been constructed as early as 1310 CE and was converted to a lighthouse in 1814.[11] By 1945, it was powered by diesel generators. The diesel shortage meant that the utility responsible for the lighthouse—Überlandwerk Nord-Hannover AG—was unable to keep it in service. Laying undersea transmission cables was deemed too costly at the time. However, because the island is also home to winds in excess of 6 m/s, the engineers at Überlandwerk Nord-Hannover AG decided that wind power was the preferred solution.

Überlandwerk Nord-Hannover AG contacted Nordwind GmbH in Porta Westfalica, a wind turbine manufacturing company recently founded by Stein. Born in St. Petersburg in 1920, Stein moved to Germany and grew up in Berlin after his family fled the Bolsheviks when he was five months old. An avid student, Stein finished his studies in electrical engineering at the Technische Hochschule in Berlin in 1943. However, after he completed his thesis, he was denied his doctorate because of his Jewish heritage and he was forced into hiding. After the war ended in the spring of 1945, he emerged from hiding determined to put his education to work. He founded Nordwind in 1945 to focus on the new field of wind power.

When Nordwind was engaged by Überlandwerk Nord-Hannover for the Neuwerk Island project, Stein agreed to design and construct the wind turbine. Stein's design called for a three-bladed 18-kW wind turbine with a 15-m rotor diameter sitting atop a 20-m tower. When complete, the wind turbine would provide enough electricity to meet 80 percent of the island's power requirements. Construction began in 1946 under the direction of Nordwind engineer Franz Villinger.[12] The wind turbine was put into operation by the end of the year.

What is interesting about the Neuwerk wind turbine is how Stein successfully incorporated concepts from the 100-kW Balaklava wind turbine built by the Russians in 1931 known as WIME D-30. It turns out that Russian-born Stein wasn't just an electrical engineering student during the war years, he was also an expert on Russian wind power developments and the author of several technical articles on wind power. To maintain constant output, the angle of the blades of the wind turbine automatically adjusted as the wind speed increased using their own centrifugal force. Just like WIME D-30, the Neuwerk wind turbine was mounted on a track with a conical wheel gear that was used to turn the wind turbine out of the wind in high winds.

Stein's wind turbine design—with its concepts borrowed from the Russians—is an example of how wind power technology progressed throughout the twentieth century. When one wind power innovator was able to incorporate design elements from fellow innovators in another country separated by time and—in this case—divided by war, wind power technology progressed. This is why incremental innovations were so important. Stein's wind power accomplishments in 1946 make him and the Neuwerk wind turbine an important part of the tapestry of twentieth century wind power.

The Neuwerk wind turbine operated successfully for twelve years. During the first four years of operation it produced 180,000 kWh of electricity.[13] This means that it operated at an average annual capacity factor of 29 percent, an impressive figure that rivals the production level of wind turbines built toward the end of the twentieth century. The wind turbine was put out of service in 1958, when an undersea transmission cable was installed.

As for Stein, he emigrated to the United States in 1947. He continued his focus on wind power and wrote technical reports on wind power development efforts. He opened an engineering consultancy in 1962 and simultaneously operated an office for the Gmelin Institute, publishers of Gmelin's Handbook of Inorganic Chemistry. In November 2008, sixty-five years after he submitted his thesis, Stein was awarded a doctorate degree from the Technical University of Berlin.

8.3 The Aesthete

In August of 1946, the Allied Control Commission (ACC) disbanded the Luftwaffe and German aviation was heavily restricted. This persisted until Germany joined the North Atlantic Treaty Organization (NATO) in 1954, at which point the Western Allies believed that the Soviet Union posed a greater threat to peace than Germany's air force. The curtailment of German's aviation industry was challenging for many German engineers who had dedicated their carriers to aviation.

Ulrich Hütter was one such German engineer. Hütter was born in Pilsen in 1910 and had a lifelong fascination with flight. He studied aircraft engineering at the technical universities of Vienna and Stuttgart. By the time he was in his twenties, he was designing and flying lightweight gliders with his brother. His gliders were manufactured by Sportflugzeugbau Schempp-Hirth in Göppingen. He continued his focus on airplanes, and during the war he led the construction at Graf Zeppelin. By 1944 he had become a chair for fluid and flight mechanics at the Technical University of Stuttgart.[14] His future was in aviation, but the ACC's restrictions after the war had interrupted his plans.

Fortunately, Hütter had another interest: wind power. In fact, he had completed his doctorate in 1942 at the University of Vienna with the thesis *A Contribution to Creating the Design Principles for Wind Turbines*. The goal of his 1942 thesis was to "determine the dimensions and shapes for this type of [wind turbine] system which result in maximum economy and to establish them in the most general form possible."[15] He had developed fundamental theoretical relations to describe the aerodynamics of a wind turbine and, from them, deduced basic design principles: a small number of blades, clean aerodynamic profiles, high rotor velocity, and extremely light construction to achieve the major design priorities of high efficiency and low weight.[16]

Hütter was also an aesthete. His sense of style and design would influence wind power development throughout the twentieth century. Hütter emphasized the design of wind turbines and the importance that wind turbine aesthetics would have on future generation. In Hütter's own words: "These (wind turbines) must for that reason in a deeper sense be of a timeless beauty, so that they do not in three or four decades hence burden a later generation with the heavy task of removing angular skeletons...."[17] He wanted wind turbines to work flawlessly, but he wanted them to look good. This is the underlying principle that would animate Hütter's approach to wind turbine development over the next forty years.

By 1948, Hütter had taken a keen interest in the work of Richard Bauer. Bauer—who had developed a successful 10-kW wind turbine in 1924—reprised his work after the war, this time for the firm Winkelstrater GmbH. Bauer built a 3-kW wind turbine that had a single blade with a rotor diameter of 8.6 m. Taken by the high power coefficient demonstrated by Bauer's single-bladed wind turbine, Hütter began the development of his own single-bladed machine at the firm Schempp-Hirth in Kirchheim in 1946. Hütter's wind turbine would work on the same principle as the British 100 kW Andreau-Enfield pneumatic wind turbine, which was under construction in England at the time.

In the pneumatic design, the wind turbine blades are hollow and have open tips. When the blades rotate, they act as a centrifugal pump and pull air through an opening at the base of the tower. The air passes through a turbine that is driven by airflow and is connected to an electric generator. Hütter began construction of a single-bladed 600-W pneumatic wind turbine project in 1946. However, Schempp-Hirth ran into financial difficulties and the project was put on an indefinite hold. Hütter was content to let the Andreau-Enfield wind turbine project prove the pneumatic concept.

However, a critical mass of interest in wind power had been reached in Germany at the time. The wind power development activities taking place in Germany, Denmark, and France inspired German wind power enthusiasts to found the Studiengesellschaft Windkraft on December 15, 1949. This academic association served as a community for information exchange among wind power experts. The annual meetings of the association brought together German representatives and other experts from Denmark, England, France, Spain, and the United States. It was the first group of its kind since Poul la Cour founded the Danish Wind Electricity Society in 1903. The association would play a key role in uniting the wind power community throughout the 1950s.

Hütter was, of course, a central figure behind the association. The primary purpose of the association was really to facilitate the construction of a large, grid-connected wind turbine in Germany. In between annual gatherings, members of the group studied the designs of existing wind turbines—they became intimately familiar with the 100-kW Russian wind turbine in Balaklava, the 1.25-MW Smith-Putnam wind turbine in the United States, and the early experiences of Denmark's Johannes Juul. All these foreign experiments had achieved positive results.

During the same period, Allgaier-Werkzeugbau GmbH (Allgaier Tool Production Corporation)—a small farm equipment company founded by the entrepreneur Erwin Allgaier from Uhingen near Göppingen in Wurttemberg—developed an interest in wind power. Allgaier had been manufacturing tractors before the war. They had great success in tractor manufacturing and sales since it became one of two worldwide licensees for Dr. Ferdinand Porsche's tractor designs. Porsche was prohibited from manufacturing his own tractors at the time because only companies that had been making tractors prior to and during the war were permitted to do so. Considering its success in building and selling the Porsche-designed tractors, Allgaier was looking for additional avenues for growth. Wind turbines seemed like a natural fit for the farm equipment company.

Wind turbines—like farm equipment—must be constructed in a manner that ensures their reliability in the face of continuous use and rugged conditions. Farm equipment manufacturers would ultimately play a key role in commercializing wind power. Twenty-five years later, another small farm equipment manufacturer in Denmark experimented with wind power before developing a 30-kW wind turbine and making it available for sale in 1979. That company—Vestas—would grow to become the top wind turbine manufacturer in the world.

Erwin Allgaier contacted Hütter in 1948 and engaged him as the senior wind turbine construction designer. Allgaier set up a test field where Hütter could conduct research on wind turbine performance. He started his work by piloting a 1.3-kW wind turbine in the fall of 1948. He quickly increased the size of the test turbine to 7.2 kW. The test systems that Hütter built during this time used blade pitch regulation, which enabled him to adjust and achieve the perfect blade angle for each wind speed.[18] Hütter used a centrifugal governor like the one used by the Jacobs brothers in the United States. He improved upon the design by arranging the centrifugal governor coaxially with the axis of rotation of the blades and of the generator. In Hütter's design, the centrifugal governor was mounted directly on the shaft of the generator and rotated at a high speed so that the adjustment mechanism operated with great force. He patented his approach in January 1951.[19]

Hütter's wind turbines performed so well that starting in 1950, Allgaier began manufacturing and selling a 10-kW version, the WE-10. Without much fanfare, this was the start of commercial wind power in Germany. This was the business that Kurt Bilau had envisioned when he opened shop as a wind turbine manufacturer in October 1921. However, Bilau was hampered by the depressed Germany economy during the interwar years and he was forced to close his business in 1926. In contrast, Allgaier benefited from the "Wirtschaftswunder," the West German economic miracle in the 1950s. Approximately 200 WE-10 wind turbines were produced and sold by Allgaier in the 1950s.

Eight WE-10s were clustered in groups at a single location in November 1953. Until this point, no one had thought of clustering multiple wind turbines in a single location to create a "farm" of wind turbines. The Water Administration Office in Meppen put the WE-10s in operation to generate power for a pumping station. The wind turbines supplied enough power to pump groundwater and rainwater for the next decade.

In the summer of 1952, the Studiengesellschaft Windkraft arranged the testing of a 1.8-kW grid-connected wind turbine that used a synchronous AC generator. The wind speed and electrical output was continuously recorded to establish relationships for the behavior of wind turbines when operating in parallel with the transmission network. The Germans discovered what the Americans and the Russians already learned—it was indeed feasible to connect a wind turbine to the transmission network. In Hütter's own words "all of these experiments had positive results. This occurred even though some specialists had predicted that parallel operation would be difficult because the wind conditions were non-uniform." The stubborn idea that wind turbines were confined to remote applications outside the reach of the transmission system was gradually fading.[20]

8.4 The Blade Breakthrough

Allgaier's WE-10 was a successful start to Hütter's postwar wind power journey. But he had bigger plans for wind power and the tests in 1952 simply reinforced his ambitions. Wind power wasn't just for use on the farm or at the water pumping station, it had the potential to transform Germany's electric power system. The first step in this transformation was the construction of a grid-connected 100-kW wind turbine that would prove the technology and the economics.

Erwin Allgaier agreed with Hütter's vision, but he was in no position to finance it. At Hütter's suggestion, the Studiengesellschaft Windkraft set up a joint venture called the Windkraft Entwicklungsgemeinschaft (WEG) in June 1954. The purpose of the WEG was to unite Germany's energy companies, suppliers, and related companies into a joint venture capable of financing a 100-kW wind turbine. The venture included seven West German public power utilities and five firms in the electrical machinery and construction business. The Federal Ministry of Economics, the Baden-Württemberg

Ministry of Economics, the Lower Saxony Ministry of Economics and Transportation, and the Deutsche Forschungsgemeinschaft participated in financing.

WEG announced a design competition inviting solicitations for the development of a 100-kW wind turbine at a rated wind speed of 8 m/s. The winner of the competition was never in doubt. Even still, wind power pioneer Richard Bauer would not go quietly. He submitted a proposal for a 100-kW version of his single-blade wind turbine with its high tip-speed ratio and power coefficient. But Bauer's machine had problems with control and aerodynamic stability because of the asymmetric loads caused by the single blade. As expected, the WEG ultimately identified Hütter's proposal as the most feasible pathway to develop a 100-kW German wind turbine.

Hütter assembled a team and began work in 1955. It would require four years of work before the 100 kW came on-line in 1959. The prospect of creating a wind turbine with ten times the power output of the WE-10 excited Hütter and his team. They were determined to build a wind turbine that was superior to the American and Russian efforts in functionally and—more importantly to Hütter—in form. According to the calculations of Hütter and team, a 100-kW machine would require a rotor diameter of 34 m. Thus, the experimental 100-kW machine would henceforth be known as the WE-34.

Hütter and team decided that the WE-34 would have two blades, just like the smaller WE-10 machine. Today, most wind turbine manufacturers have settled upon the three-bladed wind turbine design because two-bladed wind turbines put more stress on the major components. Having two blades instead of three increases the aerodynamic loads experienced by the tower, shaft, and nacelle. In one analysis, loading arising from wind shear and yaw misalignment were found to increase loads by one-third due to wind turbulence were found to increase by over 20 percent. Also, because Hütter's wind turbine was a downwind machine, the blades had to pass through the shadow of the tower as they revolved around the hub, which created additional blade stresses.

The key to overcoming these issues in Hütter's design was to give the two-bladed wind turbine a "teetering hub". A teetering hub is a hub that is mounted on a hinge, which is perpendicular to the shaft that absorbs the out-of-plane aerodynamic loads on the blades, which would otherwise result in bending movements on the shaft.

After sorting out the design approach for the rotor and the hub, Hütter and team focused their energy on the development a new type of blade. They wanted to create a blade that was lightweight, strong, aerodynamically shaped to maximize efficiency, and esthetically pleasing. Blades from existing large wind turbines were made from metal—the Russian blades on the Balaklava wind turbine were made from iron, the American Smith-Putnam blades were built on steep spars and covered with a stainless-steel skin. Putnam team chose steel blades because of the high tensile strength of steel. However, steel was not an ideal blade material because it is very heavy.[21]

Hütter and his team recognized that to build a successful wind turbine, the blade material would have to be both strong and lightweight. If a new lightweight and strong material could be used, this would unlock the future potential of wind turbines because they could become more efficient and cheaper both in terms of initial capital investment and the cost of electricity production.

Of course, the first place Hütter looked for a new blade material was the aviation industry. Hütter had been focused on wind power over the last decade, but still had an interest in flight. He examined airplane wings and gliders for inspiration on how to construct a new lightweight wind turbine blade. Could material innovations used in this field somehow be applied to wind power? The aviation industry had been on a continuous search for materials that were both lightweight and strong since its inception. When Dale Kleist—a young researcher for Corning Glass—accidently pointed a jet of compressed air at molten glass in 1932 he discovered "fiberglass" and started a chain of events that would lead to the development of glass fiber reinforced plastics (GFRP), a material that was adopted for use in airplanes.

GFRP is a composite material made of polyester resin, fiberglass, and plastic. The polyester resin was invented by Calton Ellis of DuPont in 1936, but it wasn't used with fiberglass until 1942 after German scientists refined its curing process. This occurred during World War II. British intelligence agents promptly stole the German curing process and handed it over to the Americans so that they could manufacture GFRP themselves.[22] GFRP was immediately used by the Americans for aircraft cockpit components and as a replacement for the molded plywood.

GFRP is attractive for aviation applications—and about a thousand other conceivable uses such as cars, boats, pipes, tanks, water facets, shower stalls, and bathtubs—because it has a low density and a tensile strength close to both steel and aluminum. After World War II, the suitability of GFRP in aviation applications was noted by Dr. Tsuyoshi Hayashi of the Department of Applied Mathematics and Engineering at the University of Tokyo. Dr. Hayashi pioneered studies on aircraft materials and structures during the war and took notice of American production and use of GFRP in aviation.

Hayashi would go on to become one of the world's foremost experts on composite materials. By 1954, he had constructed the world's first glider using GFRP, which he called the "Light Blue Soarer."[23] Hasashi fabricated the third version of his glider in 1955 at the same time that glider enthusiast and wind power expert Ulrich Hütter was searching the aviation industry for a new blade material for the experimental W-34 wind turbine.

Once Hütter made the connection, there was never really a question of whether GFRP was the right blade material for the W-34. He was going to use it regardless. Hütter's team tested several different variations of GFRP for the new blade before settling upon their preferred mixture. After that, they constructed a new resin dispensing machine for precise dosage of the resin quantities.[24] They worked tirelessly and carefully to craft the blades. The W-34 team shaped the blades according to airfoil profiles developed by the US National Advisory Committee for Aeronautics, the predecessor to NASA. They were extremely pleased by the flexibility that GFRP offered, according to Hütter "The manufacturing of blades of fiber-glass-reinforced plastics provides almost unlimited freedom in blade configuration and thus makes it possible to meet all known aerodynamic requirements to a large degree."[25] In February 1959, the team put the finishing touches on the two blades for the WE-34. They had created the first ever pair of fiberglass wind turbine blades. Fiberglass composites would later become the material of choice for wind turbine blades, only to be gradually displaced by carbon fiber in the twenty-first century.

A test field at Schnittlinger Berg between Schnittlingen and Stötten was set up for the WE-34. With the major components ready, assembly began in August 1959. Assembly proceeded smoothly and within a month's time the 100-kW machine was ready for operation. Once completed, the wind turbine was impressive. Its features would probably excite contemporary observers even more than those who were present at the time because WE-34 was the first wind turbine built that actually looked like a modern horizontal-axis wind turbine with long, aerodynamically shaped blades that dominated its appearance.

Its dimensions were impressive too. Each blade was seventeen meters long. It reached full output at 42 rpm. It was connected to a 100-kW generator that ran at 1,500 rpm. It stood over 24 m tall and weighed 13 tons. The switch was flipped on September 4, 1959 and WE-34 began generating electricity. Observations were made, and tests were conducted during the first three months of operation. WE-34 began to supply electricity to the German grid on December 11. On December 12, WE-34 reached its full load operation of 100 kW.

WE-34 outperformed expectations. Using output data from 1960 and 1961, Hütter would later calculate the wind turbine's actual power curve and compare it to the team's precalculated power curve based on design estimates from before construction. The actual power curve was higher across the full range of wind speeds. Based on this data, Hütter estimated that the efficiency of WE-34 was greater than 40 percent. Unlike other large-scale wind turbines such as the Smith-Putnam machine, there were no catastrophic breakdowns and dramatic failures due to mechanical problems or design flaws. Hütter's W-34 was everything he had promised.

There was, however, one very big problem, and it had nothing to do with the design, construction, or even the performance of WE-34. Between the time that WEG had awarded Hütter the contract to build a 100-kW wind turbine in 1955, and the time that the turbine was complete in 1959, Germany had lost interest in wind power. Allgaier could not see any economic potential for wind power and fully withdrew from wind turbine manufacturing in 1959.

Figure 8.1 Ulrich Hütter's 1959 100-kW WE-34 wind turbine was revolutionary in many ways. Its fibreglass composite blades would become commonplace starting in the 1980s.

WEG canceled further tests. Hütter moved on to become an associate professor in 1959, as well as head of the newly founded Department of Applied Flight Physics in Stuttgart. By 1964, even the Studiengesellschaft Windkraft ceased to exist. After nine years of intermittent operation and limited support and interest, WE-34 was shut down in 1968.

The loss of interest in wind power was driven by the plunge in energy prices starting in the late 1950s. Thus, despite Germany's wind power development efforts, and Hütter's ground breaking wind turbine design, interest in the wind power would have to wait for the next energy shortage. But with Hütter's timeliness beauty spinning in the background, there were scant signs of another energy crisis on the horizon.

Little did anyone realize at the time that an American biologist was already busy planting the seeds that would ignite a global environmental movement, which would eventually help push wind power research back to the forefront of the global energy landscape. The next part of the wind power story begins on a farm up the Allegheny River from Pittsburgh.

Notes

1 Juchem, *The Current Status of Honnef Wind Power Plants*, 1955, 9.
2 Hütter, *The Development of Wind Power Installations for Electrical Power Generation in Germany*, 1954a, 8.
3 Juchem, *The Current Status of Honnef Wind Power Plants*, 1955, 12.
4 Wikipedia, 2017b.
5 Juchem, *Are Wind-Driven Power Plants Possible?* 1953, 2–3.
6 Hütter, *The Development of Wind Power Installations for Electrical Power Generation in Germany*, 1954a, 8.
7 Hütter, *Wind Power Machines*, 1954b, 6.
8 Ibid.
9 Engdahl, 1992, 103.
10 Ibid.
11 Wikipedia, 2017a.
12 Röbke, 2010.
13 Hütter, *The Development of Wind Power Installations for Electrical Power Generation in Germany*, 1954a, 7.
14 Janzing and Oelker, 2013, 389.
15 Hütter, *Contribution to the Creation of Basic Design Concepts for Power Plants*, 1942, 7.
16 Matthias, *Signs of Hubris: The Shaping of Wind Technology Styles in Germany, Denmark, and the United States, 1940–1990*, 1998, 653.
17 Matthias, *The History of Wind Energy Utilization 1890–1990*, 1995, 92.
18 Janzing and Oelker, 2013, 391.
19 Hütter, *Wind-Driven Power Plant with Centrifugal Governor*, 1953.
20 Hütter, *The Development of Wind Power Installations for Electrical Power Generation in Germany*, 1954a, 18.
21 Material densities and tensile strengths taken from Cambridge University Engineering Department (2003).
22 FELCO Manufacturing Pty. Ltd., 2014.
23 Kimpara, 1998.
24 Janzing and Oelker, 2013, 394.
25 Hütter, *Operating Experienced Obtained with a 100-kW Wind Power Plant*, 1964, 13.

Bibliography

Burton, Tony, David Sharpe, Nick Jenkins, and Ervin Bossanyi. *Wind Energy Handbook*. West Sussex, UK: John Wiley, 2001.

Cambridge University Engineering Department. *Materials Data Book*. Cambridge, UK: Cambridge University, 2003. http://www-mdp.eng.cam.ac.uk/web/library/enginfo/cueddatabooks/materials.pdf.

Engdahl, William. *A Century of War: Anglo-American Oil Politics and the New World Order*. Ann Arbor, MI: Pluto Press, 1992.

FELCO Manufacturing Pty. Ltd. 2014. *The History of Fibreglass*. Accessed October 5, 2017. http://www.felco.net.au/fibreglass.php.

Hütter, Ulrich. "Contribution to the Creation of Basic Design Concepts for Power Plants." PhD thesis, Technische Hochschule Wien, 1942.

———. "The Development of Wind Power Installations for Electrical Power Generation in Germany." *Zeitschrift BWK* 6, no. 7 (1954a): 270–8.

———. *Wind Power Machines*. In *The Engineer's Pocketbook*, 1030–44. Vols. IIA. Twenty-eight. Berlin, Germany: Wilhelm Ernst and Sohn, 1954b.

———. Wind-Driven Power Plant with Centrifugal Governor. US Patent 2,655,604, October 13, 1953.

———. "Operating Experienced Obtained with a 100-kW Wind Power Plant." *Windkraft* 16 (1964): 333–40.

Janzing, Bernward, and Jan Oelker. *Hutter's Heritage: The Stuttgart School*. In *The Rise of Modern Wind Energy: Wind Power for the World*, edited by Preben Maegaard, Anna Krenz, and Wolfgang Palz, 387–406. Pan Stanford Series on Renewable Energy 2. Singapore: Pan Stanford Publishing, 2013.

Juchem, P. *Are Wind-Driven Power Plants Possible*. Koelnische Rundschau, April 26, 1953.

———. "The Current Status of Honnef Wind Power Plants." *Elektrotechnische Zeitschrift* 7, no. 5 (1955): 187–91.

Kimpara, I. "Overview of Professor Hayashi's Contributions to Composites R&D in Japan." In *Proceedings of the Eigth Japan-U.S. Conference on Composite Materials*, edited by Golam M. Newaz and Ronald F. Gibson, 3–6. Lancaster, PA: Technomic Publishing, 1998.

Matthias, Heymann. *The History of Wind Energy Utilization 1890–1990*. Frankfurt, Germany: Campus Verlag, 1995.

———. "Signs of Hubris: The Shaping of Wind Technology Styles in Germany, Denmark, and the United States, 1940-1990." *Technology and Culture* 39, no. 4 (1998): 641–70.

Milborrow, David. "Are three blades really better than two?" *WindPower Monthly*, August 4, 2011. Accessed October 5, 2017. http://www.windpowermonthly.com/article/1083653/three-blades-really-better-two.

Röbke, Thomas. "Energy pioneer on Neuwerk." *Die Welt*, March 1, 2010.

Wikipedia. *Neuwerk*. August 3, 2017a. Accessed October 5, 2017. https://en.wikipedia.org/wiki/Neuwerk.

———. *Rottland (Waldbröl)*. May 22, 2017b. Accessed October 5, 2017. https://de.wikipedia.org/wiki/Rottland_(Waldbröl).

9

Wind Power's Silent Decade

> *It is also an era dominated by industry, in which the right to make a dollar at whatever cost is seldom challenged.*
>
> —Rachel Carson (1962)

9.1 The Pope's Speech

The audience of scientists and entrepreneurs was rapt with anticipation as Pope John XXIII took the stage at Castel Gandolfo, Italy on a sizzling summer night in 1961. The Pope's topic was "new sources of energy" and how they relate to economic and social development. The audience was enthralled as the Pope spoke—not just with the topic and the speaker, but also with the magnificent natural beauty of Castel Gandolfo. Occupying a height on the Alban Hills overlooking Lake Albano, Castel Gandolfo has long been considered one of Italy's most scenic towns.

Against this backdrop, Pope John's speech in the summer of 1961 must have felt like a new beginning for alternative energy technologists and those who had toiled away at their development for decades. For wind power pioneers in attendance—after seventy years of incremental improvement—it was now time for the world to recognize wind power's potential role in the global energy landscape.

The Pope gave this speech during the UN Conference on New Sources of Energy, which was held in Rome from August 21 to August 31, 1961.[1] The conference was held in accordance with resolutions of the UN Economic and Social Council and on the invitation of the Government of Italy. The conference consisted of 250 experts and entrepreneurs representing the experience of twenty-nine countries in the development of wind power, geothermal energy, and solar energy. It was the first international conference of this scale focused on alternative energy technologies.

The Wind Power Story: A Century of Innovation that Reshaped the Global Energy Landscape, First Edition. Brandon N. Owens.
© 2019 by The Institute of Electrical and Electronics Engineers, Inc.
Published 2019 by John Wiley & Sons, Inc.

Wind power was prominently featured at the conference. The conference sessions on wind power featured reports from fifteen countries that had experience with wind power. America's Marcellus Jacobs, Denmark's Johannes Juul, and Germany's Ulrich Hütter, all provided insights on wind power. Other energy innovators from around the world also shared their experience. The group discussed and debated the future of wind power in both the developed and developing world. However, the picture that emerged at the conference regarding wind power was not as optimistic as wind power pioneers Jacobs, Juul, and Hütter had hoped. According to the final conference report, the discussion focused on the need for wind power to be developed in areas where "...its justification would have to be made entirely on economic grounds, such as by savings of fuels."[2] Several wind turbines in the size range of 100 kW or greater were mentioned, but further research was recommended to establish convincingly the economic viability of wind power. The Romani-BEST wind turbine in France was highlighted as an example of a wind turbine design that could be potentially economically viable at some point in the future.

In the end, the conclusion was clear: although wind resources were abundant—and some test wind turbines had been developed—more research would be needed to prove the economic viability of wind power. Governments around the world would have to embark upon additional wind power research efforts to further advance the technology. However, given the energy landscape of the 1960s, it was apparent that no government was willing to invest in wind power research. The wind power research efforts of the 1950s were either abandoned outright or gradually phased out as the 1960s progressed.

The wind power journey started with James Blyth, Charles Brush, and Poul la Cour in the nineteenth century. Brush affixed an electric generator to a multibladed American windmill in his backyard in Cleveland to create a functioning wind turbine. Brush's wind turbine was not connected to the transmission system and was used to charge batteries that powered the lights in his mansion. Denmark's Poul La Cour simultaneously conducted research and built pilot wind power projects that were based on a European Tower Mills that had fewer blades. La Cour also designed a wind turbine with a constant rotational speed and built a crude regulation mechanism.

La Cour's wind turbine design principles were used to produce the first commercially available wind turbines in Denmark during the first decades of the twentieth century. By 1908, Danish manufacturer Lykkegaard had built seventy-two wind turbines that supplied power to rural Denmark. The application of the lessons of aerodynamics from the emerging airplane industry led to the development of more efficient small wind turbines that used airplane propellers as wind turbine blades in the 1920s. By the early 1930s, small DC wind turbines were manufactured in the United States for use in remote, off-grid applications.

The first grid-connected, constant speed wind turbine was successfully developed by the Soviet Union in 1931. The 100-kW wind turbine in the Crimea fed electricity into the local transmission network and was codispatched with existing generators on that system. Russian scientists used a rudimentary mechanical control system to maintain constant rotational speed of the wind turbine. This was important for a grid-connected wind turbine because the electrical output needed to be provided at a fixed frequency to match the frequency of the transmission system. A decade later, American Palmer Putnam and his team extended wind turbine technology by developing a grid-connected 1.25-MW wind turbine. The turbine used a hydraulic motor to adjust the pitch of the wind turbine blades to maintain constant rotational speed. The Smith-Putnam wind turbine was undersized given the loads it faced and was prone to mechanical failure, but it successfully demonstrated the ability of a megawatt-scale wind turbine to feed electricity into the local transmission network.

In the postwar period, wind turbine technology advances were made primarily in Denmark, France, and Germany. In Denmark, Poul la Cour's onetime pupil—Johannes Juul—led the charge to develop the 200-kW wind turbine in Gedser that started operating in 1957. The Gedser wind turbine was a three-bladed upwind wind turbine with fixed pitch blades. To maintain constant rotational speed, Juul developed the concept of "passive stall," which relies upon the blade shape to create a natural stall at high wind speeds to regulate the rotational speed of the wind turbine. Wind turbines that use passive stalls become self-regulating, avoiding the need for mechanical control systems. Gedser's use of three blades would prove to be the optimal design for wind turbines representing the best combination of high rotational speed and minimum stress in the turbine and tower. Juul's approach would later become the basis of the "Danish Concept" design that would serve as the standard approach wind turbine design through the 1980s.

In France, Lucien Romani spearheaded an effort financed by EDF that led to the development of an 800-kW wind turbine near Nogent Le Roi west of Paris in 1958. The Romani wind turbine did not represent a technical advance in wind power as much as it was a demonstration of France's ability to use state-of-the art materials and techniques to build and operate a large wind turbine. In Germany, Ulrich Hütter pioneered the development of a 100-kW wind turbine known as WE-34 that began operating in 1959. This was the first wind turbine built that resembled modern horizontal-axis wind turbines with long, aerodynamically shaped blades that dominated its appearance. The two blades on Hütter's wind turbine were constructed of fiberglass reinforced plastic, a composite material made of polyester resin, fiberglass, and plastic. It was lightweight and strong, and was the precursor to modern wind turbine blades which are also constructed from fiberglass composites. The WE-34 outperformed even Hütter's expectations.

All of these research efforts would eventually serve as the starting point for the international wind energy research and development efforts in the 1970s.

However, these wind power advances were largely unrecognized and underappreciated in the 1960s. Romani's Nogent Le Roi wind turbine ceased operations in 1966. In Denmark, Johannes Juul's Gedser wind turbine was decommissioned in 1967. And, in Germany, in 1968—after nine years of intermittent operation and limited support—Ulrich Hütter's WE-34 was finally shut down.

The ignominious end of the wind power projects in Denmark, France, and Germany was not due to technical failure. Indeed—although more research was needed—the fundamental technical challenges associated with the production of electricity from wind turbines less than 1 MW had largely been solved. Rather, the dearth of wind power research in the 1960s was primarily the result of low global energy prices. The abundance of oil and coal in the 1960s and concomitant low fuel prices provided a disincentive to conduct alternative energy research. Who needed wind power when electricity generated from conventional technologies was so cheap?

However, just as important as the decline in coal prices, coal-fired generation technologies improved over time, which led to further declines in the cost of electricity. Thomas Edison's Pearl Street Station converted 2.5 percent of its raw input energy into electricity. The introduction of pulverized coal steam generators in 1919 improved coal combustion by allowing for bigger boilers. By 1920, state-of-the-art steam turbines had a thermal efficiency of 20 percent. Another technological boost came with the advent of once-through boiler applications and reheat steam power plants—along with the Benson steam generator—which was built in 1927. Reheat steam turbines became the norm in the 1930s, when unit ratings soared to a 300-MW output level.[3]

In the 1930s, metallurgical processes made available superheating tubing and turbine parts that enabled temperatures to be raised to close to 1,000°F.[4] The subsequent development of "superalloys" that resisted metal fatigue and cracking allowed engineers to design steam boilers with higher temperatures and pressures. This culminated in the development of "supercritical" boilers that were made available starting in 1957. These boilers have a maximum temperature of up to 1,200°F and a maximum pressure of over 4 psi.[5] By 1960, the thermal efficiency of new steam turbines in the United States was 40 percent.[6] The doubling of the thermal efficiency from coal-fired power plants between 1920 and 1960 had the same impact on the cost of electricity as cutting the fuel price in half.

Steam turbines weren't the only fossil fuel-based electricity generation technology making strides. Innovations in aircraft technology, and engineering and manufacturing advancements during both World Wars led to the development of gas turbine technology, which would later become a cornerstone of electric power systems. GE installed its first 3.5-MW gas turbine for power generation in 1949.[7] The unit used its exhaust heat to heat the feedwater for a connected steam turbine. Gas turbine technology improved rapidly and firing temperatures increased, reaching 1,300°F by 1957. By 1965, the first combined-cycle gas turbine came online. Gas turbine technology wouldn't fully take off until the 1990s, but invention and improvement through the 1960s created a new option for utilities when considering new generation investments. Fossil fuel generation technologies weren't sitting still while wind power research was underway.

Thus, the combination of declining fossil fuel prices and improving generation technology worked together to reduce the cost of electricity and increase demand. The US FPC summed up the situation nicely in its 1964 National Power Survey: "The electric utility system of the United States stands on the threshold of a new era of low-cost power for all sections of our country...Larger and larger machines are being built which can generate electricity at progressively lower costs."[8]

Furthermore, in this environment—to the extent that the need for alternative electricity generation sources was recognized at all—it was widely believed that nuclear power represented the future of electricity. The focus on nuclear power starting in the 1950s was more than just a passing fad. In the period between fiscal year 1947 and 2014, the United States government alone would spend nearly US$100 billion in nuclear power R&D. This represented roughly half of the nation's energy R&D spending over this span. All other sources together—fossil, renewable, energy efficiency, and electricity systems—accounted for the other half.[9]

By 1960, electric utilities in the United States began to take notice of nuclear power. Westinghouse and GE raced to introduce the first commercial nuclear reactors. Westinghouse designed the first fully commercial pressured water reactor (PWR) and GE designed the boiling water reactor (BWR). Westinghouse's PWR was first commercialized at the Yankee Rowe nuclear power station in Massachusetts in 1960. GE's BWR was activated at the Dresden generating station in Illinois the same year. Jersey Central Power and Light Company's commitment to the Oyster Creek nuclear power plant in December 1963 set the tone for the decade. In the United States, nuclear power plant orders gain momentum as the 1960s progressed. Utilities had ordered ten nuclear reactors by 1966. Between 1966 and 1967, that number quadrupled, as GE and Westinghouse vied with each other for turn-key contracts. US orders for nuclear power plants grew throughout the decade and peaked in 1972.[10]

The growth of nuclear power was not confined to the United States. France detonated its first atomic bomb in 1960 and immediately launched a civilian effort to harness nuclear power. The national electricity utility EDF began investing in nuclear power. EDF struggled to meet growing electricity demand in France in the 1950s and responded by increasing its hydropower, nuclear, and oil-fired electricity capacity throughout the 1960s. EDF's first move toward nuclear power started with the opening of the Chinon station in 1964. Additional reactors at Chinon were put into service in 1965 and 1966. After the commissioning of Cader Hall in 1956, the United Kingdom followed up by opening a steady stream of nuclear generating stations throughout the 1950s and 1960s. By 1968, 3,404 MW of nuclear capacity were operating in the United Kingdom alone.

9.2 Silent Spring

But what's most striking about the energy landscape of the early 1960s to present-day observers is not the low fossil fuel prices, nor the advancement in nuclear and fossil fuel technologies as wind power lagged. No, what's most striking about this era to modern eyes is the absence of any consideration of the environmental and human costs of energy production and use.

During this era, fossil fuel-fired power plants were emitting generous levels of air pollutants such as particulate matter, sulfur dioxide, and nitrogen oxides that adversely impacted human health across the globe without protest; and greenhouse gases (GHGs) such as carbon dioxide were freely emitted from electricity generators without any consideration of their contribution to climate change. In addition, concerns related to nuclear accidents and radioactive waste disposal had yet to be raised. When it came to energy production and use, the prevailing approach was to focus exclusively on economics. The cheapest source of energy won—regardless of the environmental and human consequences.

Of course, this changed significantly as the 1960s unfolded, as the consequences of energy production and use were identified and the prevailing wisdom associated with unfettered industrial growth was challenged. In fact, by the time that the first Earth Day was celebrated in 1970, the scale, scope, and power of the environmental and antinuclear movements was significant enough to permanently alter the energy landscape.

The single biggest catalyst for this change was the 1962 publication of *Silent Spring*—a book written by a biologist named Rachel Carson. Ms. Carson didn't set out to change the world, but with the publication of her book she ignited a global environmental movement that altered the course of history—and dramatically improved the fortunes of wind power and other nonpolluting electricity generation technologies.

Born in 1907 on a farm up the Allegheny River from Pittsburgh, Carson spent her early days exploring the family farm and writing stories about animals before graduating High School in 1925 at the top of her class. Her interest in animals and the natural environment led her to John Hopkins University, where she earned a master's degree in Zoology in 1932. Carson was hired as a biologist by the US Bureau of Fisheries in 1936. In this role, she conducted research and wrote educational brochures. She also submitted her writings for publication in newspapers and magazines. Her article *The World of Waters* was published in the Atlantic Monthly in 1937. She expanded the article into the book *Under the Sea Wind*, which was published by Simon and Schuster in 1941. She continued writing articles for periodicals such as *Sun Magazine*, *Nature*, and *Collier's* while maintaining her position at the Bureau. By 1948, she was chief editor of publications at the Bureau, which by that time had been renamed the Fish and Wildlife Service.

Carson began writing her second book in 1948. *The Sea Around Us* was published in 1951 by Oxford University Press. It was an enormous success. It remained on the New York Times Best Seller List for eighty-six weeks and won the 1952 National Book Award for Nonfiction. Carson was awarded two honorary doctorate degrees and licensed a documentary based on the book. The success of *The Sea Around Us* gave Carson the financial security she needed to dedicate herself to writing full-time.

Truth be told, Carson wasn't the first person to express an appreciation for the importance of the natural environment. In fact, the environmental movement in the United States and Europe started in the late nineteenth century in response to air pollution and chemical discharges from the increasing number of factories dotting urban landscapes in the late nineteenth century. In the United Kingdom, William Blake Richmond founded the Coal Abatement Society in 1898 to fight against air pollution. Richmond and others fought against the negative environmental impacts of the Industrial Revolution.

However, Richmond's movement never gained enough prominence to stand in the path of unfettered industrialization in the early twentieth century. Another thread of the early environmental movement focused more heavily on natural conservation and wildlife protection. In America, the Audubon Naturalist Society was organized in 1897 for the "protection and study of birds."[11] John Muir lobbied Congress to form Yosemite National Park and founded the Sierra Club in 1892. Conservationist principles and a belief in the inherent right of nature were the bedrock of environmentalists' beliefs in the first half of the twentieth century.

Conservationist principles grew in popularity and efforts were made to protect animals and preserve the natural environment. However, conservationist-minded environmentalists had yet to fully challenge the industrial system head on. This was mainly because beyond obvious visual evidence that foul air and polluted water were undesirable for people and the planet, there was no firm scientific evidence to back up these claims. This enabled industries and governments to largely ignore environmentalists and cast them as starry-eyed dreamers who wanted to obstruct progress.

However, starting in the 1950s, there was growing evidence that industrial activities were negatively impacting human and environmental health. The Great London Smog of 1952—which killed thousands and was caused by coal burning—served as a wake-up call. Then—in 1957—the US government began a chemical spraying program designed to eradicate specific "pest" insects such as mosquitos and ants. This involved aerial spraying of the pesticide Dichlorodiphenyltrichloroethane or DDT.

The Audubon Society began actively opposing the US government's chemical spraying program because of their concerns about DDT's potential impact on birds. They approached Rachel Carson—who had served on the board since 1948—and asked her to help them research and write about the issue. When Carson's friend—Olga Owens—wrote to the Boston Herald describing the death of birds on her property resulting from aerial DDT spraying in 1958, Carson began a four-year research project that led to the publication of *Silent Spring*.[12]

Silent Spring was first published in three serialized excerpts in the *New Yorker* in June 1962, and the book appeared in September that year. In the book, Carson meticulously describes how DDT enters the food chain and accumulates in the fatty tissues of animals—including human beings—and causes cancer and genetic damage. A single application on a crop, she wrote, killed insects for weeks and months—not only the targeted insects but countless more—and remained toxic in the environment even after it was diluted by rainwater.[13]

The overarching theme of *Silent Spring* is the negative effect that industrialization can have on the natural world. Carson argued that pesticides have harmful effects on the environment and accused the chemical industry of spreading disinformation and government officials of accepting industry claims. Decades later, the consensus was that Carson "quite self-consciously decided to write a book calling into question the paradigm of scientific progress that defined postwar American culture."[14]

The chemical industry went on the attack. The industry had too much to lose to sit by idly while Carson informed the public of the true dangers of pesticides. They accused her of being radical, disloyal, and unscientific. Furthermore, when *Silent Spring* was published—at the height of the Cold War—Carson was accused of being a Communist. The ugly specter of sexism also raised its head. Allegations that Carson was just a hysterical woman appeared in chemical and agricultural trade journals. At the time, an agricultural expert told a reporter "you're never going to satisfy organic farmers or emotional women in garden clubs."[15] And, in a letter to former President Eisenhower, Former Secretary of Agriculture Ezra Taft Benson wondered why a "spinster was so worried about genetics."[16]

The attacks on Carson are important in the context of wind power because they serve as an example of how established interests fight against scientific and technical knowledge by issuing attacks and spreading false information. As the twentieth century progressed, several environmental issues such as air quality and climate change would make wind power a progressively more attractive energy source vis-à-vis conventional alternatives. Along the way, battles had to be fought to enable scientific facts to prevail over both ignorance and the views of entrenched interests to enable wind power and other alternative energy sources to succeed. Caron's early battle with the chemical industry was a precursor to the struggles that lie ahead for alternative energy proponents. The battle continues today in the twenty-first century with wind power opponents who deny the reality of climate change and therefore the need to accelerate the adoption of low-carbon technologies.

At a personal level, Carson was stung by the attacks, but she had bigger challenges to confront at the time. She had been battling breast cancer for years. She was weakened from her treatment regimen and became ill with a respiratory virus in January 1964. The cancer spread to her liver in the spring of 1964 and she died on April 14 in her home in Maryland at the age of fifty-six.[17]

Silent Spring was successful in kick-starting the modern environmental movement that ultimately led to the control and regulation of energy technologies. The movement gathered steam throughout the decade and was aided by a continuous steam of environmental disasters. For example, in 1967, the first supertanker—Torrey Canyon—ran aground near England and leaked 120,000 tons of oil killing most of the marine life along the south coast of Britain and the Normandy shores of France. In 1969, oil spilled from an offshore well in California's Santa Barbara Channel.

Overall, it was becoming increasingly clear that limitless economic growth without taking the environment into account was leading to environmental degradation and depletion of natural resources. The environmental movement coalesced in the late 1960s. Initially an expression of leftist parties and youth movements, environmental philosophies were gradually embraced by industrial and political leaders.

This culminated in the publication of *The Limits to Growth*, a report funded by the Volkswagen Foundation and commissioned by the Club of Rome—an alliance of industrial leaders and concerned scientists. The report detailed the results of computer simulations of exponential economic and population growth assuming a finite supply of natural resources. Although the fundamental underlying assumptions were—and still are—hotly debated, *The Limits to Growth* stressed the need to curb growth rates to avoid a sudden and uncontrollable economic and population decline.

In the United States, a host of environmentally focused and litigious nonprofit organizations such as the Environmental Defense Fund (EDF) and the Natural Resources Defense Council (NRDC) were founded to advocate for science-based environmental policies. One of the earliest court victories for the environmental movement was Scenic Hudson Preservation Conference v. Federal Power Commission—decided in 1965—which helped halt the construction of a power plant on Storm King Mountain in New York State. The case gave birth to environmental litigation.[18]

By 1969, the environmental movement had been successful in encouraging the US Congress and the Nixon Administration to pass the National

Environmental Policy Act, which promoted the enhancement of the environment and established the President's Council on Environmental Quality. The law was enacted on January 1, 1970 and has been called the world's equivalent of an "environmental Magna Carta." To date, more than 100 nations around the world have enacted national environmental policies modeled after it.[19]

Furthermore, in 1970, pressure from environmental groups led to the passage of the 1970 Clean Air Act. The 1970 Act was built upon the 1955 Air Pollution Control Act—which provided funds for air pollution research—and the Clean Air Act of 1963 that authorized research into specific techniques for controlling air pollution.[20] The 1970 Act required comprehensive federal and state regulations for both industrial pollution sources and automotive vehicles. It also expanded federal enforcement. The United States Environmental Protection Agency (EPA) was established in December 1970 to consolidate environmental research and enforcement into one federal agency.[21] The era of unreported and unabated environmental emissions from power plants was over.

The FPC acknowledged the importance of the environmental movement and the challenges it had created for the US electric power industry. In their 1971 report, they conceded that "Mounting demand, sharply rising costs and changing social values have combined to place unusual stresses on the U.S. electric power industry."[22] Twelve years after the UN conference on New Sources of Energy was held in Rome, another similar UN conference convened in Sweden. The UN Conference on the Human Environment was held in Stockholm from June 5 through 16, 1972. At this conference—for the first time—representatives of multiple governments met to discuss the state of the global environment. Environmental concern had become a global issue. The impact for energy technologies would be as far-reaching because the importance of economic value versus environmental impact was recalibrated.

9.3 The Gathering Storm

By the end of the 1960s, signs were pointing to a tightening of energy markets and potential trouble for a world that had been enjoying increasing supplies of energy throughout the decade. Hints of problems occurred in the United States in the late 1960s when demand for electricity in some parts of the country exceeded the expected annual growth rate, and when some utilities could not build power plants fast enough. During the hot summer days of 1967 through 1969, some utilities on the east coast of the United States reduced voltage as one way to deal with increased demand—causing brownouts—while they asked customers to reduce power consumption.

There were signs of trouble in oil markets too. In the United States, domestic oil production had risen steadily from 174,000 bbl/day in 1900 to

9.6 million bbl/day by 1970. That was the peak; US oil production wouldn't reach that level again in the twentieth century.* US oil reserves peaked in 1966 and declined through the end of the decade. Meanwhile, oil consumption—driven by a growing number of oil-fired power plants and an increasing number of gas guzzling cars on the road—continued to rise.

The net result was a dramatic increase in oil imports in the United States and other developed nations. Between 1960 and 1970, US oil imports increased from 371 to 483 million bbl/year. As domestic oil production declined after 1970, US oil imports skyrocketed. Between 1970 and 1972, oil imports nearly doubled from 483 to 811 million bbl.[23] With a growing amount of domestic oil supplies coming from international sources, the United States was increasingly vulnerable to even the most minor disruption in oil supplies.

Collectively, the gathering clouds on the horizon pointed to the need for a different kind of energy future. A future where human health, safety, and environmental concerns were valued alongside of economics. A future where governments recognized the need for domestic energy security and chose fuel diversity as a matter of national security while they actively invested in the development of energy alternatives for the future.

Just such an energy future would begin to come to fruition starting in the 1970s in response to the gathering clouds in the late 1960s. But it would take a full-blown global energy crisis to reshape the technology landscape and launch a wind power renaissance. Thus, in hindsight, it is clear that while very little wind power development occurred during the 1960s, the events of the decade shaped the future of wind power for the remainder of the century.

Notes

1 United Nations Department of Economic and Social Affairs, 1962, 3.
2 Ibid., 7.
3 Harvey, Larson, and Patel, 2017.
4 Yeh and Rubin, 2007.
5 Ibid.
6 Ibid.
7 Harvey, Larson, and Patel, 2017.
8 Federal Power Commission, 1964, 1–3.
9 Sissine, 2014, tables 2 and 3.
10 Chater, 2005.
11 Smithsonian Institution Archives, 2011.
12 Mattheissen, 2007, 135.

---

* EIA (2018). Recent unconventional oil production increases have resulted in output levels that exceed the 1970 peak starting in 2018.

13 National Resources Defense Council (NRDC), 2015.
14 Lytle, 2007, 166–67.
15 Graham Jr., 1970, 88.
16 Lear, 1997, 429.
17 Lear, 1997.
18 Scenic Hudson, Inc., 2005.
19 Eccleston, 2008.
20 Yang, 2015.
21 Environmental Protection Agency (EPA), 2017.
22 Federal Power Commission, 1971, 1.
23 Energy Information Administration, 2012.

Bibliography

Carson, Rachel. *Silent Spring*. New York: Houghton Mifflin, 1962.
Chater, James. 2005. "Focus on Nuclear Power Generation: A History of Nuclear Power." 28–37. Accessed November 10, 2017. http://www.nuclear-exchange.com/pdf/tp_history_nuclear.pdf.
Eccleston, Charles H. *NEPA and Environmental Planning: Tools, Techniques, and Approaches for Practitioners*. Boca Raton, FL: CRC Press, 2008.
EIA. 2018. "Petroleum & Other Liquids: U.S. Oil Field Production of Crude Oil (1860-2017)." *U.S. Energy Information Administration*. April 30. Accessed May 16, 2018. https://www.eia.gov/dnav/pet/hist/LeafHandler.ashx?n=pet&s=mcrfpus2&f=a.
Energy Information Administration. *Annual Energy Review 2011*, 370. DOE/EIA-0381(2011), Energy Information Administration, U.S. Department of Energy, Washington, DC: DOE/IEA, 2012. https://www.eia.gov/totalenergy/data/annual/pdf/aer.pdf.
Environmental Protection Agency (EPA). 2017. *EPA History*. Accessed November 10, 2017. https://www.epa.gov/history.
Federal Power Commission. *National Power Survey*. Washington, DC: U.S. Government Printing Office, 1964.
———. *National Power Survey*. Washington, DC: U.S. Government Printing Office, 1971.
Graham Jr., Frank. *Since Silent Spring*. Boston, MA: Houghton Mifflin, 1970.
Harvey, Abby, Aaron Larson, and Sonal Patel. "History of Power: The Evolution of the Electric Generation Industry." *Power Magazine* 161 (2017).
Lear, Linda. *Rachel Carson: Witness for Nature*. New York: Holt, 1997.
Lytle, Mark Hamilton. *The Gentle Subversive: Rachel Carson, Silent Spring, and the Rise of the Environmental Movement*. New York: Oxford University Press, 2007.

Mattheissen, Peter. *Courage for the Earth: Writers, Scientists, and Activists Celebrate the Life and Writing of Rachel Carson*. New York: Mariner Books, 2007.

Meadows, Donella H., Dennis L. Meadows, Jørgan Randers, and William W. Behrens III. *The Limits to Growth*. New York: Potomac Associates, 1972.

National Resources Defense Council (NRDC). 2015. "The Story of Silent Spring." *NRDC*. August 13. Accessed November 10, 2017. https://www.nrdc.org/stories/story-silent-spring.

Scenic Hudson, Inc. *Scenic Hudson Collection: Records Relating to the Storm King Case, 1963-1981*. Poughkeepsie, NY: Archives and Special Collections, Marist College, 2005.

Sissine, Fred. *Renewable Energy R&D Funding History: A Comparison with Funding for Nuclear Energy, Fossil Energy, and Energy Efficiency R&D*. Government. Washington, DC: Congressional Research Service, 2014.

Smithsonian Institution Archives. *Record Unit 7294, Audubon Naturalist Society of the Central Atlantic States, REcords*. Washingon, DC: Smithsonian Institution, 2011.

United Nations Department of Economic and Social Affairs. *New Sources of Energy and Energy Development*. New York: United Nations, 1962.

Yang, Ming. *Energy Efficiency: Benefits for Environment and Society*. New York: Springer International Publishing, 2015.

Yeh, Sonia, and Edward S. Rubin. "A Centurial History of Technological Change and Learning Curves for Pulverized Coal-Fired Utility Boilers." *Energy* (Elsevier Ltd.) 32 (2007): 1996–2005.

10

America's Next Moonshot

Present energy problems stem, in large part, from the lack of a coordinated national energy R&D program over the last 20 years.
—Dr. Dixy Ray Lee, Chairman, United States
Atomic Energy Commission (1973)

10.1   The Energy Crisis

Energy was not in the public eye at the beginning of 1973, but by the year's end, countries around the world were confronted with skyrocketing gasoline prices, endless lines, and shortages at gas stations. Global energy consumers experienced a rude awakening relating to the importance of energy in their everyday lives.

The 1973 global energy crisis would ultimately benefit wind power because it led to the birth of the Federal Wind Energy Program (FWEP) in the United States, revived wind power research in Europe, and spurred actions in India and China that would eventually lead to the development of wind power in those countries as well. However, the positive effects of these efforts would not be felt until well after the crisis was over; and, in the end, there is still a lingering question about whether the government-led research programs that were initiated in the 1970s were truly effective at commercializing wind power.

The global oil market was already strained when the Organization of Petroleum Exporting Countries (OPEC) instituted an oil embargo against the United States in October 1973. The embargo was initiated by OPEC in response to America's support of Israel in the Yom Kippur War with Egypt and Syria, and was set in motion by America's visible effort to rearm the Israelis with munitions on October 14, 1973. On October 17, OPEC representatives agreed to cut oil production by 5 percent and added an additional 5 percent until Israel withdrew their forces from the Sinai Pass. On October 20, Saudi Arabia and other OPEC members halted oil deliveries to the United States.

The Wind Power Story: A Century of Innovation that Reshaped the Global Energy Landscape,
First Edition. Brandon N. Owens.
© 2019 by The Institute of Electrical and Electronics Engineers, Inc.
Published 2019 by John Wiley & Sons, Inc.

Oil prices jumped from US\$3/bbl in the summer of 1973 to US\$12/bbl by the end of the year. The era of cheap oil was over.

As sudden and dramatic as it seemed, the crisis had actually been brewing for some time. US oil production peaked in 1970 and had begun a multidecade decline. The decline in US oil production in 1970 led to a rise in oil imports, and by 1973, Americans were vulnerable to disruption from foreign sources of oil. The influence of oil reached far beyond the impact on fuel prices for the nation's cars and trucks. Oil had become increasingly influential in the electric sector in the United States and Europe as well. In the United States, the 1970 Clean Air Act placed new limits on sulfur dioxide emissions from power plants that reduced the demand for electricity from high-sulfur coal and increased the demand for electricity from oil-fired generators. Newly constructed oil-fired plants were built in the United States to meet new emission standards. Between 1963 and 1970, the share of oil-fired generation in the US power sector grew from 5.5 to 12.3 percent.[1] By 1973, the US electric sector was on its way to becoming as dependent upon foreign oil as the transportation sector already was.

In addition, the electric power industry itself was also under strain in the years leading up to the 1973 energy crisis. In November 1965, a giant blackout in the Northeastern US left thirty million people without power for 12 hours. The incident highlighted the country's growing dependence on electricity and its vulnerability to disruption. Congress later responded by calling for a research and planning organization to support the electric utility industry. The Electric Power Research Institute (EPRI) was founded in 1972 in response in order to enable the power industry to proactively address key issues like grid reliability. FPC's National Power Survey characterized the power industries' predicament at the time: "Growing demand, sharply rising costs and changing social values have combined to place unusual stresses on the US electric power industry. This is evident from the strained power supply conditions in many parts of the country and from numerous current proposals for increased rates."[2]

Prior to the oil embargo, the Nixon Administration was well aware of the potential for an energy crisis and had been focused on the need to place a greater emphasis on research to uncover new energy sources since 1971. Citing the Northeast blackout of 1965, Nixon called for a cabinet-level agency to coordinate federal energy research. In June 1971, Nixon addressed Congress on the subject of energy.[3] Options were limited for switching away from oil-fired generation.

Nuclear power was the most obvious choice. The Atomic Energy Act of 1946 established the Atomic Energy Commission (AEC), which inherited all the Manhattan Project's R&D activities. A major focus of the AEC was research on "atoms for peace," the use of nuclear energy for civilian electric power production. Over half of America's public energy R&D dollars were devoted to nuclear power between 1948 and 1978; the rest went to fossil fuel technology development.[4]

The historical records shows that some voices had tried in vain to sound the alarm about the lack of diversified energy research. Recall, Percy Thomas and the FEA explicitly asked Congress to initiate a wind research program in 1951. However, the sentiment of Congress at that time was that nuclear power was the next source of low cost energy and there was no need to look for other energy alternatives. Furthermore, having heard of emerging wind power research in Europe, US policy makers were content to let Europeans proceed with their wind power experiments and have them report back to the United States with the results. Low energy prices and shortsightedness had lulled US policy makers to sleep in the 1950s and 1960s. The October 1973 oil embargo was the wake-up call.

In the summer of 1973—prior to the October oil embargo—Nixon had called for the creation of an "Energy Research and Development Administration" to coordinate the federal energy research program. In June, Nixon asked Dixy Lee Ray—the colorful and unconventional Chairman of the AEC—to review existing federal energy research activities and make a recommendation for integrating and accelerating these programs. In the meantime—through executive action—Nixon created the Energy Policy Office and the Federal Energy Office to serve as interim offices until a formal cabinet-level agency was formed.

But by the time the Dixy Lee Ray report—*The Nation's Energy Future*—was issued in December, all hell had broken loose. Dixy Lee Ray tried to calm Washington with the report. What the country needed—according to the report—was "energy self-sufficiency."* Furthermore—according to the report—one of the most important tools to achieve energy self-sufficiency was to "promote, to the maximum extent feasible, the use of renewable energy sources and pursue the promise of fusion and central station solar power."[5]

In this manner, the report opened the door to wind power research on a scale that the country hadn't seen before. Fortunately, America wouldn't be starting completely from scratch as it ramped up its wind research program. As it turned out, with funding from the National Science Foundation (NSF), NASA had been working on a small wind research program since 1970. The program had started when Cruz Matos—Secretary of the Interior of Puerto Rico—asked NASA's Lewis Laboratory to design a wind turbine to generate electricity for the Island of Culebra.

Actually, it was more of a skunkworks program than anything else. Upon hearing of the Culebra project in 1970, Louis Divone at the NSF authorized the funds to construct and operate an experimental 100-kW wind turbine. Divone was on temporary assignment at the NSF from the California Institute

---

* An innocuous sounding phrase that would quickly become politicized and used as a political ping pong ball for decades as the United States moved further away from domestic energy independence. Interestingly, a recent surge of domestic oil and gas production coupled with the growth of renewable energy sources has driven US net imports down significantly.

of Technology's Jet Propulsion Laboratory at the time. He had initially planned to stay with the NSF for just six months, but he wound up leading the national wind research program at the Department of Energy (DOE) throughout the 1970s.

Divone was a graduate of MIT and was familiar with the work of fellow MIT alum—Palmer Putnam—who had assembled a world-class team and constructed the world's first megawatt-scale wind turbine in 1941. Divone agreed to fund NASA's wind power research effort in 1970 and he did it within the context of NSF's Research Applied to National Needs (RANN) program, which was a program that was intended to refocus NSF's research efforts on solving problems related the environment, energy, and economic productivity.

In the fall of 1973—prior to the release of the report—Dixy Lee Ray sent her team to talk to Divone about wind power. They had heard that that Divone was working with a modest research budget and wanted to get a sense of what could be accomplished with more funding. Divone's response ultimately formed the basis of the wind R&D funding recommendations contained in the report. Divone recommended an accelerated wind R&D plan that called for US$100 million in funding over a five-year period starting in 1974. Divone had a strong sense of optimism about what could be accomplished if enough money was available. Testifying before Congress in 1974, Divone stated that "there is no single problem to be overcome regarding accelerating the implementation of wind energy systems. Rather, there are a large number of unknowns to be addressed due to the dearth of activity over the past 20 years...."[6] The challenge—as Divone saw it—was the need for "cost reduction by means of improved aerodynamic performance, minimization of input loads, and advanced subsystem and system configuration."[7]

Wind power visionary and the Director of the University of Massachusetts Ocean Engineering department—William Heronemus—also played a role in convincing policy makers that wind power had a bright future. In 1971, Heronemus proposed the development of a large network of offshore wind turbines. By 1972, he was presenting his proposal for a wind power development program before the US Congress; and by 1973, he told a slack-jawed Senate Subcommittee that "It would not be foolish at all to state this this country could be totally energized by solar energy and other renewable resources by the year 2000."[8]

In the wake of the publication of the Dixy Lee Ray report, Divone and Heronemus's recommendations for wind research support were subsequently heeded by Congress and the White House. Wind power would receive US$1.8 million in annual funding in 1974. Annual funding would increase to US$60.5 million by 1980. Before the decade was out, the US FWEP—as it become known—would receive over US$150 million in funding—one and a half times the amount that Divone originally recommended to Dixy Lee Ray's staff in 1973.[9] The FWEP was funded through the US Department of Energy (DOE), which was later established in 1977, and managed by NASA's Lewis Research Center.

With money in hand—right out of the gate—NASA convened a series of workshops on wind power. The list of attendees at these events was a "Who's Who" of wind power and included the likes of Germany's Ulrich Hütter, Marcellus Jacobs, and Beauchamp Smith from the S. Morgan Smith Company. In addition to holding these meetings, Divone and NASA researchers Ronald L. Thomas and Joseph M. Savino organized a panel to assess the potential of wind and solar energy. The panel concluded that sufficient energy could be derived from US wind resources to supply up to 19 percent of America's annual electricity requirements by the year 2000.[10]

Together, Thomas and Savino worked tirelessly to become students of wind power technology. They familiarized themselves with the well-known wind turbines of the twentieth century. By examining these designs carefully, they concluded that there were basically two suitable wind turbine approaches: Ulrich Hütter's 100-kW design and Johannes Juul's 200-kW Gedser design. Hütter's wind turbine was a fast-rotating, downwind, two-bladed machine that relied upon active blade pitch to control power output. Hütter focused on aesthetics and lightweight materials to improve performance, which is why his machine had the world's first blades made out of GFRP. However, because it was a downwind turbine, the blades were subject to cyclic loads as they passed behind the tower. Hütter compensated by employing a "teetering hub" that enabled the hub to move back and forth in order to absorb these cyclic loads.

Juul's Gedser wind turbine was a slower rotating, three-bladed, upwind machine that used passive stall to control power output. The genius in Juul's design was a special blade shape that resulted in an automatic aerodynamic stall at high wind speeds. Juul focused on reliability, rather than lightweight materials, aerodynamics, or even optimal performance. The Gedser wind turbine used guy wires to support the blades. It looked like a flying mousetrap. Also, because it was an upwind machine and the oncoming wind struck the blades before the tower, there was less stress on the blades and no need for a teetering hub as in Hütter's design.

Of course, there are advantages and disadvantages to each design. Today, megawatt-scale wind turbines employ a hybrid approach that contains elements of both Juul's and Hütter's classic designs. The typical wind turbine today is a three-bladed, upwind machine with lightweight elements such as fiberglass or carbon composite blades for maximum performance. Passive stall controls worked well in the 1980s and early 1990s up to a certain size, but the disadvantages of loud noises and large blade stresses became too great as wind turbines grew beyond 1 MW starting in the late 1990s.

But in 1974, Thomas and Savino were focused on finding a single solution. After exhaustive research, they concluded that Hütter's design was the most functional wind turbine design because they were impressed with its low cut-in speed and higher performance relative to Juul's machine. Also, since they had an eye toward the development of increasingly large machines over time, they concluded that Hütter's design would scale up to the megawatt size range better. They decided that NASA's first 100-kW prototype would be similar to the German design.[11]

At the same time, NASA articulated a future pathway for FWEP, they would: (1) develop reliable wind turbine prototypes starting with a size of 100 kW; (2) learn from the smaller prototypes; and then (3) build increasingly large ones over time. An important underlying belief of the program directors was that wind power would never achieve cost-competitiveness unless the capital costs of wind turbines could be spread out across millions of kilowatt-hours. Giant machines were needed to bring costs down to competitive levels. This was a good vision for the program, but it would be essential for NASA to follow through on each step.

There was another reason NASA wanted to build giant wind turbines. After ramping up from a US$401 million budget in 1960 to nearly US$6 billion by 1966, NASA budgets had begun to shrink back down by the early 1970s. By 1974, NASA's budget had fallen back down to US$3.2 billion. The number of NASA employees was 26,000 in 1974; 5,000 lower than the 1966 level. NASA's importance in the federal budget also diminished in the early 1970s. NASA's budget accounted for nearly 4.5 percent of the federal budget in 1966, but it accounted for just 1 percent by 1973.[12] Many within NASA believed that the drive to build giant wind turbines would be the perfect vehicle to solve both the country's energy challenges and simultaneously reenergize NASA. Wind power would be NASA's next moonshot.

10.2 Back to Ohio

With Congressional funding in hand, NASA's wind power research efforts began in Ohio in 1974, just 60 miles west of the location where Charles F. Brush constructed the first American wind turbine in the backyard of his mansion in 1887. NASA's first 100-kW wind turbine—MOD-0—was built and erected at NASA's Lewis Plum Brook Station in Sandusky, Ohio. Plum Brook was an interesting choice. The site was initially constructed by the US government as an ordnance factory in 1941. On the eve of US involvement of World War II, the government took possession of 44 million acres of previously privately owned farmland in Ohio to build Plum Brook. After the war, Plum Brook was selected as a test site for a 60 MW nuclear reactor for the National Advisory Committee for Aeronautics (NACA), NASA's predecessor organization. The reactor would support NACA's effort to build a nuclear-powered airplane. The reactor was completed in 1961, just in time for the United States government to lose interest in building a nuclear-powered airplane.

The Plum Brook reactor was put into operation regardless, and refocused on conducting research in support of a nuclear-powered rocket. However, in

1973, NASA's vision for the future of the space program centered around the Space Shuttle—not a nuclear-powered rocket—and the reactor at Plum Brook was shut down. By 1974—when plans for a NASA wind research program were being assembled—Plum Brook became the preferred location for the first test wind turbine out of necessity.[13] Never mind that wind speeds averaged less than 5 m/s at the site.

MOD-0 was up and running at Plum Brook by the summer of 1975. It was a 38-m diameter downwind, two-bladed wind turbine with a 100-kW synchronous generator sitting atop a 30-m steel-truss tower. The electrical output of MOD-0 was regulated through active blade pitch control. It was like Hütter's original design except for two important aspects: first, the blades were made out of aluminum instead of GFRP because NASA couldn't find a contractor who could deliver fiberglass blades; and second, the hub did not have the ability to teeter.

From NASA's perspective, a teetering hub wasn't necessary because their computer calculations indicated that the tower would slow the oncoming wind by just 24 percent and the blades were strong enough to absorb the cyclic loads by bending. However, as soon as operations started, problems emerged. The blades began bending at the root between the blade and the hub. NASA reran its computer simulation and discovered that that the tower had retarded the wind by over 90 percent. This produced large cycling loads on the blade as they passed through the shadow of the tower. Efforts were made to modify and streamline the tower by removing the center staircase and replacing it with an elevator. This reduced—but did not eliminate—the issue.[14]

NASA conducted experiments with MOD-0 in 1975 and 1976. They experimented with a single-bladed rotor, a rotor with blades constructed of cloth-covered wood ribs, and operated it without a yaw motor. But, at the time, NASA didn't conclude—as it should have—that the downwind design was not the right approach. Instead, they pressed forward with four new wind turbines based on the MOD-0 design. MOD-0 remained an experimental wind turbine for NASA until it was dismantled in 1986. In 1985, it was equipped with upgrades that enabled it to operate at variable speed. It is important to note here, that when NASA diagnosed MOD-0 and continued to proceed with the downwind design despite the evidence that this design was not suitable, they violated their original commitment to incorporate lessons learned into subsequent wind turbine designs. The failure to incorporate feedback into their wind power research would ultimately lead to the development of a series of poorly designed large wind turbines.

Four 200-kW MOD-0A wind turbines were built and installed between 1977 and 1980 at Clayton, New Mexico, Culebra, Block Island, and Oahu Hawaii. The first was built by NASA, the subsequent three were built by Westinghouse. The wind turbines had limited success and all experienced similar challenges, including blade fatigue, stress cracking, hydraulic failures, burned out generator bearings, fluid-coupling leaks, and yaw-control system malfunctions.[15]

Figure 10.1 Erected in 1976, NASA's 100-kW MOD-0 wind turbine was the first turbine designed and constructed as part of the US Federal Wind Energy Program (FWEP). MOD-0 served as an experimental "test bed" for future NASA wind turbines.

By June 1976, NASA awarded a contract to build the giant 2-MW MOD-1 wind turbine to GE. Since the time that GE had provided support for Palmer Putnam's 1941 Smith-Putnam wind turbine, the company had focused on developing supercritical steam turbines and nuclear reactors. With the NASA contract in hand, GE was once again poised to help build the world's largest wind turbine.

The 2-MW MOD-1 would be a scaled-up version of the 100-kW MOD-0 wind turbine. The only planned difference was the use of GFRP blades. However, MOD-1's GFRP blades failed during initial testing, so GE commissioned Boeing to develop steel blades instead. The goal of FWEP was to build economically competitive wind turbines and MOD-1 was intended to be the first large-scale wind turbine designed to achieve that goal. However, at US$30 million, the final cost of MOD-1 was so high that the estimated cost of electricity was still nearly 20 ¢/kWh.[16]

MOD-1 was built on a hilltop overlooking the college town of Boone, North Carolina. The site was selected over sixty-four other possible locations after NASA measured wind speeds at each location for one year. Boone was also selected as the site for MOD-1 because of the enthusiasm of the local distribution utility—the Blue Ridge Electric Membership Company (BREMCO)—and the local population, which strongly supported the idea of alternative energy. Situated on a site known as "Howard's Knob," MOD-1 was intended to be the ideal representation of the progress and promise of wind power. Work began in July 1976, and the wind turbine was completed three years later in July 1979. The community of Boone turned out for a street party celebrating MOD-1's dedication, and then they eagerly waited for the turbine to begin producing electricity. Full operation started in February 1980.

After MOD-1 began operating, Boone residents immediately reported odd noises and interrupted television signals. Both the noises and the television signal disruptions occurred in random locations around Boone. NASA was forced to reduce the rotation of the turbine in order to reduce the signal disruptions. However, low-speed operation created a resonance in the drivetrain. NASA also halted the operation of the turbine during "prime time" television, which was also the windiest time of the day around Boone.[17] After just 130 hours of operation—in January 1981—twenty-two studs on the low-speed shaft sheared off, severing the link between the rotor and the drivetrain. The estimated US$500,000 cost for the repairs forced NASA to mothball the machine in August 1981. MOD-1 never generated at its rated 2-MW capacity. The US$30 million machine was auctioned off to a yarn company for US$50,000 in 1982.[18]

Lou Divone would later testify in front of a Congressional committee, admitting that MOD-1 was designed before the operational data was available from MOD-0. He also explained that NASA lacked the analytic tools to correctly predict and understand the stresses on a large wind turbine.[19] GE wind project manager—Reinhold Barchet—told a Congressional subcommittee that the MOD-1 experiment would never result in a "light-weight, commercially attractive, megawatt-class wind turbine."[20]

NASA funded GE to conduct a study to explore wind turbine concepts and to develop a new design that could overcome the problems of MOD-0 and MOD-1. The benefits of a teetering hub and modified blade pitch system were examined. GE recommended design changes that they believed would reduce the cost of production to 6 ¢/kWh. However, NASA never acted upon the study and never built GE's proposed MOD-1A. In fact, by 1979, the writing was on the wall regarding the lack of success and limited future of FWEP. Several previously planned wind turbines, including MOD-3, MOD-4, MOD-5A, and MOD-6H, were all cancelled or simply abandoned mid-stream.

NASA's 2.5-MW MOD-2 design did move forward. The MOD-2 wind turbines were built by Boeing and installed by the Bonneville Power Administration. MOD-2 was a downwind wind turbine with a 90-m diameter, and—at long last—a teetering hub. Three MOD-2 wind turbines were put in operation at Goodnoe Hills, Washington in April 1980. A fourth began operating at Medicine Bow, Wyoming in September 1982. However, once installed, the MOD-2 wind turbines experienced operational problems that required shutdowns and retrofits. The low-speed shaft of one of the units cracked in 1982 due to poorly placed holes drilled in it for hydraulic tubing and electrical conduits. A later review by NASA found twenty-nine areas of concern in the MOD-2 design. The MOD-2 wind turbines at Goodnoe Hills were decommissioned in 1986 and the Medicine Bow unit was sold for scape in 1987.[21]

MOD-5B was the final wind turbine built by FWEP. It was originally designed by NASA as a 7.2-MW wind turbine, but it was built as a 3.2-MW machine with a 98-m diameter rotor. It was purchased by Hawaiian Electric Company (HECO) and installed in Oahu in 1987. Ownership was passed to Makani Uwila Power Corporation (MUPC) and the machine was kept in service intermittently until late 1996. At that time, due to financial difficulties, the machine was shut down along with the rest of MUPC, and MOD-5B was conveyed to the property owner, Campbell Estates. With no prospects for continued operation, Campbell Estates decided to disassemble and scrap the machine. Prior to the decommissioning in July 1998, DOE salvaged the drivetrain gearbox and generator.[22]

Interestingly, MOD-5B was one of the first variable-speed wind turbines. In variable-speed designs, the wind turbine rotates at a variable speed depending upon the speed of the oncoming wind. MOD-5B's variable speed operation was accomplished with the assistance of a cycloconverter that converted the generator's AC output to the fixed-frequency AC required by the grid. Variable-speed wind turbines would soon catch on and become the dominant wind turbine technology by the turn of the twenty-first century. NASA was ahead of the curve.

10.3   The Turning Point

By 1981, both the nation's excitement around the prospects for wind energy and the threat of another global oil crisis had waned. The new Reagan Administration was intent on halting America's renewable energy research programs. The President informed the country that his administration would no longer seek to reduce oil imports. Further, the basic idea of government-funded energy research was the antithesis of Reagan's laissez-faire approach to the role of government. Reagan's stated goal was to eliminate the US DOE and its associated research programs, including wind power.

Reagan's strategy for eliminating renewable energy research was to simply have the Office of Management and Budget (OMB) decline to include funding for these programs in the Administration's budget request. However, because the Administration's proposed budget would have to be authorized by Congress, the Administration's request—or lack thereof—wouldn't be the final word on the matter. Thus, instead of eliminating the wind research budget immediately as desired, the Reagan Administration was forced instead to gradually reduce the budget over time. From a peak funding level of US$60.5 million in 1980, the federal wind research budget shrank to US$8.8 million by 1989.[23]

During the debate about the appropriateness of federal funding for wind power research, NASA, Boeing, GE, and other proponents of the program were caught between a rock and a hard place when confronting fiscal conservatives in Congress who wanted the research programs terminated. If proponents indicated that NASA's wind research program had been successful, then the conservative members of congress congratulated them on their success and concluded that government support was no longer needed. If they indicated that NASA's wind research programs had failed, then conservatives held up NASA's efforts as an example of government waste and incompetence.

At one point during a 1981 Congressional hearing, after Boeing and GE representatives indicated that wind power was commercially viable, representative Jim Dunn asked them to explain why they weren't willing to invest their own money in wind turbines, at which point they were forced to backtrack and admit that wind power is "too risky" for private capital.[24]

NASA wind power research program of the 1970s and 1980s have been criticized as ineffective. Writing in 1995, wind power historian Paul Gipe argued that "If the R&D funds had been used as incentives rather than misdirected to NASA and the aerospace industry, it is conceivable that the United States would not only have developed a more dynamic manufacturing industry but would also have installed more wind turbines."[25] According to Gipe "The behemoths never delivered as much energy or worked as reliable as their proponents promised...."[26] German wind historian Matthias Heyman later characterized NASA's effort to develop large wind turbines as "hubris" and wondered in retrospect "how engineers, scientists, and politicians, more or less starting from scratch, could seriously have believed in the feasibility of giant turbines with tower heights, rotor diameters, and weights exceeding the lengths, wingspan, and weight of a jumbo jet."[27]

The frustrations with government-funded wind research expressed by Gipe and Heyman were understandable. After decades of incremental wind power technology progress, the 1973 energy crisis offered a golden opportunity for wind power technology to advance to commercialization through well-funded public R&D efforts. Watching governments spend millions of dollars building machines that were destined to fail was excruciating for wind power advocates. To his credit, long-time wind proponent Paul Gipe was one of the few people willing to be frank about the promise and pitfalls of wind power at the time and he documented much of the history of wind power in his 1995 book *Wind Energy Comes of Age.*

It is true that American policy makers and scientists did believe that a big and bold solution was needed to solve the energy crisis; rather than building incrementally from existing turbine designs, FWEP raced toward the development of multi-megawatt wind turbines. This created problems because there was not enough time to incorporate the lessons learned from smaller wind turbines into the plans for much larger units. However, it must also be noted that—as it later turned out—many of the wind power technologies pioneered within the FWEP such as doubly-fed variable-speed generators, light weight tubular towers, and engineering design tools are still used in the wind industry today.[28]

Furthermore, it is now clear that wind turbine technology eventually evolved in the same direction that NASA was headed. Wind turbine nameplate capacity, hub height, and rotor diameter have all increased significantly over time. Average nameplate capacity is now 2.2 MW, average hub height is 85 m, and average rotor diameter is over 100 m.[29] It turns out that average rotor diameters are now equal to the wingspan of the largest jumbo jets. Indeed, massive wind turbines rule the day in the twenty-first century.

Thus, in hindsight, it is more accurate to say that while NASA wasn't necessarily moving in the wrong direction, they were trying to get there too quickly. As wind turbine technology history demonstrates, you cannot simply upsize from a 100-kW to a 1-MW wind turbine and maintain acceptable levels of reliability and performance. Instead, wind researchers—both public and private—would need to build and test successively larger wind turbines over a period of decades—not years—to reach the 1-MW level in the late 1990s. However, neither the American public, nor US policy makers had the wherewithal to support wind power research over such a protracted period, particularly after oil prices collapsed in 1986 and much of the original justification for the program was eroded.

The American FWEP revived government interest in wind power research around the world. Germany started building its own giant wind turbine— Growian—largely in response to US efforts. In Denmark, NASA funded the repair of Juul's Gedser wind turbine and helped re-ignite Danish wind power development efforts.

However, the most impactful wind power innovations during this period would occur out of the spotlight, far away from these national research efforts. While NASA was building its fragile giants, a Danish carpenter and a blacksmith were busy reinventing wind power from the confines of their small workshop in Denmark. It is here that wind power was reinvented to meet the needs of the global electric power system. These bottoms-up research efforts enabled Danish companies to eventually emerge as global leaders in wind turbine manufacturing. The next part of the wind power story takes place in a carpenter's workshop on Denmark's Jutland peninsula.

Notes

1 Energy Information Administration, 2012.
2 Federal Power Commission, 1971, 1.
3 Nixon, 1971.
4 Sissine, 2014, table I.
5 Ray, 1973, vii.
6 U.S. House of Representatives, Committee on Science and Astronautics, 1974, 116.
7 Ibid.
8 Heronemus, 1973.
9 U.S. Department of Energy, 1990.
10 Thomas and Savino, 1973, 1.
11 R. L. Thomas, 1976, 2.
12 Office of Management and Budget (OMB), 2016, table 9.8.
13 Bowles and Arrighi, 2004, 9–129.
14 R. L. Thomas, 1976, 3.
15 Schefter, 1982, 54.
16 Poor, 1979, 42.
17 Schefter, 1982, 64.
18 Serchuk, 1995, 148–9.
19 U.S. House of Representatives, Committee on Science and Technology, Subcommittee on Energy Devleopment and Applications, 1979, 74–5.
20 Ibid., 118.
21 Serchuk, 1995, 170.
22 Carlin, Laxson, and Muljadi, 2001, 23.
23 U.S. Department of Energy, 1990.
24 U.S. House of Representatives, 1981, 498–9.
25 Gipe, 1995, 74.
26 Ibid., 96.
27 Matthias, 1998, 668.
28 Spera, 1994.
29 Wiser and Bolinger, 2017.

Bibliography

Bowles, Mark D., and Robert S. Arrighi. *NASA's Nuclear Frontier: The Plum Brook Reactor Facility*. Monograph in Aerospace History. Washington, DC: NASA History Division, 2004.
Carlin, P.W., A.S. Laxson, and E.B. Muljadi. *The History and State of the Art of Variable-Speed Wind Turbine Technology*. Technical Report. Golden, CO: National Renewable Energy Laboratory, 2001.

Energy Information Administration. *Annual Energy Review 2011*. DOE/IEA, 370. DOE/EIA-0381(2011). Washington, DC: US Energy Information Administration, US Department of Energy, 2012. Accessed November 29, 2017. https://www.eia.gov/totalenergy/data/annual/pdf/aer.pdf.

Federal Power Commission. *The 1970 National Power Survey*. Washington, DC: US Government Printing Office, 1971.

Gipe, Paul. *Wind Energy Comes of Age*. New York: John Wiley, 1995.

Heronemus, William E. "The Possible Role of Unconventional Energy Sources in the 1972–2000." Testimony, Committee on Science and Astronautics, Subcommittee on Science, Research and Development, U.S. House of Representatives, 92nd Congress, Washington, DC, June, 1973.

Karnøe, Peter. "Technological Innovation and Industrial Organization in the Danish Wind Industry." *Entrepreneurship and Regional Development* 2, no. 2 (1990): 105–24.

Matthias, Heymann. "Signs of Hubris: The Shaping of Wind Technology Styles in Germany, Denmark, and the United States, 1940–1990." *Technology and Culture* 39, no. 4 (1998): 641–70.

Nixon, Richard M. "Special Message to the Congress on Energy Resources." In *Public Papers of the Presidents: Richard M. Nixon*, 703–14. Washington, DC: US Government Printing Office, 1971.

Office of Management and Budget (OMB). *Fiscal Year 2017 Historical Tables, Budget of the United States*. Washington, DC: Executive Office of the President of the United States, 2016.

Poor, R.H. "The General Electric MOD-1 Wind Turbine Generator program." In *Large Wind Turbine Design Characteristics and R&D Requirements*, edited by Lieblein, S., 35–61. Cleveland, OH: NASA Lewis Research Center, 1979.

Ray, Dixy Lee. *The Nation's Energy Future : A Report to Richard M. Nixon, President of the United States*. Washington, DC: United States Atomic Energy Commission, 1973.

Schefter, James L. *Capturing Energy from the Wind*. Washington, DC: NASA, 1982.

Serchuk, Adam Harris. "Federal Giants and Wind Energy Entrepreneurs: Utility-Scale Windpower in America, 1970–1990." PhD thesis, Virginia Polytechnic Institute and State University, 1995.

Sissine, Fred. *Renewable Energy R&D Funding History: A Comparison with Funding for Nuclear Energy, Fossil Energy, and Energy Efficiency R&D*. Washington, DC: Congressional Research Service, 2014.

Spera, David A., ed. *Wind Turbine Technology: Fundamental Concepts of Wind Turbine Engineering*. New York: American Society of Mechanical Engineers (ASME) Press, 1994.

Thomas, Ronald L. *Large Experimental Wind Turbines – Where We Are Now*. Technical Memorandum. Cleveland, OH: NASA Lewis Research Center, 1976.

Thomas, Ronald L., and Joseph M. Savino. *Status of Wind-Energy Conversion*. Technical Memo. Washington, DC: NASA, 1973.

U.S. Department of Enregy. *Renewable Energy Budget History.* Internal Document. Washington, DC: Department of Energy (DOE), 1990.

U.S. House of Representatives, Committee on Science and Astronautics. *Hearing on Wind Energy Before the Subcommittee on Energy.* Washington, DC: U.S. House of Representatives, 1974.

———. *Fiscal Year 1982 Department of Energy Authorization – Fossil, Solar, and Geothermal Energy and Basic Sources.* Washington, DC: U.S. Congress, 1981.

U.S. House of Representatives, Committee on Science and Technology, Subcommittee on Energy Devleopment and Applications. *Oversight: Wind Enery Program.* Washington, DC: U.S. Congress, 1979.

Wiser, Ryan, and Mark Bolinger. *2016 Wind Technologies Market Report.* Washington, DC: U.S. Department of Energy, Wind Energy Technologies Office, 2017.

11

Denmark Reinvents Wind Power

Wind power was re-invented by local blacksmiths, students, independent researchers and other innovative persons.

—Benny Christensen (2013)

11.1 Feared to Freeze

The 1973 global energy crisis had a significant negative impact on countries around the world. However, in some countries more than others, the oil crisis posed an existential threat. Gasoline shortages were tough for consumers in the United States; but in Denmark, the crisis posed a real threat to human life because oil was widely used for space heating. In the words of Danish renewable energy pioneer, author, and expert—Preben Maegaard—when the crisis hit in the winter of 1973 "there was a real concern in this country on how we are going to get through the winter because people were relying on heating their homes using oil ... and when you live in a cold climate here where we have these cold winters, we really feared to freeze."[1]

Ninety-four percent of Denmark's energy supply consisted of imported oil at the time; the remaining 6 percent consisted of foreign coal.[2] When it comes to energy security, Denmark had been in this precarious position throughout the twentieth century and had not taken much action to correct it. In fact, one of the reasons that Poul la Cour embarked upon wind energy research at the turn of the century was because of his concern over Denmark's energy dependency. In 1900, as he considered a government research grant that he received, he wrote "[O]ne may find it surprising that a small country like Denmark, taking a bold step ahead of any others, is willing to devote such large sums of money to research on the use of wind power. Nothing could be more natural, however, considering that this country, which has no waterfalls or coal, is

The Wind Power Story: A Century of Innovation that Reshaped the Global Energy Landscape, First Edition. Brandon N. Owens.
© 2019 by The Institute of Electrical and Electronics Engineers, Inc.
Published 2019 by John Wiley & Sons, Inc.

exposed to the risk of paying increasingly high rates for the purchase of fuel from foreign countries...."[3]

Yet, the country hadn't yet heeded la Cour's call for energy independence.[4] By 1936, coal and oil together accounted for 95 percent of electric power generation. Wind amounted to less than 1 percent of total electricity production.[5] Later—in 1947—after the fossil fuel shortages of World War II abated, utility engineer and wind power pioneer Johannes Juul made an argument for wind power deployment because he believed that wind power was necessary to "...ensure Denmark's electricity supply especially in the event of critical times, such as we have lived through twice in one generation."[6] By the time the 1973 crisis hit, Denmark was still desperately reliant on foreign energy supplies.

However, the 1973 oil crisis was a turning point for Denmark. It marked the beginning of a widespread, permanent recognition in Denmark that energy independence was of vital importance. The actions of government policy makers, electric utilities, corporations, energy entrepreneurs, and the general public since 1973 have reflected this belief. In Maegaard's words: "Whereas, the interest in wind energy formerly had vanished, as soon as wars and periods of energy shortage were over, there was now a new realization that the resource shortage and environmental problems had come to stay. It was a crisis without an end."[7]

Despite the agreed upon urgency of the energy problem, no one could agree upon on the right technology solution to address it. The national government in Copenhagen sprang into action and developed a national energy plan. When the plan was published in 1976, it didn't mention wind or other alternative power sources. Instead, the focus was on coal, natural gas, and nuclear power. According to the plan, by 1985, nuclear power was expected to account for 6 percent of Denmark's electricity production and by 1995, a full quarter of the nation's energy supply was expected to come from nuclear power plants.[8] The first policies intended to support wind power weren't enacted until 1979, and even then, wind power was viewed as a small part of the solution. It wasn't until 1985 that the political majority in the Danish Parliament decided to cancel the plans for the use of nuclear power.

Not everyone in Denmark saw things in the same light as the government and the electric utilities. Denmark had been engaged in an ongoing social discussion about the role of nuclear power in its society since the 1950s. By the time the crisis hit in 1973, there was a strong and vocal antinuclear movement. The Organization for Information on Nuclear Power was one of leading voices against nuclear power and they actively distributed information across the country in opposition to it. In addition, in 1974, a small team of engineers, blacksmiths, and teachers from a local technical school formed the North-Western Jutland Institute for Renewable Energy (NIVE). NIVE's stated aim was to seek the "optimal use of local human and technological resources" through the use of renewable energy.[9] Starting with a DKK 50,000 grant from The United Nations Educational, Scientific and Cultural Organization (UNESCO), NIVE focused its efforts on providing information on small-scale distributed energy technologies including wind turbines.[10] In addition, in September 1975, The Danish Organisation for Renewable Energy (OVE) was established. OVE advanced renewable energy along with F.L. Smidth & Company (FLS), the Danish industrial company that manufactured small wind turbines during World War II.

In 1974, FLS received a loan from Copenhagen to develop plans for a new, larger wind turbine. By September, the company unveiled its plans to create a new 1.2-MW wind turbine. They postulated that five hundred 1.2-MW wind turbines could be erected along the Western coast of Jutland to replace one nuclear power plant. FLS continued to advocate for wind power throughout 1974, but by 1975 they stopped their wind power efforts because of electric utility and government resistance to the idea. FLS CEO Benned Hansen said at the time "we cannot manage this task on our own." FLS felt that government and electric utility support would be needed to develop wind power.[11]

In the meantime, the Danish Academy of Technical Sciences (ATV) began looking into alternative energy options. In 1975, ATV's Wind Energy Commission concluded that wind resources were plentiful in Denmark and about DKK 50 million would be enough to establish a technology base to start the production of wind turbines.[12] According to ATV, in a decade these wind turbines would be able to provide 10 percent of the energy needed in Denmark at competitive prices.

The period between 1973 and 1975 was the beginning of a new national discussion in Denmark on the potential role of wind power. Copenhagen would eventually set up a wind power research program. It was funded to the tune of DKK 41.4 million and operated from 1977 to 1981. Of this amount, DKK 36.4 million was used for large wind turbine research and development. The program was built in the image of the US Federal Wind Energy Program. The Danish nuclear laboratory—Risø—and the Technical University of Denmark (DTU) were both tapped to conduct wind power research for the national research program.

Risø was founded in 1956 by Nobel laureate Niels Bohr, who had spent time at Los Alamos National Laboratory in the United States during World War II and was one of the fathers of the atomic bomb. Bohr supported the founding of Risø to advance peaceful uses for atomic energy. However, opposition to nuclear power starting in the early 1970s had tempered enthusiasm for nuclear power research and—much like NASA and its laboratories at the same time—Risø was looking for its next mission. Wind power fit the bill. Risø would eventually transform itself into a national laboratory for sustainable energy before it was eventually made an institute of DTU in 2008.

11.2 More Than a Carpenter

Denmark's government-led research activities started at Risø until 1978 with the establishment of a test station for small wind turbines. A year later, the test station was named the official authority for certification of wind turbines. Risø's certification role would become a key part of the eventual success of Danish wind turbine manufacturers. However, at the time it was initiated, the government-led wind research program was seen by many as a way for the Danish government and electric utilities to delay wind power development under the guise of a long-term research program.

A growing segment of the Danish population had come to believe that what was needed instead in the years immediately following the 1973 crisis was a strong—almost religious—commitment from individuals, entrepreneurs, students, and communities to rapidly develop alternative energy solutions. And that's exactly what happened in Denmark between 1974 and 1979. This period reinvention enabled Danish companies to eventually emerge as global leaders in wind turbine manufacturing starting in the 1980s.

A prime example of wind power's reinvention is the work of a carpenter from Jutland named Christian Riisager. Riisager teamed up with master blacksmith Erik Nielsen to design and build small wind turbines in 1974. When Christian Riisager set out to build his own wind turbine, he needed a starting point. After a little digging, he discovered the proceedings from the United Nation's conference on *New Sources of Energy and Energy Development*. The conference was held in Rome in August 1961 and the proceedings of the conference were published in 1962. The proceedings contained an entire chapter on wind power that included forty articles from authors such as American small wind turbine pioneer Marcellus Jacobs, Germany's Ulrich Hütter, and—of course—Denmark's Johannes Juul. The proceedings would become a roadmap for Danish wind power development in the 1970s. The only problem for Riisager was that the proceedings were written in English. Riisager had to rely upon his son to translate the English texts to Danish.[13]

With Juul's plans in hand, Riisager set about his task of building a reliable small wind turbine. Riisager's approach was similar to that of the Jacobs brothers in the United States in the 1920s. Starting in 1922, the Jacob brothers built a series of small wind turbines on their farm in Montana using whatever parts they could come by as long as it worked, including the rear axle of a Ford Model-T. Riisager was similarly resourceful and used a secondhand heavy-duty rear shaft from a British military vehicle. He connected the shaft to the electric generator using a simple chain and built the nacelle and blades out of waterproof plywood.

In 1975, Riisager built a small 7-kW wind turbine, placed it in the garden at his home in Skærbæk and—without permission from the local utility—connected it to the grid. Riisager's wife—Boe—watched as the electric meter turned backwards.

This was the first instance of what would later become known as "net metering." By early 1976, Riisager had built and sold a larger 22-kW wind turbine. It cost less to produce than the 7-kW model and it worked better.[14]

Riisager's accomplishment might have gone unnoticed if not for the fact that he sold his first turbine to journalist Torgny Møller. Møller wrote about Riisager's wind turbine in the Danish newspaper *Information*. Møller and his wife had purchased a farmhouse in 1976 with no insulation and had high heating costs. Møller had his eye on an electricity-generating wind turbine. He purchased Riisager's 22-kW model and received permission to interconnect it with the local electricity network.

Møller wrote about the wind turbine in *Information* and provided evidence of its reliability. In fact, he was so enthusiastic about wind power that eleven months after Riisager's turbine was connected to the electricity network, Møller and eleven others formed the Danish Windmill Association, which has since become the Danish Wind Industry Association (DWIA). In 1978, Møller began publishing a newsletter that focused on wind energy called *Naturlig Energi* (Natural Energy). Four years later, Møller started *Windpower Monthly*, which remains one of the primary sources of information on the global wind power industry.

After this, orders poured in for Riisager's wind turbines. His design—based on Juul's original—became the standard for Danish wind turbines in the mid-1970s. He upgraded his wind turbines as more information and better components became available. Eventually competitors emerged and Riisager's component suppliers were able to provide them with ready-made parts based on similar Juul-inspired wind turbine designs. After a decade of success, Riisager stopped production in 1985, yielding to the new class of larger industrial wind turbine manufacturers that had emerged by that time.

Riisager wasn't the only amateur wind turbine manufacturer to reinvent wind power in Denmark during this period; he's just the most well-known due to Møller's promotion of the Riisager wind turbine. For example, in the summer of 1976, Claus Nybro and Rio Ordell designed a 5-kW wind turbine on the island of Endelave. The turbine began operating in early 1978. By November 1979, Nybro and Ordell—along with new partners Hans Dollerup and Flemming Allerslev—erected a 22-kW test turbine at Kolding Højskole on the Jutland mainland. Rio Ordell would later become one of the first CEOs of the Danish Wind Turbine Manufacturers Organisation. In 1975, Jens Erik Madsen built a Johannes Juul wind turbine clone that had a 10-m diameter and a 14-m hub height. Although now inoperable, the turbine still stands today.[15]

While Riisager and others helped ignite a Danish reinvention of wind power in the image of Johannes Juul—at the same time across the Atlantic—NASA was forging ahead with its plans to build large wind turbines based on the work of Germany's Ulrich Hütter. Recall, Hütter's 100-kW wind turbine was a fast rotating, downwind, two-bladed machine that relied upon active blade pitch to control power output. Hütter had focused on aesthetics and making the wind turbine lightweight to improve performance, which is why he used the world's first GFRP wind turbine blades. However, because it was a two-bladed downwind turbine, the blades were subject to cyclic loads as they passed behind the tower. Hütter compensated by employing a teetering hub that enabled the hub to move back and forth to absorb these cyclic loads. NASA—in its hurry to build megawatt-scale wind turbines—neglected to include the teetering hub. As a result of this omission, NASA's 100-kW MOD-0 and 2-MW Mod-1 wind turbines experienced vibration problems as the blades passed through the wind shadow behind the tower and created unexpected harmonics.

In 1974, when NASA's Thomas and Savino were conducting research to determine the design of their first wind turbines, they met with some Danish wind power pioneers including Neils I. Meyer, professor at the DTU of Copenhagen, and Helge Claudi Westh and Jean Fischer from FLS, developers of the World War II era Aeromotors. Their correspondence lasted several years and the Danish trio tried unsuccessfully to convince NASA to adopt a wind turbine configuration that was like Juul's 200-kW Gedser wind turbine. NASA paid the costs to repair and restore the Gedser wind turbine in 1978. The restored Gedser wind turbine ran from 1977 to 1979. After decommissioning it became a museum piece in the Danish Electricity Museum in Bjerringbro, Denmark.

However, the NASA researchers weren't the only wind power innovators interested in the work of Ulrich Hütter. Mogens Amdi Petersen—the son of a school principal with long hair, a beard, and an ever-present Icelandic sweater—founded an alternative Danish school system for troubled youths called Tvind. Fresh off a six-month prison term for throwing rocks at German police during a 1969 riot, Petersen created Tvind to change society for the better and eventually he built a vast network of schools and commercial ventures. By the end of the 1970s, Petersen's approach to education led to scandal and legal troubles because of an alleged money-laundering network within the group's commercial and nonprofit ventures. But back in the wake of the 1973 oil crisis, Petersen and his students were focused on how to make their school energy self-sufficient.

The Tvind teachers and students decided to take up the cause of large wind power by building their own 2-MW machine from scratch. Their efforts eventually led to the successful installation of the world's largest operating wind turbine in 1978. The successful design and development of the Tvind wind turbine demonstrated the power of amateurs to bring wind power to life at a grand scale, but it didn't lead to the successful commercialization of any wind turbines. The Tvind team originally examined both wind and solar as potential options to make their school energy self-sufficient. They settled on wind power because of the strong winds at the school campus in Western Jutland.[16] Neither the teachers nor the students had any knowledge of wind power at the time—rather, the project would represent the collective efforts of the entire Tvind community.

The first thing they did was enlist the help of Ulrich Hütter who was at the University of Stuttgart at the time. Tvind enlisted Hütter as a consultant on the project and piled into a bus to make the trip from Denmark to Germany to learn from Hütter directly. The group was struck by Hütter's design and they were particularly impressed with the GFRP blades. In particular, they were impressed by the way the fiberglass strands of the blade root wound around the bolt holes so that the blade could be connected easily to the hub. This design approach avoided the need for guy wires to support the blades like those used in Johannes Juul's Gedser wind turbine. The Tvind team recognized the elegance of Hütter's machine and committed to adopting the same blade design.

The design process of the Tvind turbine was a collective effort of Tvind High School teachers, a group of students, interested community members, engineers, and affiliates. A team of up to twenty-one students was actively working on the turbine development at any given time, and they held regular meetings where the two or three engineers hired by the school for the project would provide different potential solutions for design aspects of the turbine and these would then be discussed at length until a unanimous decision was reached on the final turbine design.[17]

Figure 11.1 Denmark's 2-MW Tvind wind turbine. The Tvind wind turbine was put into operation in 1978. It was the product of a collective development effort at the Tvind High School. The wind turbine was largely based on German Ulrich Hütter's 1950s-era design.

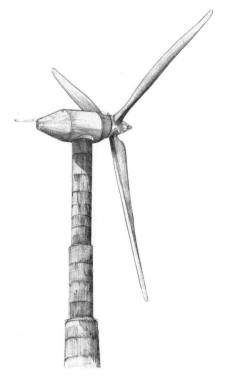

After a comprehensive research effort, the Tvind team eventually settled upon a downwind design—just like Hütter's—but the Tvind machine would operate at variable speeds and would be connected to an induction generator with a two-phase converter that converted the wild AC to 50-Hz AC as needed to interconnect to the transmission network. The turbine would be controlled by an active pitch system and had an active yaw system. The hub had a 54-m rotor diameter that was affixed atop a 50-m concrete cylindrical tower. Construction started in May 1975 and the completed wind turbine was put into operation in March 1978.

The Tvind wind turbine was not without its problems. Since Tvind was a downwind machine, when in operation, the wind turbine experienced vibration problems when the blades passed behind the tower. In order to avoid dangerous blade vibrations, the Tvind team initially limited the speed of the turbine to a lower level. As a result, the wind turbine reached less than half of the stated generator capacity. Thus, the 2-MW Tvind wind turbine actually only produced 900 kW of output. Still it was a leap forward, but because of their attachment to Hütter's downwind design, Tvind followed NASA's footsteps by adopting a design that ultimately wasn't suitable for large-scale wind power. The Tvind wind turbine is still in operation today with all of the original parts, except the blade and blade bearings. In 2018, its fortieth year of successful operation was celebrated.

11.3   Let 100 Windmills Bloom

The Tvind wind turbine was a source of inspiration and innovation for Denmark. Up to 100,000 people visited Tvind as it was being built, and a large number of Danish wind power pioneers became hooked on the idea.[18] The Tvind slogan for the project was "Let 100 windmills bloom." In this spirit, all of the technology and experience from the project was made publicly available— it was the world's first "open source" wind turbine. Inspired by the development of the Tvind wind turbine—in 1977—a group of five craftsman got busy building a small wind turbine using the same technology and design principles as the large Tvind turbine. The group included Arne Friis, Leif Nielsen, Henry Jørgensen, Svend Adolfsen, and Torben Andersen.

The blades on the smaller wind turbine developed by this group were 4.5-m GFRP blades. They were downsized versions of the Tvind wind turbine blades. The Tvind students had created the blades and the accompanying mold as a prototype prior to the construction of the larger 2-MW wind turbine. The Tvind students had subsequently made the molds available to anyone who wanted to use it. A small fiberglass workshop—TV Glassfiber—was the first to use the molds for a wind turbine built by Arne Friis. Erik Grove-Neilsen subsequently used the molds to make wind turbine blades in a farm in øaker in Middle

Jutland.[19] By making the wind turbine molds available, Tvind set in motion a series of events that would lead to a blossoming of Danish wind turbine designs.

Grove-Neilsen's blades quickly became a standard component for small Danish wind turbines. Grove-Neilsen's first set of øaker blades were used by Svend Adolfsen. Adolfsen's wind turbine worked reliably and he subsequently started serial production and started competing with the Riisager wind turbines. In 1979, Adolfsen's wind turbine with Grove-Neilsen's øaker blades became the first wind turbine design to achieve official certification at the Risø test station for small wind turbines.

During this period, Henrik Stiesdal partnered with blacksmith Karl-Erik Jørgensen to develop a 15-kW wind turbine through a company called Herborg Vind Kraft (HVK). They followed up with a 22-kW wind turbine and—later—a 30-kW version. The 30-kW version was called HVK-10 because of its 10-m rotor diameter. It was an upwind, three-bladed wind turbine that used Erik Grove-Neilsen's øaker blades. Although the HVK-10 wind turbine performed extremely well, Stiesdal and Jørgensen had financed the project with DKK 50,000 from a public fund for investors and they didn't have money to build any additional wind turbines. They began making the rounds, trying to find investors to support the commercial production of their wind turbines.

As Stiesdal and Jørgensen attempted to move forward with their small commercial wind turbines, Denmark's government-led wind research effort also progressed. One of the first actions was to sign on to the International Energy Agency's Agreement for Co-operation in the Development of Large-Scale Wind Energy Conversion Systems. Denmark, along with the United States, Germany, and Sweden joined the agreement in 1977. The United Kingdom and Canada also joined subsequently. A condition for joining the agreement was that each participating country undertake at least one large-scale national wind power project.

The Danish government had been working on the design and construction of two 630-kW wind turbines at the time. Both wind turbines were three-bladed, upwind models with 40-m rotors and 45-m hub heights affixed to asynchronous or induction generators. The models differed only in their control systems—Model A used a passive stall like Johannes Juul's Gedser wind turbine, and Model B used an active pitch-control system. Experiments based on these differences ultimately concluded that full blade-pitch control yielded smoother electrical output and offered greater rotor-thrust reduction at high wind speeds.

These nationally funded wind turbines became known as Nibe-A and Nibe-B because they were in the town of Nibe, on the Jutland Peninsula in northern Denmark. The wind turbine designs were developed by the Danish national wind research program through a yearlong design study in 1977. In February 1978, final specifications were sent to part manufactures FLS, Thrige Titan, and Frichs. Construction was completed in 1979 and government researchers began testing the wind turbines under the supervision of the Danish utility ELSAM.[20] Unfortunately, the tests didn't go smoothly. "Tests and trial operations followed and revealed a lot of problems that caused delays of a few days to several months," the project leaders later reported.[21] The biggest problem was metal fatigue in the steel parts of the rotor blades. Ultimately, Nibe B—the pitch-controlled turbine—outperformed Nibe A, having completed more than 18,000 hours of operation by the fall of 1988. The Nibe wind turbines were transferred to ELSAM in 1989. They resumed operation after an overhaul, though their performance never matched those of the smaller commercially manufactured Danish turbines from Riisager and others.[22]

The limited success of the Nibe-A and Nibe-B wind turbines provided ammunition to supporters of Riisager, Stiesdal–Jørgensen, NIVE, and others who believed that Denmark's government-led wind research efforts were ineffective. Just like the United States federal research program, Denmark's government-led wind research effort was frequently criticized for its slow moving, top-down approach to wind power innovation. It was difficult for government-led research efforts to innovate at the pace necessary to make significant advances in wind power technology. Smaller, less formal contributors were able to learn more quickly and develop improved solutions. Peter Karnøe—a Danish political scientist—has argued that Danish manufacturers' bottoms-up strategy was better than the top-down approaches of science-oriented research efforts like the American and German federal wind research programs.[23]

11.4 From Blacksmiths to Businessmen

In the fall of 1977, the Social Democrats were leading a minority government in Denmark. They wanted to pass an economic stimulus package due to the ongoing economic recession. To get the necessary votes for the stimulus package, the Social Democrats needed support of the Radical Liberal party. The Radical Liberal party offered their votes for the economic stimulus in exchange for the inclusion of a 30 percent subsidy for renewable energy investments. In addition, the government agreed to provide guaranteed loans for Danish renewable energy equipment exporters. These incentives were intended to support Danish industry and create jobs. It wasn't a preplanned policy but—rather—it emerged out of a political negotiation that was later described as a "garbage can type process."[24] Regardless of the nature of the process, the incentives were effective at jumpstarting Denmark's infant wind power industry.

The new incentives provided a direct signal to Danish firms that wind turbine manufacturing and exporting could be a profitable business. One company on the Jutland peninsula that was already trying to break into the wind turbine manufacturing business at the time was Vestas. Vestas's roots can be traced back to 1898 when twenty-two year-old Hand Smith Hansen stepped off a train in the farming town of Lem in Middle-Jutland. Hansen bought the local blacksmith shop and established himself as the town blacksmith. For the next forty years, Hansen built his business along with his son Peder. Hansen created a thriving business creating steel window frames until the outbreak of World War II and the Nazi occupation. After the war ended, Hansen and a handful of colleagues founded Vestas and focused on manufacturing household appliances and later agricultural equipment such as milk urn coolers.

By 1968, Vestas was manufacturing and exporting hydraulic cranes for light trucks to sixty-five countries across the globe.[25] When the 1973 energy crisis hit, Vestas began thinking about how to become a manufacturer of alternative energy technologies. Hansen privately experimented with Darrieus-type vertical-axis wind turbines through 1978 but failed to develop a prototype that could reliably and economically produce energy. Vestas's alternative energy efforts seemed to have led to a dead end.

Then their phone rang. Henrik Stiesdal and Karl-Erik Jørgensen from HVK were on the other end of the line. They were looking for a corporate partner to commercialize their HVK-10 wind turbine. Stiesdal and Jørgensen called Vestas to gauge their interest in manufacturing horizontal wind turbines instead of vertical ones. Vestas engineer Birger Madsen invited the pair to visit the company so that the Vestas management team could kick the tires on the Stiesdal–Jørgensen wind turbine design. After meeting with Stiesdal and Jørgensen, Vestas approved of the design and agreed to license it.

By 1979, Vestas began manufacturing its first wind turbine model—the 30-kW V-10—which was essentially just the production version of the HVK-10. Within a year after the release of V-10, Vestas began mass producing the model. By 1981, they released their second model—the 55-kW V-15. The V-15 was the first wind turbine to be considered a "standard product" from a factory.[26]

Karl Erik Jørgensen battled cancer during his partnership with Stiesdal. He lost his battle in October 1982. Henrik Stiesdal worked for Vestas as a consultant until 1986. In 1987, he joined the Danish wind turbine manufacturer Bonus. In 2004, when Bonus was acquired by Siemens, Stiesdal became the Chief Technology Officer (CTO) of Siemens Wind Power. He remained in this position until he "retired" in 2014. Since 2014, Stiesdal has devoted much of his time to working on floating offshore wind and energy storage. He is now focused on overcoming the challenges of scaling wind turbines to 15 MW and beyond. He remains one of the most enduring and respected voices in the wind industry.

Johannes Juul's work was the foundation of the small wind turbine efforts of Riisager and Stiesdal–Jørgensen, which lead to the development of Vestas's

V-10. Vestas's V-10 was an upwind rotor design with three fixed blades based on Stiesdal–Jørgensen's HVK-10 wind turbine, which itself was designed after Juul's. This design approach was known as the "Danish Concept." However, one of the key innovations of the HVK-10 was the addition of fiberglass blades produced by øaker. These blades were produced using the blade molds made by Tvind, who modeled their wind turbine after Ulrich Hütter's. Hütter was the first one to introduce fiberglass composite blades to wind turbines, whereas Juul's blades were made of wood covered with aluminum plates. Thus, the lineage of the Danish Concept comes from both Juul and Hütter.

Vestas wasn't the only wind turbine manufacturer to emerge from Denmark during this era. The 1979 incentives played a key role in launching the Danish wind turbine manufacturing industry, and by 1980—along with Vestas—there were nine other firms active and registered in the sales of wind turbines in Denmark. Nordtank—a manufacturer of road tankers for oil companies—used its knowledge of rolling and welding steel tank sections to introduce wind turbines with tubular steel towers.[27] Nordtank's towers would be adopted by other wind turbine manufacturers and would eventually displace traditional steel lattice towers. Nordtank Energy Group (NEG) would eventually merge with Moerup Industrial Windmill Construction Company (MICON) to form NEG-MICON in 1997. NEG-MICON itself merged with Vestas in 2004.

Another Danish firm—Danregn—entered the wind turbine manufacturing market in 1981. Danregn—which means "the rain,"—originally built mobile irrigation systems for farms. Given the seasonal nature of the business, their manufacturing plant peaked from April through June and went underutilized the rest of the year. Danregn was owned by Søren Sørensen. Sørensen's son, Peter, and Danregn sales consultant Egon Kristensen, approached Søren with the idea to manufacture wind turbines. Although Søren thought it was a "brainless idea," he provided the boys with DKK one million to launch Danregn Vindkraft. Søren was eager for a solution to his seasonal problem, but he half expecting the money to be wasted. Peter and Egon had already spent a good deal of money on projects that were not realized.[28]

Peter Sørensen and Egon quickly got up to speed on wind turbines, connecting with Risø, Tvind, Preben Maegaard of OVE, and others. Danregn Vindkraft was established as an independent enterprise in 1981, and its first wind turbine went in service in 1982. Danregn played the role of fast follower and developed 20- and 55-kW wind turbines in Vestas's wake. In 1983, they changed their name to Bonus Energy, and—by 1984—they had more than 200 wind turbines installed worldwide.[29] Bonus was later acquired by Siemens of Germany in 2004 and renamed Siemens Wind Power. Siemens Wind Power merged with Spanish wind turbine manufacturer and developer Gamesa in 2017 to form Siemens Gamesa, one of the largest wind turbine manufacturers in the world.

Today, two of the top five wind turbine manufacturers in the world can be traced back to Denmark's efforts to reinvent wind power in the wake of the 1973 energy crisis. Make no mistake—the global wind power industry was born in Denmark with the release of Vestas's V-10 model in 1979. However, it would take some time for the rest of the world to catch up; and it would require a dramatic increase in the global demand for wind turbines for the new Danish wind power manufacturers to survive. Fortunately, as if on cue, state and federal policy makers in the United States were in the process of creating the conditions that would enable Danish wind turbine manufacturers to thrive in the first half of the 1980s.

Notes

1 Dykes, 2013.
2 Danish Energy Agency, 2012, 6.
3 Champly, 1933.
4 Stein, 1942, 358.
5 No Author, 1943, 107.
6 Juul, 1947.
7 Maegaard, 2013.
8 Danish Ministry of Industry and Commerce, 1976.
9 Dykes, 2013, 128.
10 Ibid.
11 Maegaard, 2013.
12 Karnøe, 1990.
13 Maegaard, 2013.
14 Ibid.
15 Grove-Neilsen, 2017.
16 Dykes, 2013.
17 Ibid.
18 Backwell, 2018.
19 Christensen, 2013, 74.
20 The International Energy Agency (IEA), 1984, 23–7.
21 Matthias, 1998, 670.
22 Ibid.
23 See Karnøe, *Technological Innovation and Industrial Organization in the Danish Wind Industry*, 1990 and *Wind-Powering Denmark: A Surprising Journey from Nothing to Dominance*, 2005.
24 Karnøe, *Wind-Powering Denmark: A Surprising Journey from Nothing to Dominance*, 2005, 20.
25 Vestas, 2017.

26 Christensen, 2013, 75.
27 Ibid., 76.
28 Kristensen, 2013, 283–4.
29 Ibid., 288.

Bibliography

Backwell, Ben. *Wind Power: Struggle for Control of a New Global Industry.* Abingdon, UK: Routledge, 2018.

Champly, R. *Wind Motors: Theory, Construction, Assembly and Use in Drawing Water and Generating Electricity.* Translated by Leo Kanner Associates Sponsored by NASA. Paris, France: Dunod Publishers, 1933.

Christensen, Benny. "History of Danish Wind Power." In *The Rise of Modern Wind Energy: Wind Power for the World*, edited by 642. Singapore: Pan Stanford Publishing, 2013.

Danish Energy Agency. *Energy Policy in Denmark.* Copenhagen, Denmark: Danish Energy Agency, 2012.

Danish Ministry of Industry and Commerce. *Danish Energy Policy.* Copenhagen, Denmark: Danish Ministry of Industry and Commerce, 1976.

Dykes, Katherine. "Networks of Wind Energy Enthusiasts and the Development of the 'Danish Concept'." In *The Rise of Modern Wind Energy: Wind Power for the World*, edited by Preben Maegaard, Anna Krenz, and Wolfgang Palz, 115–62. Singapore: Pan Stanford Publishing, 2013.

Grove-Nielsen, Erik. 2017. *Winds of Change.* Accessed November 27, 2017. http://windsofchange.dk.

Juul, Johannes. "The Application of Wind Power to Rational Generation of Electricity." *Electro-Technician* 43 (1947): 137–48.

Karnøe, Peter. Technological Innovation and Industrial Organization in the Danish Wind Industry. *Entrepreneurship and Regional Development* 2, no. 2 (1990): 105–24.

———. *Wind-powering Denmark: A Surprising Jouney from Nothing to Dominance.* Copenhagen, Denmark: Draft, 2005.

Kristensen, Egon. "From Danregn to Bonus." In *The Rise of Modern Wind Energy: Wind Power for the World*, edited by Preben Maegaard, Anna Krenz, and Wolfgang Palz, 282–9. Singapore: Pan Stanford Publishing, 2013.

Maegaard, Preben. "From Energy Crisis to Industrial Adventure: A Chronicle." In *The Rise of Modern Wind Energy: Wind Power for the World*, edited by Preben Maegaard, Anna Krenz, and Wolfgang Palz, 181–248. Singapore: Pan Stanford Publishing, 2013.

Matthias, Heymann. "Signs of Hubris: The Shaping of Wind Technology Styles in Germany, Denmark, and the United States, 1940–1990." *Technology and Culture* 39, no. 4 (1998): 641–70.

No Author. "Utilization of Wind Energy in Denmark." *La Technologie Moderne* 35, no. 13–14 (1943): 106–9.

Stein, Dimitry. "Importance and Progress of Wind Power Utilization in Denmark." *Elektrizititswirtschaft* 41, no. 17 (1942): 390–2.

The International Energy Agency (IEA). *Co-operation in the Development of Large-Scale Wind Energy Systems: Annual Report 1983*. Paris, France: IEA LS WECS Executive Committee, 1984.

Vestas. 2017. *Vestas History*. Accessed November 17, 2017. https://www.vestas.com/en/about/profile#!history.

12

The Wind King

Wind power needs to be developed at a steady and appropriate pace, but the free market capitalistic system that we hold so dear will not do the job.
—William E. Heronemus (1999)

12.1   Heronemus's Dream

William Heronemus was one of America's earliest wind energy visionaries and he is sometimes called the "Wind King" or "father or modern wind power."[1] Indeed, Heronemus was among the few who saw the energy crisis coming in the late 1960s and advocated for an alternative that offered an escape from power and pollution. He was the first American to advocate for wind and solar power on a massive scale. Heronemus's teachings led to the development of a core group of wind energy professionals that would go on to lead the early wind power industry. His ideas were used as the inspiration behind the founding of the world's largest wind power manufacturing and development company in the world in the 1970s and 1980s.[2] As such, any discussion of the rise of wind power in the wake of the 1973 global energy crisis must include Mr. Heronemus.

Originally from a dairy farming family in Wisconsin, Heronemus attended the US Naval Academy at Annapolis and was commissioned Ensign in 1941. He was a gunnery officer aboard the USS Woodworth and fought in the Pacific during World War II. After the war, he attended MIT and received an MS in Naval Architecture. Starting in 1948, he spent seventeen years working on the design and construction of the US nuclear submarine fleet. Heronemus was deeply impacted by the sinking of the nation's first nuclear attack submarine— the USS Thresher. The Thresher failed a deep-diving test in April 1963 and broke apart on the bottom of the Atlantic, killing 129 people. He was critical of

The Wind Power Story: A Century of Innovation that Reshaped the Global Energy Landscape,
First Edition. Brandon N. Owens.
© 2019 by The Institute of Electrical and Electronics Engineers, Inc.
Published 2019 by John Wiley & Sons, Inc.

the sub's design and helped establish a more rigorous submarine safety program before he retired from active duty at the rank of Captain in 1965.[3] In 1967, he was offered the opportunity to start the Ocean Engineering department at the University of Massachusetts. At this time, he began to distance himself from his previous work with nuclear power and started focusing on the potential of alternative energy.

It is from this perch at the University of Massachusetts that Heronemus began to think seriously about America's energy future. In 1968—four years before the 1972 US power crunch and a full five years before the 1973 global energy crisis—Heronemus predicted that "in the immediate future, we can expect the 'energy gap' to result in a series of crises as peak loads are not met. The East Coast will be dependent on foreign sources for most of its oil and gas. The environment will continue to deteriorate despite ever-increasing severity of controls."[4] In response, he proposed the development of a large network of offshore wind turbines. By 1972, he was presenting his proposal for a wind power development program before the US Congress. And—by 1973—he told a stunned Senate Subcommittee that "It would not be foolish at all to state that this country could be totally energized by solar energy and other renewable resources by the year 2000."[5]

Heronemus's recommendations for wind research support were subsequently heeded by Congress. Wind power would receive US$1.8 million in annual funding in 1974 and funding would increase to US$60.5 million by 1980. Before the decade was out, the US federal wind power research program would receive over US$150 million in funding.[6] Heronemus went on to found the Wind Power Group at the University of Massachusetts, which trained the first generation of wind engineers and scientists. He continued to testify before Congress and still believed that large offshore wind farms were the solution to America's energy challenges. His "Grand Scale Offshore Wind Power System" was showcased in the December 1975 issue of the *National Geographic* magazine.[7] Heronemus had a dream—and he wasn't going to stop sharing his vision of the future until it came to fruition. Just like Hermann Honnef—who envisioned large offshore wind turbines in Germany in the 1930s—Heronemus was ahead of his time.

Thinking along more practical lines, Heronemus applied for a grant from the NSF to design and build a wind turbine at the University of Massachusetts. With NSF funding in hand, Heronemus and his students built and designed a small wind turbine—the WF-1—as a showcase prototype. The WF-1 was an upwind 25-kW wind turbine with three pitch-regulated blades. In addition to designing and constructing the WF-1, Heronemus continued to spread the wind power gospel—even within his own family. Heronemus's daughter was a converted believer. In fact—while attending Wheaton College—she spread the word to the daughter of an engineer named Russell Wolfe. Wolfe's daughter was left wondering why her dad wasn't involved in a more worthwhile line of work. As described in author Peter Asmus's account of the "California Wind Rush", in his book *Reaping the Wind* two weeks later, Wolfe showed up at Heronemus's office to discuss the matter.[8]

Russell Wolfe himself was a trained scientist and left Heronemus's office intrigued by the potential of wind power and the commercial viability of the WF-1 wind turbine. Wolfe immediately called his friend Stanley Charren. Charren had training in aerodynamics and had done work on airfoil configurations and propellers in graduate school. Charren was also a technology entrepreneur always on the lookout for the next new opportunity. After sitting down with Heronemus directly, Charren huddled with Wolfe and the two men decided to form a new company called U.S. Windpower (USW). They offered to employ Heronemus directly, but Heronemus felt it was more important to continue to advocate for wind power and train future industry leaders. Instead, he directed Charren and Wolfe to his two top students—Forrest Stoddard and Ted Van Dusen. Stoddard and Van Dusen subsequently joined USW to help build the company's first commercial wind turbine.[9]

Despite NASA's initial focus on large, multi-megawatt-wind turbines, an increasing amount of funding was available to support small wind turbine development during this period. According to the estimate of Federal Wind Energy Program (FWEP) managers, between 1973 and 1985, US$60 million or 15 percent of total wind research expenditures had been spent on the development of small wind turbines.[10] Part of this funding went toward the establishment of a small wind turbine test site at Rocky Flats, in between Boulder and Golden in Colorado. Rocky Flats was a nuclear weapons production facility that operated from 1952 to 1992. It was originally under the control of the United States Atomic Energy Commission (AEC) and later the United States Department of Energy (DOE). DOE allocated a large plot of land on the western edge of Rocky Flats as the wind turbine test site.

Between 1976 and 1982, fifty-four different small wind turbines from an array of manufacturers were tested at the site.[11] Although the site has a modest average wind speed which kept annual wind turbine generation figures low, it was also home to periodic wind guests that challenged the reliability of small wind turbines. Several of the designs tested at Rocky Flats went on to achieve commercial success. Carter Wind Energy's 25-kW wind turbine with flexible blades and a flexible tower became legendary for its ability to withstand the harsh winds at Rocky Flats. Five Rocky Flats employees left the test site in 1980 to form Energy Sciences, Inc., which eventually developed a commercially successful 50-kW wind turbine. Overall, two-thirds of the small wind turbines tested at Rocky Flats resulted in commercial products.[12]

During the same period, a national trade association called the American Wind Energy Association (AWEA) was formed. AWEA brought together wind turbine manufacturers and others who had an economic interest in wind power. AWEA was founded by wind power equipment seller Allen O'Shea in the basement of a Detroit police station, which offered space across the street from the 1974 World Energy Congress. The following year, AWEA held its first conference at the University of Colorado in Boulder.[13] By 1979, AWEA represented 1,200 members. AWEA was funded by DOE to conduct research on the wind energy market. Most importantly—perhaps—AWEA became a political voice for the emerging wind industry. Much to the chagrin of federal wind R&D program manager Lou Divone, AWEA's initial focus was on the advancement of small wind turbines instead of the large turbines being built by the government.[14]

Small kilowatt-sized wind turbines were technically proven by the 1970s and had always been part of the background of the American energy system. For example, from their Minneapolis factory, the Jacobs Wind Electric Company manufactured 2.5- and 3-kW DC wind turbines between 1933 and 1956. At their peak, they employed 260 workers that could produce eight to ten small wind turbines a day working one shift. According to Marcellus's own account, they ultimately built and sold approximately US$50 million worth of small wind turbines in 25 years.[15] Until his death in 1985, founder Marcellus Jacobs told everyone within earshot that the challenge of harnessing wind energy had already been solved decades ago. Marcellus was right of course, for small wind turbines, but his insistence that Jacobs wind turbines represented the final evolutionary step for wind power alienated him from the wind energy entrepreneurs of the 1970s and 1980s, some of whom would go on to lead the global wind power industry.

Truth was, although small wind turbines were technically mature and reliable, they hadn't been scaled much beyond the 10-kW range yet and they generally weren't designed to be interconnected to utility transmission systems. This is the part that FWEP was assisting with through the Rocky Flats site. FWEP provided US$2 million in grants to another small wind turbine manufacturer—Enertech—to work on scaling up its 4-kW machine to 40 kW. Enertech was a small wind turbine manufacturer based in New Hampshire. It was founded by Bob Sherwin who taught an energy conservation course at Rutgers University. Sherman launched the company in 1975 with a friend who had worked for NASA. They initially imported wind turbines and sold them domestically, but later they began designing their own wind turbines.

By 1978, Enertech had a commercially established 4-kW wind turbine that was generating US$200 million in annual sales. It was billed as the only utility-grade wind turbine on the market. That's when Russell Wolfe reached out to Bob Sherwin to place an order. USW was in the process of developing its first wind turbine and they wanted to learn from the best-in-class small wind turbine manufacturer. With Heronemus's prize students on staff—and Sherwin's wind turbine in hand—USW was ready to make a serious run at this wind power thing.

In 1978, Charren and Wolfe put Stoddard and Van Dusen to work designing a 25-kW wind turbine, just like they had done in Heronemus's class. Stoddard was given instructions that the turbine needed to be ready by the end of 1979. In the meantime, Charren and Wolfe focused their efforts on raising money for USW. Norman Moore, Herbert Weiss, and Alvin Duskin joined USW to help. Moore was installed as president and Weiss was put in charge of engineering. Duskin—an entrepreneur with limited energy experience—led marketing and government relations. Throughout 1979, Stoddard was challenged to develop a reliable wind turbine design. The prototype units experienced numerous blade failures throughout the year. Stoddard and Van Dusen expressed their concerns, but they went unheeded. Soon after, Stoddard and Van Dusen were asked to leave the company.

USW needed functioning wind turbines for its first big project: the world's first "windplant." Instead of erecting just one wind turbine at a time, Charren had conceived the idea of clustering wind turbines together in a collection on a contiguous plot of land. This way the wind turbines can be operated and controlled in unison like a single power plant. The idea was to maximize production and to minimize downtime. When a single turbine had to be fixed, it could be taken out of service with minimum impact on the collective windplant. Charren also described wind power as a new way of farming. "Wind farming will be a second source of income for farmers and landowners," he predicted.[16] Charren had big plans, "we're talking thousands at one site."[16]

The term windplant never stuck, but the phrase "wind farm" did. From this point forward, nearly all wind turbines would be built in clusters called wind farms. Wolfe and Charren found the perfect site for the world's first wind farm on Crotched Mountain in New Hampshire. It had a high average annual wind speed and easy access to the utility transmission system. USW selected Crotched Mountain and began a public relations blitz to grab headlines. They touted their project as a source of electricity for a nearby institute for disabled children and claimed that it would create jobs for nearby mill workers. Crotched Mountain generated positive local publicity and USW moved forward with the project.

The Greenfield Wind Farm on Crotched Mountain was constructed in 1980. It consisted of twenty 30-kW wind turbines designed by USW. The wind farm fed power into the New England Electric Service system. Within a year, USW retrofitted the 30-kW turbines, replacing them with 50-kW models. The Greenfield Wind Farm was intended to be a research site and was launched to provide USW with enough positive publicity to position the company for even greater things. Both of these goals were accomplished. Unnoticed by the public was the fact that the blades soon started snapping off the wind turbines at the Crotched Mountain site. USW hadn't solved the riddle of reliability yet. Creating a wind turbine that could be used to demonstrate wind power was

Figure 12.1 Built in 1980, U.S. Windpower's Greenfield Wind Farm on Crotched Mountain in New Hampshire was the world's first. It was originally equipped with twenty 30-kW wind turbines. Because of the unreliability of the wind turbines, within one year, all of the wind turbines on Crotched Mountain were destroyed and the entire site was bulldozed over.

one thing, building one that produced electricity was quite another. Within one year all of the wind turbines on Crotched Mountain were destroyed and the entire site was bulldozed over.[17] USW was ready to move on to bigger and better things.

12.2 California's Soft Energy Path

Between 1942 and 1972, electricity use in California increased by 8 percent annually.[18] Through 1971, the state's two major electric utilities—Pacific Gas & Electric (PG&E) and Southern California Edison (SCE)—had been able to keep up with growing demand by installing a portfolio of coal steam plants and hydropower stations with a handful of gas turbines and nuclear plants. PG&E and SCE planners expected demand to continue to grow at 8 percent per year and put together plans to build even more power plants to meet rising demand.

Nuclear power was the preferred choice of PG&E and SCE decision makers to meet growing demand. However, the state's increasingly stringent air quality standards and siting restrictions limited the types of plants that could be constructed and stretched out the timing of planned capacity additions. By 1973, SCE was issuing warnings that predicted brownouts in Los Angeles unless power plant siting and permitting processes were streamlined.[19] At the same time, several grassroots groups began to have an increasing impact on the thinking of state policy makers. Citizens and environmental groups emerged that favored energy efficiency and alternative energy options as the preferred solutions to California's energy future instead of large hydropower, fossil fuel, or nuclear power plants. California's grassroot activists were attempting to alter the regulatory landscape and policy environment in order to enable alternatives to thrive.

Environmentalists, citizen groups, and entrepreneurs pushed for the creation of a state-sanctioned energy resource planning group that adopted a holistic approach to energy planning instead of rubber stamping utility plans like many believed the California Public Utilities Commission (CPUC) was doing at the time. In the wake of the October 1973 global energy crisis, the environmental movement gained increased traction in California. In 1974, conservative governor Ronald Reagan signed Assembly Bill (AB) 1575 which created the California Energy Resources Conservation and Development Commission, known after 1980 as the California Energy Commission (CEC). The CEC would be staffed by incoming governor Jerry Brown.

In 1976, energy policy analyst and advocate Amory Lovins coined the term "soft energy path" when he published an article in *Foreign Affairs* called "Energy Strategy: The Road Not Taken?"[20] Lovins outlined two possibilities: one future involved the continued development of nuclear power and reliance on fossil fuels, and the other centered upon energy efficiency and renewable energy technologies. Brown was a fan of Lovins' soft energy path concept and staffed the CEC with likeminded energy and environmental individuals. In recognition of some of the limitations of the federal wind energy program, CEC initiated a wind energy research program that received US$800,000 in funding starting in 1978.[21]

Brown also created an Office of Appropriate Technology (OAT) that focused on new energy technology research and policy. Tyrone Cashman was the self-assigned OAT wind technology director. Starting in 1977, Cashman grew frustrated with the results of the federal wind research program and started devising ways that California could create the conditions to enable more wind power innovation. He eventually settled on tax credits as the preferred policy approach. Through OAT, Cashman advanced a bill through the State Legislature proposing a 25 percent energy tax credit (ETC) for wind power plants in California between 1981 and 1985. The legislature passed Cashman's bill in 1979. The state of California would ultimately provide US$570 million in tax credits to wind developers as a result of the state ETC.[22] During the same year, the annual AWEA conference was held in San Francisco. After Governor Brown backed out of his keynote speech, Cashman delivered the address and shared his vision for California's wind power future. The next day he was elected president of the AWEA board.[23]

To provide additional support to wind developers and investors, the CEC conducted thirteen separate regional studies of wind speeds in California between 1980 and 1985. CEC employed a meteorologist who traveled throughout the state in a helicopter. Providing developers with wind speed data for siting wind farms would save them the trouble of traversing the state looking for the right location to build their wind turbines. Through this effort, CEC discovered that California had the potential to develop 13,000 MW of wind power and identified the three most favorable regions—Altamont Pass in Northern California, the Tehachapi Mountains in the south, and the San Gorgonio Pass in the San Bernardino Mountain Range.

The state's policies created the right conditions to entice investors and entrepreneurs to develop wind projects in California. The federal government also provided incentives to stimulate wind power investment and changed federal law to force utilities to purchase the electricity generated by wind farms. Jimmy Carter was elected to the presidency in 1976. Carter strongly believed that an energy crisis threatened the country. In 1977, he appeared on television to caution Americans that the current energy situation is the "greatest challenge our country will face during our lifetimes." He warned that the country was running out of oil and gas and said that the nation must prepare for the permanent use of "renewable energy sources."[24] He sent his proposed energy strategy to Congress in the spring of 1977. Congress spent nearly eighteen months arguing over the bill, but eventually, in October 1978, they passed the National Energy Act of 1978 (NEA).

The NEA focused on a set of three related goals: reducing the consumption of imported oil; encouraging energy conservation; and diversifying the nation's energy supply by developing alternative energy technologies. NEA included the Public Utility Regulatory Policies Act (PURPA), the Energy Tax Act (ETA), the National Energy Conservation Policy Act, the Power Plant and Industrial Fuel Use Act, and the Natural Gas Policy Act. NEA was followed up with the Energy Security Act, which Carter signed into law in 1980. The most consequential Acts for wind power turned out to be ETA and PURPA.

ETA provided wind, solar, and other alternative energy project investors with a 20 percent investment tax credit. The credit was added to an already existing 10 percent credit. Two years later—in 1980—the Crude Oil Windfall Profits Tax Act extended the credit through 1985 and added another 5 percent, bringing the total federal credit to 25 percent. In addition, the Economic Recovery Tax Act of 1981 allowed investors to depreciation alternative equipment at an accelerated five-year rate. Collectively—through 1985—these tax credits would cost the US Treasury US$342 million.[25]

For investors that leveraged their equity with debt, the value of the tax benefits was greater than their initial investment. Other states offered wind incentives too, but only California offered incentives so generous that investors could receive a guaranteed profit just by investing in wind power projects without having to deliver any electricity. It was a no-brainer for individuals and corporations with large tax burdens to invest in wind power. The federal and state tax incentives turned California wind farms into massive tax shelters. Several years later, *Forbes* magazine called it "The great windmill tax dodge."[26]

As attractive as the state and federal tax incentives were for wind investors, PURPA and the CPUC's implementation were just as important. Unbeknownst to even its authors at the time, PURPA would unleash forces that would eventually lead to a wholesale transformation of the United States electricity system. Along the way, it would create profitable opportunities for wind power generators by requiring utilities to buy the output from wind farms. Nonetheless, many wind power advocates believed that state regulators could have been even more aggressive in their interpretation and implementation of PURPA.

To understand the role and impact of PURPA, it is useful to consider the origins of the US power system. In the early twentieth century, electric production and distribution companies were formed to serve new markets. Because the largest companies could realize the greatest cost reductions via economies of scale in the central station model, electricity production and distribution quickly became the domain of large "electric holding companies," each of which owned and operated a portfolio of electric companies. Monopolies had been outlawed by the Sherman Antitrust Act of 1890, so cities, and eventually states, regulated the electricity industry through the formation of Public Utility Commissions. In 1907, Wisconsin and New York became the first states to regulate electric utilities. By 1914, forty-three additional states had followed suit.[27] This was also the beginning of a political bargain between the electric holding companies and regulators. Electric utilities were sheltered from competition and provided with a guaranteed rate of return in exchange for providing electricity at "reasonable" prices and an obligation to serve all customers with minimal interruptions. Reasonable prices were defined as prices that covered the cost of electricity production and delivery plus a predetermined rate of return.[28]

This bargain established electric utilities as the gatekeepers to the electrical system. It enabled them to select only those generation technologies that best suited their purposes—such as large coal-fired steam turbines—while erecting barriers to technologies that didn't fit with the central station model, such as wind power. This is basically how US electric utilities operated until PURPA came along. The main thrust of PURPA was electric retail rate reform. However, tucked into a few short pages of Title II, Sections 201 and 210 encourage the use of smaller, "nonutility generators." PURPA defines "small power production facilities" as ones that produce electricity using biomass, waste, renewable resources, and whose total electric capacity is less than 80 MW. This described every wind turbine or farm in existence in 1978. PURPA required utilities to purchase power from these nonutility generators at a "just and reasonable" rate, while not discriminating against them in any way. Furthermore, PURPA required state utility commissions to determine exactly what "just and reasonable" meant and implement such rates within one year.

Thus—in a few small passages—PURPA changed the face of the US electric power system. Utilities would be required to purchase power from small generators and provide them with a fair price for their power. Not only that, but utilities wouldn't be allowed to provide discriminatory access to the grid any longer. Anyone could now start a small power production facility and the local utility would be required to connect them into the system and purchase power form them. Electric utilities fought PURPA tooth and nail, ultimately taking their case to the Supreme Court. The Supreme Court eventually upheld the federal government's authority to regulate state utility commissions. The utilities' only hope was that state utility commissions would adopt a conservative approach to PURPA when they defined "just and reasonable" rates.

By 1979, the Brown Administration had filled the CPUC with like-minded progressives that were interested in helping renewable resources to thrive in California. The relationship between CPUC and California's major utilities— PG&E and SCE—grew increasingly contentious. The CPUC had made a commitment to deciphering the utilities' "black box" of ratemaking and pricing. The CPUC adopted the principle of rate transparency. The key concept at the time was "avoided cost." The phrase originally coined by FERC in the context of PURPA was intended to represent the cost that an electricity utility would avoid by purchasing power from a small nonutility generator instead of having go built its own new power plant.

When it came to determine what "fair and reasonable" meant in the context of utility power purchases from nonutility generators under PURPA, CPUC convened a series of meetings to hammer out a standard contract that the state's utilities could offer to small power generators. The CPUC worked to put small independent producers on a level playing field during the standard offer negotiations. The small power producers were represented by Jan Hamrin, the head of a new organization called the Independent Energy Producers Association (IEPA). IEPA represented dozens of small power producers that wanted to receive a standard offer from SCE or PG&E.

The process was contentious and IEPA used the opportunity to hold the utilities accountable for the difficulties they have provided small power producers when they had previously attempted to negotiate contracts. The transparency exposed some of the utilities' bullying tactics. Prior to the development of a standard offer, when wind farm developers attempted to get power purchase agreements from the utilities they simply had to take what the utility was offering, even if they didn't fully understand the rationale behind the terms and conditions. The utility black box was impenetrable, until CPUC, CEC, PG&E, SCE, and IEPA got into a room together and started negotiating the standard offer in the context of PURPA.

In the spring of 1983, all of the parties involved agreed to the Interim Standard Offer #4 contract (ISO4). The CPUC approved the contract in September 1983 and planned on reviewing it one year later to evaluate its success. However—by this time—wind development had already accelerated in California, but the existence of ISO4 would facilitate future project development, create financial certainty for wind investors, and streamline the project development process. At the time, none of the parties believed that the contract represented a very good solution from their perspective—just a reasonable compromise. No one felt like they had won the day.[29]

ISO4 was a thirty-year contract. In the first ten years of the contract, the specific prices that utilities would pay small power producers for their electricity were explicitly provided in ISO4. These prices were based on the utilities' expectations of the avoided cost of power at the time. In the early 1980s, oil prices were high and they were expected to stay high by most observers for the foreseeable future. Since oil-fired generation was used as the avoided cost benchmark plant, the ten-year prices in the ISO4 contract were extremely attractive for the small electricity producers. For example, the price schedule obligated PG&E to pay 5.5 ¢/kWh in 1985, rising to 11.4 ¢/kWh by 1994.[30] After the initial ten-year contract period, the prices reverted to a new short-term avoided costs calculation based on the energy prices of the day. Everyone agreed that oil prices would be astronomical ten years hence and no one foresaw the 1986 oil glut that would drive oil prices down from US\$35/bbl in 1980 to below \$10/bbl by 1986.[31] Given how energy markets evolved in the 1980s, ISO4 turned out to be the deal of a lifetime for California's wind farm investors. If Heronemus's dream of a national network of wind turbines was ever going to be realized—it would have to start in California because all of the pieces were now in place for a spectacular wind power boom.

12.3 This Is the Place!

In 1980, USW's Charren and Wolfe were looking for the next big thing after "successfully" developing a wind farm at Crotched Mountain. Although still largely unproven, they believed that they had a commercial-ready 50-kW wind turbine that could be the basis for wind farms up to 100 MW. USW purchased land rights in Montana, Wyoming, and Washington. They also dispatched their Chief Marketing Officer—Alvin Duskin—to California to look for more land.

In the 1970s, Duskin had become an antinuclear crusader and held benefit rock concerts. By 1977, he had burned out on dealing with rock stars and moved to Washington, DC to write a book about the dangers of nuclear proliferation.[32] While there, he worked for Senator James Abourezk from South Dakota and ended up writing an amendment to the Crude Oil Windfall Profits Tax Act that expanded the 10 percent business tax credit to 25 percent for alternative energy technologies, including wind power. It passed Congress and Duskin was thus partially responsible for helping to create the conditions that made wind power so attractive in California in 1980.

Duskin returned to California in 1980 representing USW. He reached out to an old friend, State Resources Secretary Huey Johnson. Duskin knew that Johnson had jurisdiction over the state Department of Water Resources (DWR), which was a large purchaser of electricity and a significant California landholder. Like the rest of the Brown Administration, Secretary Johnson was enthusiastic about the potential for wind power and had consulted with meteorologists from Oregon State University to survey wind speeds on state lands. They identified particularly strong winds at Pacheco Pass, one of DWR's land holdings. With a little convincing from Duskin, Johnson offered to give USW 5,000 acres at Pacheco Pass and have DWR sign a contract for 100 MW of wind power at 5 ¢/kWh. This was a big deal. Until that point, the largest wind farm contract had been at Crotched Mountain for 2 MW.[33]

With the DWR contract in hand, USW was able to line up investors for its new California wind farm. When word got out that USW had secured land and a contract for a new 100-MW wind farm, AWEA hurriedly relocated its annual conference from Massachusetts to San Francisco to celebrate USW's accomplishment. Johnson himself helped to spread the news that DWR had signed a major contract with USW. In the meantime, USW engaged its own meteorologists to assess the wind at the Pacheco Pass site. Unfortunately, they found that the wind resources were overstated by 40 percent by Oregon State. Duskin needed to find another site.

After consulting with a state meteorologist, Duskin learned that the Altamont Pass region had some of the best wind resources in California. After confirming the strength of the winds at Altamont Pass by photocopying a leaked draft of a wind resource study that CEC was finalizing, Duskin rang Charren with the news. "This is the place," he told him. USW would have to build its first California wind farm in Altamont Pass, not Pacheco Pass. Charren had already lined US investors based on the Pacheco Pass site and they already had the DWR contract and was reluctant to switch horses at this point, but he relented after Duskin threatened to quit.[34]

Duskin then signed up half of the land rights on Altamont's windiest ridges for pennies per acre by promising local ranchers future royalties based on actual wind production. His bigger challenge was negotiating a new power purchase agreement with PG&E. This was before the PURPA ISO4 contract had been established, so he expected it to be tough sledding. With little to lose, Duskin explained the whole situation to PG&E executives. He told them that USW had the contract with DWR, but couldn't use it because the winds had been overstated, and now USW wanted to build at Altamont Pass in PG&E's service territory instead. Would they be willing to provide USW with a power purchase agreement?

Unbeknownst to Duskin, PG&E was more than eager to appease wind power developers and receive the positive press. They needed something to offset the negative publicity that they were receiving as a result of their attempts to obtain an operating license for their new nuclear power plant, Diablo Canyon. Deemed safe when construction started in 1968, by 1981, PG&E had to update its plans and add structural support after a seismic fault line had been discovered nearby. The public soured on the plant and over a two-week period in 1981, 1,900 activists were arrested and sent to jail for protesting at the construction site. In this environment, Duskin and USW had more leverage than they realized. PG&E agreed to sign a 100-MW contract for USW's wind farm at a whopping 9 ¢/kWh. Furthermore, PG&E offered help with interconnection and accelerated the delivery time for new transformers. This kind of cooperation by an electric utility was previously unheard off. The contract was signed in April 1981. Construction began in July, and the first USW wind-generated electricity hit the PG&E grid on the last day of 1981.

USW's Altamont Pass wind farm was the joint product of visionaries, innovators, entrepreneurs, dogged opportunities, utility decision makers, and state and federal policy makers. It would open the door to a flood of wind power development over the next five years. California would be the proving ground for wind power to show that it belonged as a reliable and economically competitive source of electricity for the nation and the world. Wind power's future looked bright on New Year's Day in 1982 as the celebratory ball dropped back east in Times Square.

Nobody could have imaged what was in store in the coming year. Wind power was about to turn California back into the Wild West.

Notes

1 Chernow, 1978.
2 Stoddard, 2014.
3 Fialka, 2017.
4 Heronemus, *A National Network of Pollution-Free Energy Sources*, 1971.
5 Heronemus, *The Possible Role of Unconventional Energy Sources in the 1972–2000*, 1973.
6 U.S. Department of Energy, 1990.
7 Hamilton, 1975.
8 Asmus, 2001, 57.
9 Ibid., 57–8.
10 Serchuk, 1995, 124.
11 Ibid., 111.
12 Ibid., 112.
13 American Wind Energy Association, 2017.

14 Serchuk, 1995, 116.
15 Jacobs, 1973.
16 Asmus, 2001, 61.
17 Ibid.
18 Serchuk, 1995, 196.
19 Pryor, 1973.
20 Lovins, 1976.
21 Serchuk, 1995, 206.
22 Starrs, 1988, 140.
23 Asmus, 2001, 74.
24 Carter, 1977.
25 Starrs, 1988, 140.
26 Paris, 1984.
27 Smithsonian Institution, 2014.
28 Hein, 2003.
29 Serchuk, 1995, 268.
30 Ibid.
31 Energy Information Administration, 2012.
32 Asmus, 2001, 83.
33 Duskin, 2014.
34 Asmus, 2001, 84.

Bibliography

American Wind Energy Association. 2017. "History of Wind." *American Wind Energy Association.* Accessed November 29, 2017. https://www.awea.org/history-of-wind.

Asmus, Peter. *Reaping the Wind.* Washington, DC: Island Press, 2001.

Carter, Jimmy. "The President's Proposed Energy Policy." In *Vital Speeches of the Day* XXXXIII, 418–20. Arlington, VA: Public Broadcasting System, 1977.

Chernow, Ron. "Heronemus the Wind King." *Quest*, January/February (1978): 13–20.

Duskin, Alvin, Interview by Huey Johnson. 2014. *Launching the Wind Industry.* Forces of Nature: Environmental Elders Speak, June 26. Accessed November 30, 2017. http://theforcesofnature.com/movies/alvin-duskin-2.

Energy Information Administration. 2012. *Annual Energy Review 2011.* DOE/IEA 370. DOE/EIA-0381(2011). Washington, DC: Energy Information Administration, U.S. Department of Energy. https://www.eia.gov/totalenergy/data/annual/pdf/aer.pdf.

Fialka, John. 2017. "Meet 'The captain', the Father of Modern Turbines." *E&E News*, November 28. Accessed November 28, 2017. https://www.eenews.net/stories/1060067379.

Hamilton, Roger. 1975. "Can we Harness the Wind?" *National Geographic,* December.

Hein, Jeff. 2003. *Shining Light on the Utility Industry's Earliest Foundings.* Lakewood, CO: Western Area Power Administration.

Heronemus, William E. 1971. "A National Network of Pollution-Free Energy Sources." Proposal, Research Applied to National Needs (RANN), National Science Foundation (NSF), Washington, DC.

———. 1973. "The Possible Role of Unconventional Energy Sources in the 1972–2000." Testimony, Committee on Science and Astronautics, Subcommittee on Science, Research and Development, U.S. House of Representatives, 92nd Congress, Washington, DC, June 1973.

———. 1999. "Lifetime Achievement Award Acceptance Speech." American Wind Energy Association WINDPOWER '99 Conference, Burlington, VT, June 22, 1999.

Jacobs, Marcellus L. "Experience with Jacobs Wind-Driven Electric Generating Plant, 1931-1957." In *Wind Energy Conversion Systems,* edited by Joseph M. Savino, 155–8. Washington, DC: NSF/NASA, 1973.

Lovins, Amory. "Energy Strategy: The Road Not Taken?" *Foreign Affairs* 55 (1976): 65–96.

Paris, Ellen. 1984. "The Great Windmill Tax Dodge." *Forbes,* March 12, 39–40.

Pryor, Larry. 1973. "Energy Crisis – Hot Water for Legislature." *Los Angeles Times,* March 4, A1.

Serchuk, Adam Harris. "Federal Giants and Wind Energy Entrepreneurs: Utility-Scale Windpower in America, 1970–1990." PhD thesis, Virginia Polytechnic Institute and State University, 1995.

Smithsonian Institution. 2014. *Powering a Generation of Change.* July. Accessed August 23, 2017. http://americanhistory.si.edu/powering.

Starrs, Thomas A. "Legislative Incentives and Energy Technologies: Government's Role in the Development of the California Wind Energy Industry." *Ecology Law Quarterly* 15, no. 103 (1988): 103–58.

Stoddard, Forrest. *The Life and Work of Bill Heronemus, Wind Engineering Pioneer.* Amherst, MA: University of Massachusetts Wind Energy Center, 2014.

U.S. Department of Energy. *Renewable Energy Budget History.* Internal Document, Washington, DC: Department of Energy (DOE), 1990.

13

The California Wind Rush

Industry observers and participants both agree that many wind developers have overstated their capabilities and provided projections that were not achievable.

—The California Energy Commission (1986)

13.1 The Deluge

USW's Altamont Pass project marked the beginning of a flood of wind power projects in California. USW's entry into Altamont Pass was mirrored by a new wind power project development in the Tehachapi Mountains in the south; and wind power would soon spread to San Gorgonio Pass in the San Bernardino Mountain Range. With California's regulatory and policy landscape supportive of wind power, investors and entrepreneurs flocked to California to build wind turbines and operate wind farms. The flood of wind farms would later become known as the "California Wind Rush," named after a similar frenzy of gold mining activity that started in 1849 when gold nuggets were discovered in the Sacramento Valley.

Back in the nineteenth century, the California Gold Rush peaked in 1852 after US$2 billion in gold had been extracted.[1] The California Wind Rush would peak too, but not before statewide wind energy capacity reached 1.2 GW in 1986. Over 14,600 wind turbines were installed by the end of 1986 and they were generating 1,217 million kWh of electricity annually at their peak.[2] Before the end of the California Wind Rush, 96 percent of the world's existing wind power capacity was in California.[3]

USW's Altamont Pass project provided the company with the credibility it needed to attract investors like Merrill Lynch and gain leverage with its utility partner PG&E. Furthermore, having partners like Merrill Lunch and

The Wind Power Story: A Century of Innovation that Reshaped the Global Energy Landscape,
First Edition. Brandon N. Owens.
© 2019 by The Institute of Electrical and Electronics Engineers, Inc.
Published 2019 by John Wiley & Sons, Inc.

PG&E provided USW with access to additional sources of capital. USW continued to develop additional wind farms in California using its own wind turbine technology and achieved increasing levels of success. By 1985, the company was earning US$90 million in revenue and making US$6 million in profits per year.[4]

USW relied upon its latest wind turbine, the 56-50. It was a 50-kW downwind wind turbine that moved freely in response to shifting winds. The 56-50 was a constant-speed machine that used variable pitch blades to maintain constant speed. It was based on the original 56-30 30-kW model that engineers Forrest Stoddard and Ted Van Dusen had designed for the USW Crotched Mountain wind farm. Of course, Stoddard and Van Dusen had already been let go by USW allegedly for raising concerns with the readiness of USW's wind turbine technology. But as of 1983, USW's wind turbine represented the very best that America had to offer.

An interesting aspect of the 56-50 model—and a harbinger of things to come—was its use of digital technology to control its variable pitch blades. The wind turbine used software and a microprocessor to control the pitch of the blades at different wind speeds. The blade started moving once the wind hit 5 m/s and rotated at an increasing rate until the wind speed hit 13 m/s. Between 13 and 20 m/s, the software changed the angle of the blades to maintain a constant rotational speed. If the wind speed exceeded 20 m/s, the blades went parallel to the oncoming wind and the wind turbine returned to a resting position as the wind sailed by the feathered blades.

USW use of a digital control system to adjust blade pitch was unique. The Danish models at the time didn't use software to regulate blade pitch because they were passive stall machines that used wind speed and blade shape to create enough turbulence to stall the blades to maintain a constant output. Digital control systems would grow increasingly complex over time and assume more responsibilities as the sophistication of information technologies improved and costs declined. USW use of a digital control system for the 56-50 was the first commercial implementation of a digital control system, and it improved the performance of USW turbines relative to other manufactures. USW wind turbines would lead the pack in California achieving 73 percent of their expected production.[5]

USW remained the top wind turbine manufacturer and wind farm operator in California throughout the 1980s. By 1986, USW installed 245 MW of wind power capacity in and around Altamont Pass. They had over 2,700 wind turbines in operation by the end of 1986, feeding 280 million kWh of electricity into the PG&E transmission system.[6] The wind farms of Altamont Pass—including USW's—accounted for half of the world's wind-generated electricity by 1985.[7] This was quite an accomplishment for a company that started when USW co-founder Russell Wolfe visited Bill Heronemus's office back at the University of Massachusetts.

However, USW's wind turbines—like almost all of the wind turbines installed in California at the time—experienced severe reliability problems. And that's putting it mildly. Blades flew off and struck whatever—or whoever—happened to be in range. By 1984, 179 of 557 56-50 wind turbines had broken down. An additional 352 were shut down due to the significant risk of damage. Although USW's 56-100 100-kW wind turbine would fare better, wind turbine reliability would continue to be the Achilles Heel of USW, and indeed, the entire California wind energy experiment. After everyone finished congratulating themselves over the USW-PG&E wind project in Altamont Pass, it soon became clear that although California might have been ready for wind power, wind wasn't quite ready for California.

James Dehlsen would try to change that. Born in Guadalajara, Mexico, Dehlsen had been operating in risky and uncertain environments his entire life. The son of a Danish businessman and an engineer, Dehlsen joined the US Air Force at a young age to see the world. After his military service was complete, he briefly studied engineering at San Diego State University before dropping out to sail around the world. Later, he put his engineering studies to good use and invented a spray-on lubricant called Tri-Flon. Tri-Flon used micron-sized Teflon particles that acted like tiny bearings on any surface. By 1978, Dehlsen was selling cans of Tri-Flon lubricant for US$5.95 each out of his business in Costa Mesa, California. Dehlsen made good money when he sold the company in 1980.[8]

Now Dehlsen needed a new adventure, and renewable energy intrigued him. In late 1980—using money from the sale of Tri-Flon—Dehlsen founded a new company called Zond to take advantage of the new renewable energy technology wave and business opportunity. He was in the mood for an adventure—and the renewable energy business environment reminded him of his days in the wild west of Guadalajara. He started by purchasing a Santa Barbara-based solar and wind manufacturer that specialized in small off-grid technologies. However, after attending a wind conference in Palm Springs, he became convinced that wind power was the answer and let go of the solar business.

Dehlsen needed just two things to get serious about wind power: wind turbines and a place to build them. He asked around and discovered that manufacturers of reliable wind turbines were in short supply. He contacted a California wind turbine designer named Ed Salter who had founded Wind Power Systems in San Diego in order to develop a 40-kW wind turbine called the "Storm Master." The Storm Master was modeled after Ulrich Hütter's 1950s era design—it was a lightweight and slender downwind machine with long fiberglass blades. Dehlsen would eventually purchase 450 Wind Master wind turbines and buy the manufacturing rights. The only problem is that they hadn't actually been tested in the field yet, so there was no telling if they would actually work.

But where would Zond place the Storm Masters? Dehlsen and his small band of Zond employees familiarized themselves with CEC's recent "Wind Power Potential in California" report that identified three favorable wind resource areas in California: Altamont Pass in the north, the Tehachapi Mountains and San Gorgonio Pass in the south.[9] Dehlsen's stomping ground was around Los Angeles since his Tri-Flon business had been operating out of an office in Costa Mesa just south of LA, so he focused initially on nearby Tehachapi. While he negotiated with Salter for the Storm Master wind turbines, Dehlsen and his wife purchased 750 acres on a ridge of the Tehachapi Mountains.[10]

With the turbines and the land, Zond began the effort of designing their new wind farm in 1981. Locating in Tehachapi meant the Zond would have to sell their wind farm output to Southern California Edison. The standard power purchase agreement for nonutility generators—ISO4—hadn't yet been established in 1981, so Dehlsen would have to negotiate directly with SCE. This wouldn't be an easy task, particularly considering Dehlsen's limited experience in electric power. Zond initiated the conversation and then raced to install the Storm Master wind turbines on their land in the Tehachapi Mountains. To qualify for the state and federal tax credits, the wind turbines would need to be installed and delivering power to the SCE system by the end of the year. Zond started installing the wind turbines and hoped to have a contract with SCE by the end of the year. In Dehlsen's own words, "it was insane."[11]

After months of negotiations—by mid-December—Zond had gotten SCE to agree to purchase the electricity generated by their first wind farm, which he dubbed Victory Garden. A mad rush ensued to get as many Wind Master's installed and operating as possible by the end of the month. They succeeded in getting fifteen wind turbines up and running and fed power into the SCE grid on Christmas Day. Victory Garden had 600 kW of capacity operating by year end and thus become the first company to sell wind power to SCE. It was as if an alien from the planet Zond had just touched down in Southern California with the gift of wind-generated electricity and the promise of a clean energy future.

13.2 The Danish Invasion

Zond would increase its Tehachapi wind capacity to 32 MW by the end of 1982 and Dehlsen would raise US$8 million in capital for its projects in that year. However, just as Dehlsen and his cohorts cheered their success at Victory Garden, they ran into big problems. The Wind Master wind turbines didn't work. "As soon as we turned them on they started disintegrating," Dehlsen later admitted. "The next day we picked up the pieces and concluded that we'd better get a better technology pretty damn quick."[12]

Dehlsen launched a second search for a reliable wind turbine; this time they'd have to look beyond the borders of California. The search led him to Europe, which was known to have built hardy, small wind turbines for farm use since the 1940s. Dehlsen along with Zond executive Bob Gates—and new CFO Ken Karas—ended up in Amsterdam in discussions with a Dutch wind turbine manufacturer. However, When Vestas's Finn Hansen got word that Dehlsen was in Holland getting ready to buy Dutch wind turbines, he hopped in his small airplane, picked up Dehlsen, Gates, and Karas, and brought them back to Denmark to visit Vestas. Vestas had just rolled out the 55-kW V-15 wind turbine in the summer of 1981 and they were still looking for their first big customer.

Dehlsen examined Vestas's V-15 wind turbine. It looked far sturdier than anything the Zond team had seen before. Dehlsen purchased two on the spot and followed up an order of 150 when he got back to his office in California. Vestas would end up selling 1,784 wind turbines to project developers like Zond in California through the end of 1986. This was a turning point for both Zond and Vestas. Zond had found a more reliable wind turbine that it could use to fuel the California Wind Rush, and Vestas found a high-volume customer that it could use to scale up its manufacturing operations. The number of Vestas employees grew from 200 to 870 during this period.[13]

Led by Dehlsen—and powered by Vestas's wind turbines—Zond would continue to expand over the next several years. By 1985, the company was producing 205 million kWh of wind generated electricity annually, second only to USW's 280 million kWh.[14] And Vestas wasn't the only Danish manufacturer to get into the act. By 1986, Danish manufacturers—primarily Bonus, Micon, Nordtank, Vestas, and Windmatic—accounted for 38 percent of the total installed capacity in California. In 1986 alone, a full 68 percent of new wind capacity in California came from Denmark.

Although the Danish manufacturers learned tough lessons from California's harsh winds—overall—the Danish wind turbines performed better than US-built wind turbines. By 1986, although Danish wind turbines accounted for 38 percent of installed capacity but were generating 48 percent of the electricity from California wind farms. Furthermore, all Danish wind turbines achieved an average capacity factor of 14 percent in 1986. By comparison, the US machines were only operating at an average capacity factor of 9 percent, which meant that the Danish wind turbines were outperforming the US ones by 50 percent.[15]

Vestas used the California Wind Rush as an opportunity to refine its wind turbine designs. By 1985, they had introduced their first pitch-regulated, constant-speed turbine, as opposed to the traditional Danish passive-stall approach original pioneered by Johannes Juul and later adopted as part of the Danish Concept. Passive stall relies upon aerodynamically shaped blades that create increasing turbulence at higher wind speeds causing the blades to stall, thereby maintaining a constant rotational speed. This approach requires

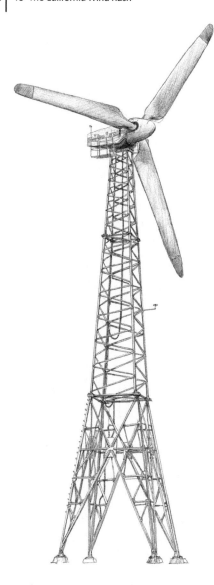

Figure 13.1 Vestas's 65-kW V-15 wind turbine became the workhorse of the California Wind Rush. Based off the original Danish HVK-10 wind turbine by Henrik Stiesdal and Karl Erik Jørgensen, the V-15 was the first wind turbine that was manufactured as a "standard product" from a factory.

wind turbines to absorb elevated levels of torque that is hard on mechanical components such as gearboxes. This leads to more expensive construction costs—particularly at larger sizes—as passive-stall wind turbines are fortified to absorb higher loads.

Vestas's pitch-regulated wind turbine incorporated an active pitch control system that combined elements of both pitch control and passive stall. It was operated by a digital control system that smoothed power output and reduced structural loads under turbulent wind conditions. Besting USW's digital technologies, Vestas's digital control system would set the bar for other wind turbine manufacturers and would become a key feature of their wind turbines moving forward.

Despite the early wind power activity at Altamont Pass and the Tehachapi Mountains, it was actually San Gorgonio Pass that the CEC identified as having the best wind resources in the state. However, wind power development was slow to take off in San Gorgonio Pass for a couple of reasons. First, SCE had concentrated its wind energy procurement efforts in Tehachapi and wasn't too interested in opening a can of worms with another set of developers in a whole new geographic area. California's response to PURPA, the ISO4 contract wasn't available until 1983. This meant that wind developers that wanted to build in San Gorgonio had to negotiate directly with SCE.

Second, the residents of nearby Palm Springs weren't too keen on having their desert surroundings littered with wind turbines. This was before wind turbines adopted a more standardized and sleek look and feel with modern wind blades and tubular towers. Many of the turbines built in California in the early 1980s looked and operated like losing entries in a children's Erector Set competition. Palm Springs residents revolted and filed suit against wind developers in 1985. Palm Springs wouldn't end its fight against San Gorgonio wind farms until the end of the 1980s after reaching an agreement with the wind industry on wind farm siting guidelines. The crusade against wind turbines in Palm Springs was led by former pop singer turned mayor Sonny Bono, who would later reverse course entirely and embrace wind farms as a way to increase the tax base in Palm Springs.

Third, while USW was focused on Altamont, and Zond was focused on Tehachapi, there was no dedicated and determined wind developer with its sights set on San Gorgonio until 1983 when San Diego-based California Wind Energy Systems arrived. California Wind Energy Systems—which changed its name to SeaWest in 1984—started developing wind projects in Altamont in 1982, but moved to San Gorgonio starting in 1983 after arranging financing from Kidder Peabody. Peabody had been convinced that wind power was real after observing USW and Zond's successful development efforts up north.

SeaWest followed Zond's strategy by focusing just on wind farm development, and not on wind turbine manufacturing. They sourced their wind turbines originally from Bob Sherwin's Enertech, which was one of the design inspirations behind USW's wind turbine, and later from Danish wind turbine manufacturers. Where SeaWest differed from Zond was their focus on wind farm operations. This operational focus would serve them well because it enabled them to improve the performance of their wind turbines and increase the electrical output of their wind farms. And also, because it provided them with steady business later by fixing all of California's broken wind turbines after the California Wind Rush ended.

SeaWest was ultimately successful in taking advantage of the superior wind resources of San Gorgonio. By the end of 1986, SeaWest had 125-MW of installed capacity operating. Collectively, 1,500 SeaWest wind turbines generated 137 million kWh of electricity per year by 1986—third in California behind USW and Zond.[16] Overall, wind turbines installed in Gorgonio performed slightly better than their counterparts in Altamont and Tehachapi. The average capacity factor of the San Gorgonio wind turbines was 13 percent in 1986, compared to 12 percent in both Altamont and Tehachapi, indicating that the San Gorgonio wind turbines performed 8 percent better. This was due to the high wind speeds in San Gorgonio and also because the San Gorgonio wind turbines were installed later and therefore benefited from some of the early learnings at Altamont and Tehachapi.

Nonetheless, the wind turbines of San Gorgonio did experience many of the same reliability challenges as those in Altamont Pass and Tehachapi. Even more so in some instances because the desert sand sandblasted the wind turbines and wreaked havoc on the gears. Even the Danish wind turbines experienced problems and many of the blades on these machines needed to be replaced.[17] Beyond the poor performance of many of the wind turbines during the California Wind Rush—and the local resistance in Palm Springs because of the visual blight caused by early wind farms—there was an additional problem. As it turned out—in some instances—wind turbines and birds didn't mix well. Or rather, they sometimes mixed too frequently, and when they did, the birds came out on the losing end.

After wind power development ramped up in Altamont, Tehachapi, and San Gorgonio, reports began emerging from birdwatchers and locals that the wind turbines were killing birds in record numbers. The CEC jumped in and began tallying the numbers. They found that as many as 160 birds had been killed near California's wind turbines from 1984 to 1988. CEC documented ninety-nine dead birds near Altamont wind turbines, nine near Tehachapi wind turbines, and forty in the San Gorgonio pass. Of particular concern was the endangered Golden Eagle, which is protected under federal and state law. CEC commissioned a follow-up study that estimated that facilities in the Altamont pass were killing up to 400 birds per year, including forty golden eagles.[18]

CEC's research and the associated uproar about the prospect of wind turbines killing Golden Eagles focused attention on the issue and created a national "bird problem" for the fledgling wind power industry. USW hired a biologist to study the issue and he concluded that although bird deaths due to wind turbines represent a very small fraction of bird populations, "some species, such as Golden Eagles may be more vulnerable to impacts form wind turbines...."[19] Now the wind industry had a protected species problem.

As it turned out, Altamont Pass is unique, and wind power's bird problem was vastly overstated. Had an adequate environmental impact assessment been conducted, it would have shown that Altamont Pass is a migration route and wintering place for many species of raptors. Its landscape and canyons make it an ideal setting for birds of prey. Properly siting the wind turbines in areas with low bird populations would have dramatically reduced collision rates in Altamont. Even today, more raptors are killed in Altamont Pass than any other wind farm area in the United States. But it is now common practice for a bird study to be conducted before wind farm construction begins.[20]

Wind turbine designs also matter. The lattice towers of the early California wind turbines contributed to bird deaths and recent studies have found that—all other factors being equal—modern monopole towers result in fewer bird deaths.[21] Furthermore, modern larger wind turbines with slower rotational speeds and higher capacity ratings have been found to result in lower collision rates. The Altamont Pass experience forced the wind industry to adapt over time by taking bird populations into account when considering prospective wind farm locations and wind turbine designs.

This bird issue also changed the perception of wind power. Initially viewed as a completely begin source of energy, there was now a greater awareness of the impact that wind has on the environment. Even in some circles today, wind power is viewed by the uninformed simply as a technology that "kills all your birds."[22] However, the US National Wind Coordinating Committee (NWCC) was the first organization to put wind turbine bird kills into perspective. In a 2001 study, they estimated that up to 40,000 birds per year were killed by wind turbines in the United States at the time. They also found that up to eighty million were killed by moving vehicles, up to 980 million were killed when they ran into buildings and flew into windows and up to fifty million were killed when they flew into communication towers every year.[23] In other words, flying is dangerous business in general, and wind turbines are just one of many hazards to be avoided by birds.

13.3 Brand New Day

Wind power's rapid journey from a promising energy alternative to the star of California's energy revolution ended abruptly. The clock struck midnight with the passage of the Tax Reform Act of 1986. The Act terminated the federal investment tax credit for wind power. Attempts to extend the state's wind energy tax credit past their scheduled termination at the end of 1986 also failed.

But the writing was on the wall as early as 1985 after California wind farms gained a national reputation as a tax dodge. President Reagan promised to simplify the tax code and eliminate tax fraud, including loopholes like "jojoba bean shelters, windmills, race horse write-offs, and Cayman Island trusts."[24] Some elements of the wind power community had fallen in with the wrong crowd and the entire industry would pay the price.

Wind power additions tapered off between 1985 and 1987. In 1985, when the California Wind Rush was in full swing, approximately 400 MW of new wind capacity was added to the California grid. In 1986, when the tax credits were still available, but there was considerable uncertainty about their future, wind capacity additions dropped to 275 MW. A year later—in 1987—only 150 MW of new wind power was added to the grid.[25] Of the forty developers building wind farms in California between 1982 and 1985, only a handful remained in 1987. One developer—Fayette—was the first to sell all its assets to pay off its multi-million debt of back taxes after its wind farm projects were found to violate tax laws. USW, Zond, and SeaWest were still active, albeit at a reduced level. By that point, the only American-made wind turbines being installed in California were those made by USW—all of the other surviving wind developers were installing Danish machines.

The only thing that held up California's wind energy market after the expiration of the tax credits at the end of 1986 were the ISO4 contracts which provided wind developers with guaranteed fixed power purchase prices from PG&E and SCE for ten years. Wind developers who opted for the ISO4 contracts were insulated from falling oil prices for a decade.

World oil prices were in the range of US$40/bbl in 1981, just before the PURPA negotiations started. But by 1986, an economic slowdown in the developed world and ongoing conservation measures in the face of the 1973 and 1979 oil crisis had created an oil glut. In addition, new sources of oil from Siberia, Alaska, the North Sea, and the Gulf of Mexico flowed into the market in the early 1980s. OPEC decreased its production annually to keep prices high. However, at the end of 1985, swing producer Saudi Arabia got fed up when OPEC partner countries didn't stick to their agreed upon production quotas. They began producing at their full capacity and oil prices promptly fell to US$10/bbl.

The drop in oil prices eroded much of the justification for providing tax credits for alternative energy technologies. Furthermore, it made the ISO4 contract prices appear artificially inflated and created a future day of reckoning after the ten-year window expired. Even those wind power developers that survived the expiration of the tax credits in 1986 were facing into the future prospect of much lower revenue when the fixed prices reverted to a floating price based on the new price of oil.

Both the California wind developers and the Danish wind turbine manufacturers went into survival mode. USW took a two-pronged approach. They attempted to leverage their capabilities as a wind turbine manufacturer and began to market their turbines internationally. That's why they changed their name to Kenetch in 1988 to facilitate the transformation from a US wind developer to an international wind turbine manufacturer. At the same time, they continued to develop domestic wind farms where they could. This meant that they needed to initiate a relentless effort to cut costs and improve wind turbine productivity however possible. They hired electric utility executive Dale Osborn to help facilitate the transition. They faced into the 1990s with an uncertain future.

SeaWest didn't have the burden of being a wind turbine manufacturer with inventory to sell. Instead they focused on improving the performance of existing wind farms. They made a business out of retrofitting older American and Danish wind turbines. They salvaged existing turbines and implemented performance upgrades that increased their output and kept them running. SeaWest would develop wind farms selectively and cautiously for the reminder of the 1980s and throughout the 1990s. It was the only wind development company to thrive during the California Wind Rush, survive the post-tax credit era, and enter the twenty-first century intact. By the end of 2004, SeaWest operated over 500 MW of wind facilities in California, Wyoming, and Oregon and had 1,800 MW of development sites in ten states in the Western United States. SeaWest was eventually acquired by US independent power producer AES Corporation in 2005.

Zond continued its focus on wind farm development even after the tax credits expired. Dehlsen needed another showcase project to demonstrate that wind power was alive and well. He chose the town of Gorman, an isolated community on the boundary between Los Angeles and Kern countries. Its proximity to LA meant that the power could be fed into the SCE transmission network and serve the large LA population. It would be a visible demonstration of wind power's ability to provide pollution-free electricity to one of the countries must polluted cities. The project was supported by Ralph Nader's non-profit group, Public Citizen, because of its potential to offset 10 million bbl of oil during its lifetime.

Unfortunately for Zond, the project attracted a lot of negative attention because of its potential to kill condors, an endangered species that was going to be released from the LA Zoo. A local group gathered 1,000 signatures and created strong local opposition to the project. The opposition enlisted the help of the Sierra Club and Audubon Society to help sink the project. Opponents were able to successfully convince the LA County Board of Supervisors to deny an operations permit for the wind farm.

After the Gorman project rejection, a frustrated Dehlsen almost threw in the towel and sold Zond to Kenetech.[26] But Dehlsen eventually soldiered on. For the next several years, the only progress Zond made was by reengineering virtually every aspect of the wind projects it had built, with the goal of increasing revenue through greater productivity. They succeeded in increasing output by 22 percent from 1986 to 1989.[27]

Vestas had a near death experience after the tax credits expired at the end of 1986 and the California Wind Rush ended abruptly. The company was scheduled to deliver 1,200 wind turbines to Zond in December 1985. But the shipment was delayed because the shipping company went bankrupt and when it arrived Zond wasn't able to buy the turbines anymore. Vestas found itself with a large stockpile of wind turbines that it couldn't sell. The company suspended payments to all of its creditors at the end of 1986. Nordtank and Micon found themselves in the same position and filed for bankruptcy. In 1987, Vestas was resurrected as a smaller company—Vestas Wind Systems A/S—that focused exclusively on wind power. Vestas Wind Systems focused on supplying wind turbines to other, "easier" markets than the United States— like India and China.

The California Wind Rush has a special place in wind energy lore. As such, it has become customary for those writing about the California Wind Rush to cast a suspicious eye toward wind farm developers, wind turbine manufacturers, tax investors, policy makers, politicians, or all the above. However, with the benefit of over thirty years of hindsight—and insight on the interim evolution of wind power technology—what's perhaps more useful at this stage is to better understand how the mis-steps in California were subsequently corrected in a manner that enabled wind power to eventually flourish in the twenty first century. In this context, the biggest topic for discussion is the role of the energy tax credits.

In total, the CEC has reported that the federal and California state treasuries lost approximately US$630 million and US$770 million, respectively, which would have been paid as taxes but for the implementation of the energy tax credit programs.[28] This is the amount that was collected in avoided taxes by wind farm investors. As a result of the tax credits—by the beginning of 1987 just after the credits expired—1.2 GW of wind power capacity was operating in California generating over 1,700 million kWh of electricity annually.

Whether or not tax payers paid a fair price for this wind power unknowable. Some have compared wind energy incentives to those received by fossil fuels and noted the disparity. A DOE study demonstrated that as late as 1999, emerging renewable energy sources were receiving just 17 percent of total US energy subsidies, while oil, gas, coal, and nuclear power technologies were still receiving 41 percent of all US energy subsidies.[29] Even given this, however, it is apparent that the credit created inefficiencies and waste by incentivizing the development of wind farms built with turbines that were unable to produce much electricity. Their poor design and abuse of their income sheltering properties led to their elimination. It is also clear that the sudden expiration of the tax credits pulled the rug out from under wind energy investors in such an abrupt manner that prospective future investors were scared off.

Both of these tax credit issues were spotted early on and remedies were proposed. In fact—in 1985—in the US House of Representatives, a bill was proposed that changed the investment credit to a "production credit," that provided a tax credit for electricity generation instead of wind farm investment.[30] The bill also proposed tying the credit value to the price of oil and providing an orderly credit phase out if oil prices increased. This bill didn't become law, but the Energy Policy Act of 1992 (EPACT) did. EPACT provided a 1.5 ¢/kWh tax credit for wind-generated electricity, which helped fuel the return of wind power in the United States starting in the late 1990s.

Would a properly structured tax credit have saved taxpayers money and still yielded an enormous 1.2 GW of wind capacity in a five-year period? Hard to tell. More likely, the California Wind Rush was a singular event tied to a specific time, place, and group of people who did their part to get wind power off the sidelines and into the game. Wind power wouldn't be where it is today without them. Truth is, everyone who benefits from the presence of wind power in the twenty-first century global energy landscape owes a debt of gratitude to the wind power pioneers that struck gold during the California Wind Rush.

Notes

1 History, 2017.
2 CEC, 1987, 26.
3 Righter, 1996, 215.
4 Asmus, 2001, 95.
5 Smith, 1987.
6 CEC, 1987, 23.
7 Smith, 1987.
8 Dehlsen, 2003.
9 Miller and Simon, 1978.
10 Asmus, 2001, 105.
11 Ibid., 107.
12 Dehlsen as quoted in Yergin (2011, 596).
13 Vestas, 2017.
14 CEC, 1987, 30.
15 Data derived from the summary data table of CEC (1987, 30).
16 Ibid.
17 Asmus, 2001, 122.
18 Orloff and Flannery, 1992.
19 Howell, 1991, 32.
20 Distefano, 2007.
21 Loss, Will, and Marra, 2013.
22 Cama, 2016.
23 Erickson et al., 2001.
24 Reagan, 1985.
25 CEC, *Results from the Wind Project Performance Reporting System: 1987 Annual Report*, 1988.

26 Asmus, 2001, 142.
27 Dehlsen, 2003.
28 Starrs, 1988, 140.
29 EIA, 2008, xxi, table ES1.
30 Starrs, 1988, 143.

Bibliography

Asmus, Peter. *Reaping the Wind*. Washington, DC: Island Press, 2001.

Cama, Timothy. 2016. "Trump: Wind Power 'Kills All Your Birds.'" *The Hill*, August 2, 2016. http://thehill.com/policy/energy-environment/290093-trump-wind-power-kills-all-your-birds.

CEC. *Results from the Wind Project Performance Reporting System: 1985 Annual Report*. San Francisco, CA: California Energy Commission, 1986.

———. *Results from the Wind Project Performance Reporting System: 1986 Annual Report*. San Francisco, CA: California Energy Commission, 1987.

———. *Results from the Wind Project Performance Reporting System: 1987 Annual Report*. San Francisco, CA: California Energy Commission, 1988.

Dehlsen, James, interview by U.S. Department of Energy. 2003. *From Zond to Enron Wind to GE Wind*. WINDExchange.

Distefano, Michael. "The Truth About Wind Turbines and Avian Mortality." *Sustainable Development and Law Policy* 8, no. 1 (2007): 10–11.

EIA. *Federal Financial Interventions and Subsidies in Energy Markets 2007*. Washington, DC: U.S. Department of Energy, Energy Information Administration, Office of Coal, Nuclear, Electric, and Alternative Fuels, 2008.

Erickson, Wallace P., Gregory D. Johnson, M. Dale Stickland, David P. Young, Jr., Karyn J. Sernka, and Rhett E. Good. *Avian Collisions with Wind Turbines*. Resource Document. Washington, DC: National Wind Coordinating Committee (NWCC), 2001.

History. 2017. "The Gold Rush of 1849." *History*. Accessed December 8, 2017. http://www.history.com/topics/gold-rush-of-1849.

Howell, Judd. *Assessment of Avian Use and Mortality Related to Wind Turbine Operations*. Livermore, CA: U.S. Windpower, 1991.

Loss, Scott R., Tom Will, and Peter P. Marra. "Estimates of Bird Collision Mortality at Wind Facilities in the Contiguous United States." *Biological Conservation* 168 (2013): 201–9.

Miller, A, and R Simon. *Wind Power Potential in California*. San Francisco, CA: California Energy Commission, 1978.

Orloff, Susan, and Anne Flannery. *Wind Turbine Effects on Avian Activity, Habitat Use, and Mortality in Altamont Pass and Solano County Wind Resource Areas*. San Francisco, CA: California Energy Commission, 1992.

Reagan, Ronald. "Message to the Congress Transmitting Proposed Legislation." In *Weekly Compiliation of Presidential Documents*, 21, 707–8. Washington, DC: Office of the Federal Register, National Archives and Records Administration, 1985.

Righter, Robert W. *Wind Energy in America: A History*. Norman, OK: University of Oklahoma Press, 1996.

Serchuk, Adam Harris. "Federal Giants and Wind Energy Entrepreneurs: Utility-Scale Windpower in America, 1970–1990." PhD thesis, Virginia Polytechnic Institute and State University, 1995.

Smith, Don. "The Wind Farms of the Altamont Pass Area." *Annual Energy Review* 12 (1987): 145–83.

Starrs, Thomas A. "Legislative Incentives and Energy Technologies: Government's Role in the Development of the California Wind Energy Industry." *Ecology Law Quarterly* 15, no. 103 (1988): 103–58.

Vestas. 2017. *Vestas History*. Accessed November 17, 2017. https://www.vestas.com/en/about/profile#!history.

Yergin, Daniel. *The Quest: Energy, Security, and the Remaking of the Modern World*. New York: Penguin Books, 2011.

14

Germany's Giant

The biggest wind power station in the world was to be built in Germany by all means.

—Bernward Janzing and Jan Oelker (2013)

14.1   Ulrich Hütter Returns

When the global oil crisis struck in October 1973, West Germany was impacted along with the rest of the world. At the time, 75 percent of Germany's oil came from Arab countries.[1] However, the West German government in Bonn didn't impose restrictions or price controls in order to contain high oil prices. Instead, they let German citizens buy as much gasoline and heating oil as they wanted if they were willing to pay the high prices. In response, many Germans cut back on their overall spending and the German economy softened. Energy demand fell and this took some steam out of the oil price rise. By December 1973, German oil demand was down by 6 percent. It was an effective, free market strategy.

The oil crisis spurred talk of energy alternatives within Germany. Industrial factories and power plants began a gradual movement toward coal. Coal had been the traditional source of electricity supply in Germany, but oil had made inroads starting in the late 1960s. By the time the crisis struck, oil-fired generators accounted for 30 percent of total power plant additions. This trend promptly reversed after the 1973 crisis, and oil-fired power plant additions all but disappeared by the end of the 1970s.

That wasn't the only type of fuel switching that was being considered. After a twenty-year hiatus, wind power was back on the table as an alternative energy option. In 1974 Bonn created a federal energy research program in the image of the United States Federal Wind Energy Program (FWEP). The German Ministry

The Wind Power Story: A Century of Innovation that Reshaped the Global Energy Landscape,
First Edition. Brandon N. Owens.
© 2019 by The Institute of Electrical and Electronics Engineers, Inc.
Published 2019 by John Wiley & Sons, Inc.

of Research and Technology (BMFT)—led by Hans Matthöfer—prepared a four-year research plan that was designed to explore "new energy sources for large-scale technological application."

BMFT also created the German Energy Research Program (GERP) at this time.[2] The plan was to start where Ulrich Hütter had left off in the late 1950s. Recall, Hütter teamed up with farm equipment manufacturer Allgaier-Werkzeugbau in the 1950s and began selling a 10-kW wind turbine. With financing from a consortium of energy companies, Hütter then began the development of a 100-kW wind turbine in 1955. This led to the creation of a 100-kW wind turbine known as WE-34 in 1959. The defining feature of the WE-34 was its two blades made from GFRP, a composite material made of polyester resin, fiberglass, and plastic.

WE-34's performance exceeded expectations, but by the late 1950s oil prices had fallen and Germany lost interest in wind power. After nine years of intermittent operation and limited support, WE-34 was shut down in 1968. The blades from WE-34 were salvaged and later ended up as a showpiece in front of a building at the Institute of Aircraft Design (IFB) in Stuttgart. Hütter joined IFB and taught aircraft design and also chaired the Institute of Structure and Design (DLR) at the German Test and Research Institute for Aviation and Space Flight (DFVLR).

Although forgotten in the 1960s, Hütter had since become a legend in wind power circles because of his development of the WE-34 wind turbine. In fact, after the 1973 energy crisis, Hütter expected the German government to contact him to restart wind power development. And—to be sure—Hütter's phone did indeed ring at the outbreak of the 1973 energy crisis. However, the voice on the other end of the line didn't speak German. In the wake of the crisis, the American space agency—NASA—had thoroughly researched wind power and combed through decades of technical literature on the subject. They identified W-34 as the premier wind turbine design and wanted to buy the plans from Hütter. Hütter obliged, earning US$55,000 for IFB.[3] Fact was, NASA also needed some hand-holding as they got up to speed on wind power and enlisted Hütter as a consultant.

During this period, the first German wind power association was formed. It was called the "Verein für Windenergieforschung- und Anwendung" (Association for Wind Energy Research and Applications). It was later renamed "Deutsche Gesellschaft für Windenergie" (German Association for Wind Energy). Momentum was gradually building in Germany for the support of wind power.

BMFT's Matthöfer did eventually reach out to Hütter in 1974. He invited him to Bonn to discuss Germany's wind research program. The two hit it off at the meeting in Bonn and a decision was made to recreate WE-34 but on an even larger scale. BMFT was thinking big, just like their counterparts at NASA. Hütter agreed—in theory—and suggested that wind turbines could hypothetically be built as large as 3 MW if further research was first done to successfully scale it up.

But Matthöfer and the rest of the BMFT researchers were thinking even larger, and during their initial meeting, they asked Hütter why his proposed 3-MW wind turbine was "so little."[4] BMFT then commissioned Hütter to write a report entitled *Energy Sources for Tomorrow?*, that studied the feasibility of building a megawatt-sized wind turbine based on the WE-34 experience.

Hütter and his research team at the University of Stuttgart got to work and assembled the feasibility study. The team looked at the feasibility of two different wind turbines: a 1-MW machine with an 80-m diameter rotor, and a 3-MW version with a 113-m rotor. The team ultimately recommended that Germany start with the 1-MW machine by noting that "it appears to be appropriate to maintain continuity, to proceed with a smaller step in researching the problems of and control of a large wind turbine. For this reason, the one-megawatt turbine with a tower height of fifty-two meters and a rotor diameter of eighty meters should be constructed in the short term."[5]

But wires had been crossed. BMFT thought that Hütter had already agreed to build a wind turbine that was at least 3 MW in size. Their eyes were on NASA across the Atlantic, which had already launched a large wind turbine research program to build megawatt-scale machines. The German government wanted to ensure that Germany didn't fall behind in the new space race. Plus, BMFT had already contracted with MAN in Munich to build the big machine. Hütter eventually reluctantly relented. As long as it was built in the image of WE-34 with GFRP blades, Hütter would scale it up to 3 MW. Hütter's design philosophy was based on lightweight, strong materials, which he believed were the key to successfully building megawatt-sized wind turbines.

Hütter and his team at Stuttgart wanted the blades to be moulded from GFRP. However, their industrial partner MAN didn't have any experience with this material, so they made the blades with a steel frame and then shrouded it with a shell of fibreglass compound. Hütter and his team were aghast. The blades would be heavier now, which required a complete redesign of the turbine. At this stage in the project, Hütter stepped back and was no longer involved in its daily activities. He simultaneously retired as the chair of DFVLR and focused on his professorship at IFB in Stuttgart. From his perspective, he had done his part to guide the German government and its industry partner in the right direction. Whether or not they followed his guidance and ultimately succeeded would be up to them.

MAN pushed forward and constructed the prototype blades, which it then ran through rigorous tests. MAN later issued a report indicating that the blades were structurally sound and suitable for use in the large wind turbine. Under the instruction of BMFT, MAN prepared a final report on the wind turbine in advance of the actual construction.[6] The report green-lighted the design and recommended construction of the wind turbine. In the spirit of research diversification—during this period—BMFT also funded another industrial partner—Messerschmidt Bölko Blohm (MBB) of Munich—to build a 400-kW downwind, single-bladed wind turbine with a 30-m rotor. The wind turbine—named Monopteros—began operating in 1984 and experienced a massive failure in 1985 when part of the blade broke off.[7]

Construction began for the BMFT-sponsored 3-MW wind turbine developed by MAN, which came to be known as Growian, an acronym for GRoße WIndenergie Anlage. The wind machine test station at Kaiser-Wilhelm-Koog in northern Germany was chosen for the installation location. The nacelle was finally raised to the top of the tower in October 1982. Testing began during the summer of 1983. Because of a delay caused by modifications of the locking bolt of the rotor hub, the rotor brake, and a bearing in the teetering hub, the wind turbine was not connected to the grid until 1984.[8]

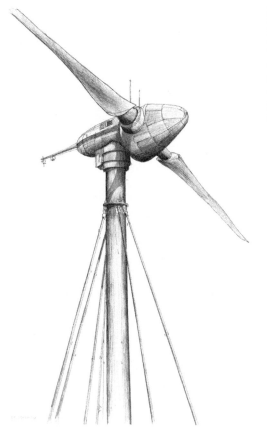

Figure 14.1 Although it only ran briefly starting in 1984, Germany's 3-MW "Growian" wind turbine was the world's first grid-connected variable-speed wind turbine. It used a variable-speed generator configuration that is now favored by twenty-first century wind turbine manufacturers.

Upon completion, Growian was a two-bladed, 3-MW, variable-speed, downwind machine with a teetering hub that sat atop a 100-m concrete tower. The nacelle mass was 240 tons and its orientation was actively controlled. From the cut-in wind speed of 6.3 m/s up to the rated wind speed of 11.8 m/s, the digital control system maintained the optimum tip-speed ratio. Full-span pitch allowed power output to reach three megawatts for winds at the cut-out speed of 24 m/s.[9]

The most innovative aspect of Growian was its variable-speed design. In a typical wind turbine design, the generator is connected to the rotor via a gearbox, so the generator's rotational speed and the frequency of its output is determined by the rotational speed of the rotor. Recall that up until this point, grid-connected wind turbines employed a constant-speed design. Within this approach, keeping the rotational speed of the rotor fixed ensures that the electrical output is provided at a fixed frequency and voltage as required by the electrical grid. However, the drawback of the constant speed approach is that all the additional energy contained in the wind is wasted at higher wind speeds. The rotor speed had to be throttled back at higher wind speeds either by changing the blade pitch (i.e., active pitch control) or by designing the blades in such a manner that turbulence causes the blades to slow (i.e., passive stall).

In contrast, variable-speed approaches—like that used for Growian—enable the wind turbine rotor to vary freely based on the wind speed. This enables the energy contained in the wind at higher speeds to be captured and converted to electricity. The net result is that variable-speed machines capture 10 percent more energy than constant-speed machines.[10] In addition, the peak mechanical stresses caused by wind gusts is reduced by limiting the torque acted on the wind turbine by the generator and allowing the wind turbine to speed up in response to wind gusts.

However, variable-speed approaches have their drawbacks too. Namely, variable-speed machines produce electrical output with varying frequency and voltage. If the rotor speed is variable because of changes in the wind speed, then so is the generator's rotational speed, and the resultant output will not be suitable for the grid. To address this problem, variable-speed configurations use power electronics to convert the wild AC to fixed-frequency AC. While technically feasible, power conversion technologies had been large, inefficient, and costly. However, advances in solid-state electronics drove down the size and cost of power electronics dramatically starting in the 1970s. This meant that variable-speed approaches started to become increasingly economically feasible starting in the mid-1980s.

The designers of Growian decided that their machine represented an opportunity to finally implement a variable-speed approach. The Growian designers considered three options: (1) create a constant-speed wind turbine that uses a synchronous generator; (2) build a variable-speed wind turbine that uses direct-in-line power conversion; or (3) develop a variable speed wind turbine that uses a doubly-fed induction generator (DFIG). They chose the third option, a variable-speed wind turbine using a DFIG. Thus, Growian was the first variable-speed, grid-connected large wind turbine.

In a DFIG configuration, both the rotor and the stator are separately connected to equipment outside the generator. This is why the machine is called "doubly-fed" because power is fed to the grid independently from the rotor and the stator. One winding is connected directly to the grid and produces AC power at the required grid frequency. The other is connected to AC power at variable frequency. Input power is then adjusted in frequency to compensate for changes in speed of the wind turbine. Adjusting the frequency requires a converter. Because of the doubly-fed configuration, even though the wind turbine operates at a variable speed, only about one-third of the power needs to be fed through a converter.

This means that the power conditioning system only needs be one-third as large as those used in direct-in-line conversion, so the DFIG configuration reduces power conversion needs, and associated capital costs and efficiency losses, by two-thirds compared to direct in-line converters that must condition all of the power before it is exported to the grid. This is why variable-speed wind turbines that use a DFIG configuration are the preferred technology solution for large, grid-connected wind turbines in the twenty-first century.

However, despite these impressive innovations—or perhaps because of them—Growian was riddled with problems that limited its operation. From the very beginning severe fatigue problems, faulty bearings and brakes, and frost damage hampered operations. In 1988, after only 420 hours of operation over a four-year period, the machine was dismantled. Growian project engineers quickly tried to understand what went wrong with the wind turbine and incorporated insights into their plans for a new 1.2-MW wind turbine—WindKraftAnlage-60 (WEKA-60). WEKA-60 was later installed on Helgoland Island in the North Sea in 1990. However, neither WEKA-60, nor any of its successor units experienced commercial or technical success.

The German national wind research program took a beating in the German press and this ultimately led to the widespread impression that wind power wasn't ready for prime time. Some criticized the German federal research program as a NASA-like folly that produced a single large turbine that had a very limited operational period and didn't result in any immediate commercialization. In fact, Growian has been characterized as "a disaster" and Germany's "most spectacular aeronautical failure."[11]

While it is true that the process was fraught with error, in hindsight it is also clear that Growian engineers and scientists had a firm grasp on the future technology pathway for wind power. Growian was more of a sign post to the future rather than the beginning of a commercial-ready wind turbine. Germany's experience here was like that of the United States. In America, federal research led to a series of costly and unreliable large wind turbines that failed to produce commercial wind turbines. However, many of the technology solutions employed in NASA's machines were precursors to twenty-first century wind power technology.

After exiting the Growian program, Ulrich Hütter wasn't done putting his stamp on wind power. To better advance his vision for large-scale wind power, Hütter teamed up Voith GmbH & Co. KGaA. Voith was a German family-owned multinational engineering corporation in Heidenheim. Under the direction of managing director Hugo Rupf, Voith was looking for ways to diversify its operations and expand into new industries. Voith's Viktor Kaplan had invented the Kaplan turbine in 1922 and Voith had since developed, built, and installed hydraulic machines for the development of hydropower. During this time, Voith Hydro had contributed significantly to the advancement of many types of hydro turbines and wind energy seemed like a natural extension of this work.

Together Hütter and Voith designed a high-speed 265-kW two-bladed, downwind wind turbine that embodied Hütter's lightweight design philosophy. The rotor had a diameter of 52 m, so the wind turbine was given the name WEC-52. WEC-52 was innovative because it included two gearboxes, one in the tower and one in the base. Energy was transmitted through the tower-mounted gearbox and down to the base of the tower through a shaft. The tower base contained the second gearbox and electric generator. In this manner, the gearbox and the generator were more easily accessible, and the load on each gearbox was reduced. WEC-52 was constructed near Stuttgart in 1981. When the wind turbine went into operation in October 1981, it was the largest in Germany.[12]

The Voith wind turbine was tested between 1982 and 1988. Overall, the operation of WEC-52 was acceptable, but the wind turbine experienced its fair share of challenges. Among other problems, the long transmission shift through the tower oscillated at high wind speeds.[13] Furthermore, the blades bent at the tip speed during high winds.[14] An engineering team studied these problems and concluded that Hütter's blade designs would need to be refined. But by 1988, Voith had lost patience and terminated its wind energy efforts. Because of the uniqueness of WEC-52, the Mannheim Museum for Engineering and Work purchased the machine for display.[15] Voith later re-reentered the wind power industry in 2007 when the company introduced a hydraulic drive for wind turbines that is used to convert the energy from a variable-speed wind rotor into constant-speed. Their technology solution is situated between the wind rotor and the gearbox and therefore eliminates the need for a DFIG configuration or any other solution designed to contend with wind speed variability.

Ulrich Hütter died in 1990 after a lifetime of wind power innovation and is one of the most notable figures in the history of wind power. Hütter's blade design went on to heavily influence the direction of wind power. His turbine and blade designs were first used by a group of students and teachers at a Danish school named Tvind. The Tvind students made the blade moulds available to anyone who wanted to use them and a small fiberglass workshop—TV Glassfiber—was the first to use the moulds for a wind turbine built by Arne Friis. Erik Grove-Neilsen subsequently purchased the moulds and he began making wind turbine blades in a farm in øaker in Middle-Jutland, Denmark.[16] The Danish wind turbines developed during this period embodied what became known as the "Danish Concept" and one of the key components of the Danish Concept were the fiberglass composite blades produced by øaker.

To be sure, Germany and America weren't the only countries seduced by the grandeur of building megawatt-sized wind turbines during this time. Sweden, the United Kingdom, Italy, Spain, and the Netherlands all built government-funded wind turbines in the megawatt range in the 1970s and 1980s. A record of this research can be found in the proceedings of the Annual International Symposium on Wind Energy Systems, which started in 1976 when the first meeting was held in Cambridge, England. The symposium included written contributions from nine countries including Tanzania and Singapore.

Swedish participants at the symposium described the Swedish government's ten-year program to study wind energy. The program was managed by the National Swedish Board for Energy Source Development (NE; in Swedish: Nämnden för energiproduktionsforskning) and was focused on building state-of-the-art wind turbines. The Swedish government produced a 1976 study that analyzed the potential for large wind turbines in the range of one to 5 MW to meet the country's energy requirements. They found that wind power could meet up to 20 percent of the country's needs using just 1 percent of the land area.

As a practical first step, Sweden started with a 60-kW wind turbine that was installed in Älvkarleby, Uppsala County. The machine was built by NE and Saab-Scania AB and was a two-bladed, downwind turbine with an 18-m rotor. It sat atop a 25-m concrete tower and was put into operation in April 1977.[17] It used GFRP blades and was built in the spirit of Hütter's lightweight approach with a teetering hub. However, the wind turbine was destroyed in mobile crane incident in 1980.

The Swedish program continued with the development of two large wind turbines. The 3-MW WTS-3 was installed by Swedish utility Sydkraft in 1983 at Maglarp on the west coast of Sweden. It was built through a partnership with American company Hamilton Standard and Swedish company Karlskronavarvet. It was one of the most successful megawatt-sized wind turbines of the 1980s. The wind turbine recorded 34 million kWh of generation by 1993, which was a world record at the time, and was kept in operation without support from the Swedish government.[18]

The second megawatt-sized wind turbine built in Sweden was the WTS-75, located at Näsudden on the southwest cost of the island of Gotland. The 2-MW wind turbine was manufactured by Kamewa and it had a 75-m rotor and a 77-m hub height. The turbine started operation in 1984 and ran until 1990. In 1993, an improved nacelle was installed on the same tower as part of a German-Swedish research partnership between Swedish Kvaerner and German utility Preussen Elektra.[19] The US$34 million wind turbine was privately funded by Kvaerner and Preussen Elektra.[20] By the time it was put out of operation due to a gearbox failure in 2006, it held the world record for electricity generation from a wind turbine.

There were other efforts in Europe to develop large wind turbines in the 1980s. These included the 3-MW Orkney-60 wind turbine in the United Kingdom. European governments were collectively promoting the development of megawatt-scale wind turbines through the European Economic Community's (EEC) Wind Energie Grösse Anlagen (WEGA) program. This program resulted in the development of the Tjaereborg 2-MW wind turbine in Denmark, the 1-MW Richborough wind turbine in the United Kingdom, the 1.2-MW AWEC-60 wind turbine in Spain, and the 2-MW Gamma wind turbine in Italy. However, like the German and Swedish efforts, these large wind turbines didn't lead directly to commercialization.

14.2 Bottoms Up

As important as Growian and the other large wind turbines were for Germany's wind power research efforts, there was a parallel effort underway to develop smaller wind turbines that would eventually bear greater commercial fruit. It started when BMFT asked its industrial partner MAN to build a smaller 11-kW wind turbine. The resultant wind turbine—Aeroman 11—performed very well and was used to build a wind farm on the Greek island of Kythnos in 1982. A second 20-kW model—Aeroman 20—was deployed at demonstration sites throughout Germany. Jacobs Energie of Heide in Northern Germany purchased the license to the Aeroman design in 1988.* Jacobs increased the capacity to 33 kW and continued to build successively larger wind turbines, eventually reaching 1.5 MW in 1998.

Jacobs Energie would later become part of Repower, which was formed in 2001 when Jacobs merged with Husumer Schiffswerft (HSW), BWU, and pro+pro energy systems. Repower began producing its 1.5-MW ProTec wind turbine in 2001. The company was subsequently acquired by India's Suzlon Energy in 2007 as part of its global expansion. Repower—which was renamed Senvion—was later resold to Centerbridge Partners in 2015.

---

* The Germany company Jacobs Energie is not to be confused with the American small wind turbine manufacturer Jacobs Wind Electric Company, which was founded by brothers Marcellus and Joe Jacobs in the 1920s.

HSW itself was a boat manufacturer from Husam in northern Germany that decided to enter the wind industry in 1986. They started with a 30-kW two-bladed wind turbine and later built a 200-kW version. They employed the Danish approach and featured a stall-regulated rotor. In 1988, they introduced serial production of their 250-kW model. They continued to operate as a domestic wind power manufacturer and developed a 750-kW model in 1993. They successfully navigated the German wind market throughout the 1990s, but eventually filed for bankruptcy in 1999. HSW was purchased out of bankruptcy in 2000 by Jacobs Energie and later became part of Repower.[21]

Furthermore—in 1985—wind turbine designer Dr. Günter Wagner developed several prototypes and was searching for a partner to help him begin commercial production. He found a willing and able partner in Franz Tacke. In 1960, Tacke had taken over the Tacke KG in Rhein, a factory that his grandfather founded in 1886. By the mid-1980s Tacke was looking to expand his manufacturing business. He purchased Wagner's wind turbine designs and began producing 150-kW wind turbines that were successfully installed in California's Tehachapi area as part of the California Wind Rush. In 1990, Franz Tacke founded Tacke Windtechnik with his sons to focus exclusively on wind power.

By 1996, Tacke had sold several hundred 600-kW wind turbines and had become the second largest German wind turbine manufacturer.[22] In 1997, Tacke was purchased by United States energy company Enron. Enron had already purchased America's Zond earlier in the year and the engineers at Zond urged their parent company to purchase Tacke because they respected its technology. When Enron went bankrupt at the end of 2001, Enron's wind power assets—including Tacke and Zond—were later purchased for US$358 million by GE at a bankruptcy auction 2002. GE Wind Energy would go on to become one of the largest wind turbine manufacturers in the world. Today it is known as GE Renewable Energy and it is headquartered in Paris, France.

Aloys Wobben was also a very important German wind turbine designer during this period. Wobben was born 1952 in Lower Saxony, Rastdorf. As a child, he loved playing and exploring in the natural environment. He studied electrical engineering at the Technical University in Braunschweig and he specialized in electrical control systems and inverter technologies.[23] In 1981, Wobben purchased a set of Danish økær blades and built a 15-kW, three-bladed wind turbine in his backyard. He later founded Enercon in 1984 to sell wind energy equipment.

The company initially focused just on wind turbine control systems, but in 1985 he had developed a 55-kW wind turbine called the Enercon-15 or E15. The E15 was a three-bladed, upwind wind turbine that sat atop a lattice tower. It was a constant-speed machine that used a synchronous generator to provide fixed-frequency power to the German grid. The first unit was set up in Aurich in northern Germany.[24] By 1988, Enercon was manufacturing an 80-kW wind turbine—the E17—and a 100-kW machine, the E32. The company then focused on manufacturing its own blades and built its own production facility in 1991. Enercon's wind turbines attracted attention from the beginning because of their quality design and construction. Enercon's massive contributions to the global wind power industry were just beginning.

Figure 14.2 Enercon's E15 wind turbine developed in 1985. It was a three-bladed, upwind wind turbine that sat atop a lattice tower. It was a constant-speed machine that used a synchronous generator that provided fixed-frequency power to the German grid.

14.3   Our Common Future

As discussed, the modern environmental movement started in earnest in the 1960s after the publication of Rachel Carson's *Silent Spring*. Carson highlighted the negative environmental impacts of pesticides, which led to a reassessment of the costs of rapid industrialization. By the end of the 1960s, the movement had coalesced and gained important political power that resulted in the implementation of environmental policies in the 1970s. The 1972 United Nations Conference on the Human Environment was a watershed movement when nations of the world came to together to identify and address global environmental challenges. The global energy crises of 1973 and 1979 further underscored the importance of environmental stewardship and highlighted the benefits of domestic renewable energy resources.

However, by the mid-1980s, oil prices had collapsed, and environmentalism was no longer a global priority. Budgets for national wind power research programs were flat or declining. In the United States, the conservative Reagan Administration gutted American wind research efforts. However, the world's environmental challenges were not yet solved. In fact, in terms of the impact on the environment, these challenges had grown. The 1986 Chernobyl nuclear power catastrophe underscored the world's global energy challenges and highlighted the need for less risky energy options. It also amplified public questions surrounding the role of nuclear energy.[25]

It is in this context that the Secretary-General of the United Nations—Javier Pérez de Cuéllar—asked the Prime Minister of Norway—Gro Harlem Brundtland—to create an organization independent of the UN to focus on environmental challenges and solutions. This new organization was the Brundtland Commission—or more formally—the World Commission on Environment and Development. The Brundtland Commission was first headed by Gro Harlem Brundtland as Chairman and Mansour Khalid as Vice-Chairman.

The organization aimed to create a united international community with shared sustainability goals by identifying environmental problems worldwide, raising awareness about them, and suggesting the implementation of solutions. In 1987, the Brundtland Commission published the first volume of *Our Common Future*, also known informally as the Brundtland Report. The report provided a clear definition of "sustainable development" as "development that meets the needs of the present without compromising the ability of future generations to meet their own needs."[26] It also called for carbon dioxide emissions reduction targets.

The report had a strong influence on governments across the globe, particularly in Europe. In Denmark, Minister of Energy Jens Bilgrav-Nielsen was in the process of developing the third Danish energy plan, Energy 2000. Inspired by the Brundtland report and its focus on sustainable development, Bilgrav-Nielsen adopted the recommendation of the Brundtland report related to the reduction of energy consumption and carbon dioxide emissions and set an ambitious goal of a 20 percent reduction in Danish carbon dioxide emissions by 2005. Wind power would help meet the goal by providing 10 percent of the nation's electricity consumption by 2005.[27] A tariff for wind energy was introduced in 1993 that required electric utilities to pay wind generators a price equal to 85 percent of the utilities production and distribution costs.

Germany too was influenced by the Brundtland Report. The German Enquete Commission on Preventive Measures to Protect the Earth's Atmosphere was established in 1987, and came out with two additional influential reports, one in 1989 and one in 1990. This led to the beginning of the "Energiewende," Germany's energy transition. In December 1990, the country passed the Electricity Feed Act (Stromeinspeisungsgesetz, StrEG). The law was passed when three members of the parliament with different party affiliations formed an alliance to finalize its passage. The Act required utilities to purchase electricity from small renewable energy generators and obligated them to pay favorable prices for the power. The contract price was set by the federal law. This was the first implementation of a national "feed-in tariff" (FIT), which would later become the preferred mechanism for driving renewable energy supply into the electricity network throughout Europe.

For wind power, the German law required utilities to purchase power at 90 percent of the average retail rate.[†] To the electric utility industry in Germany, this was a shock to the system. A subsequent battle ensued between German utilities and small independent power producers that wanted access to the German market. German electric utilities eventually acquiesced considering the relatively small quantities of renewable power that were added to the systems at the beginning of the program.

Federal payments under the StrEG FIT were supplemented by additional tax credits made available through the 100-MW wind power initiative that was created in 1989. This initiative provided a tax credit of 8 Pf/kWh. Taken together—the FIT and the tax credit—made wind power economically viable in high wind areas along Germany's costs. As a result, these incentives were strong enough to encourage the growth of Germany's wind market. By 1990, 71 million kWh of wind power was generated nationally. That's equivalent to about 40 MW of capacity operating at an average annual capacity factor of 20 percent.

To recap, Germany responded to the 1973 energy crisis by enacting a federal wind energy research program that focused primarily on building

---

[†] Germany's electric power system was later deregulated, and the retail price of electricity declined. This meant that the FIT—as implemented in StrEG—also declined. As the FIT price declined, the nascent wind market in Germany threatened to stall. It would require another renewable energy law, the 2000 Renewable Energy Sources Act (Erneuerbare-Energien-Gestz, EEG), to correct this problem.

megawatt-scale wind turbines. These efforts—including the development of the Growian wind turbine—produced technological insights that would help shape the future of the wind power industry. Growian's variable-speed configuration and implementation of a DFIG was a sign of things to come for wind power technology. The concurrent efforts of individuals and small corporations to develop commercial-ready wind turbines led to the creation of several companies—Jacobs, Tacke, and Enercon—that would eventually have a big impact on the global wind power industry.

Still, at the dawn of the 1990s, wind power had a long way to go before it would become meaningful part of the global electric power system. The technology was still small-scale and unreliable. From the vantage point of 1990, it was still unclear whether the wind power progress of the 1970s and 1980s represented another false start or whether this was this a new era that provided wind power with the opportunity to permanently make its mark on the global energy landscape.

The next decade would prove to be pivotal for wind energy. The future of wind power would begin in a packed conference room in Madrid in 1990. Tempers flared at the European Economic Community's wind energy conference where an old wind power schism reared its ugly head, and the fate of Europe's wind energy research efforts hung in the balance.

Notes

1 Whitney, 1974.
2 Janzing and Oelker, 2013, 399.
3 Ibid.
4 Ibid., 400.
5 Hau, 2013, 25.
6 Thiele, 1983.
7 Jaeger, 2013, 413.
8 Carlin, Laxson, and Muljadi 2001, 21.
9 Ibid.
10 Varrone, 2013.
11 See Karnøe (1989), Gipe (1995), and Matthias (1998).
12 Janzing and Oelker, 2013, 402.
13 Hau, 2013, 310.
14 Janzing and Oelker, 2013, 403.
15 Jaeger, 2013, 412.
16 Christensen, 2013, 74.
17 Jaeger, 2013, 495.
18 Gipe, 1995, 109.
19 Jaeger, 2013, 496.
20 Gipe, 1995, 109.

21 Jaeger, 2013, 422.
22 Ibid., 417.
23 SuccessStory.Com, 2008.
24 Jaeger, 2013, 420.
25 Palz, 2013, 26.
26 Ibid., 27.
27 Christensen, 2013, 82.

Bibliography

Carlin, P. W., A. S. Laxson, and E. B. Muljadi. *The History and State of the Art of Variable-Speed Wind Turbine Technology*. Technical Report. Golden, CO: National Renewable Energy Laboratory, 2001.

Christensen, Benny. "History of Danish Wind Power." In *The Rise of Modern Wind Energy: Wind Power for the World*, edited by Preben Maegaard, Anna Krenz, and Wolfgang Palz, 642. Singapore: Pan Stanford Publishing, 2013.

Gipe, Paul. *Wind Energy Comes of Age*. New York: John Wiley, 1995.

GWEC. *Global Wind Energy Report 2006*. Brussels, Belgium: Global Wind Energy Council, 2007.

Harrison, Robert, Erich Hau, and Herman Snel. *Large Wind Turbines: Design and Economics*. New York: John Wiley, 2000.

Hau, Erich. *Wind Turbines: Fundamentals, Technologies, Appications, Economics*. Berlin; Heidelberg, Germany: Springer-Verlag, 2013.

Jaeger, Arne. "Overview of German Wind Industry Roots." In *The Rise of Modern Wind Power: Windd Power for the World*, edited by Preben Maegaard, Anna Krenz, and Wolfgang Palz, 407–44. Singapore: Pan Stanford Publishing, 2013.

Janzing, Bernward, and Jan Oelker. "Hütter's Heritage: The Stuttgart School." In *The Rise of Modern Wind Energy: Wind Power for the World*, edited by Preben Maegaard, Anna Krenz, and Wolfgang Palz, 387–406. Singapore: Pan Stanford Publishing, 2013.

Karnøe, Peter. "Technological Innovation and Industrial Organization in the Danish Wind Industry." *Entrepreneurship and Regional Development* 2, no. 2 (1990): 105–24.

Matthias, Heymann. "Signs of Hubris: The Shaping of Wind Technology Styles in Germany, Denmark, and the United States, 1940–1990." *Technology and Culture* 39, no. 4 (1998): 641–70.

Morey, Matthew J., and Laurence D. Kirsch. *German Experience with Promotion of Renewable Energy*. Madison, WI: Christensen Associates Energy Consulting, LLC, 2014.

Palz, Wolfgang. "Wind Power Development in the European Union." In *Wind Power for the World, Interntional Reviews and Developments*, edited by Preben Maegaard, Anna Krenz, and Wolfgang Palz, 23–38. Singapore: Pan Stanford, 2013.

SuccessStory.Com. 2008. *Aloys Wobben Success Story*. Accessed December 18, 2017. https://successstory.com/people/aloys-wobben.

Thiele, Hans Martin. *Growian Rotor Blades: Production, Development, Construction, and Test*, 76. Bonn, Germany: BMFT, 1983.

Varrone, Chris. "The Selfish Invention." *Renewable Energy Focus* 14, no. 2 (2013): 34–7.

Whitney, Craig R. 1974. "West Germans, at a Price, Avoid Oil Crisis." *New York Times*, Janurary 24.

15

Spain's Wind Power Miracle

Why reinvent the wheel?
          —Lopez Gandasegui, CEO of Gamesa from 1995 to 2005

15.1   A Hot Day in Madrid

By 1990, nearly two decades of government-led wind power research in Denmark and Germany had failed to produce commercially viable wind turbines. National wind turbine research was derided in some camps as an example of bureaucratic waste and ineptitude. Even those wind turbine manufacturers that did exist in Denmark and Germany were struggling to find success in the global marketplace after the collapse of the California market in 1986. Vestas—Denmark's flagship manufacturer—recently emerged from its 1987 bankruptcy. Enercon—Germany's homegrown wind turbine manufacturer—had yet to experience sustained success. The future of wind power was very much up in the air at this juncture.

It is against this backdrop that attendees of an invitation-only meeting at the European Economic Community's (EEC) 1990 wind conference gathered in Madrid, Spain. The EEC had been supporting wind power since 1979, and—in the 1980s—the EEC developed a research program to support large, megawatt-scale wind turbine development called the Wind Energie Grösse Anlagen (WEGA). The purpose of the program was to build experimental wind turbines to test the feasibility of further size increases. The program was conducted in two phases, WEGA I started in 1985 and supported the development of large wind turbines in the United Kingdom, Spain, and Denmark. WEGA II was scheduled to start in 1990, based on the outcome of the Madrid meeting.

WEGA I produced a 2-MW Danish wind turbine called Tjaereborg. Tjaereborg had a 60-m rotor diameter and it was installed at Esbjerg in Jutland.

The Wind Power Story: A Century of Innovation that Reshaped the Global Energy Landscape,
First Edition. Brandon N. Owens.
© 2019 by The Institute of Electrical and Electronics Engineers, Inc.
Published 2019 by John Wiley & Sons, Inc.

WEGA I also built a Spanish 1.2-MW wind turbine in Cabo Villano on the northwestern corner of the Iberian Peninsula. It had a 60-m rotor diameter. WEGA I was also responsible for a 1-MW wind turbine installed at Richborough in Kent in England. It had a 55-m rotor diameter and it sported wood epoxy composite blades. All three wind turbines started producing power in 1989. However, despite coordination and communication among European scientists and engineers, none of these wind turbines experienced a large degree of technical success and they didn't lead to any commercialization.

A second phase of WEGA was about to commence and there was uncertainty regarding the future direction of the program. Should EEC continue along the multi-megawatt path, or pivot back toward smaller wind turbines? WEGA program director Wolfgang Palz convened the Madrid meeting to chart the course for WEGA II. Palz—a PhD physicist who had previously led power systems development for the French National Space Agency—was a well-respected renewable energy expert. In 1973, he co-organized the UNESCO Congress—*The Sun in Service of Mankind*—in Paris. Between 1976 and 1978 he wrote the book, *Solar Electricity: An Economic Approach to Solar Energy*, which was published by UNESCO in seven languages.[1] Now he was in Brussels, managing the EEC's renewable energy R&D program, of which WEGA was a key component.[2]

Palz introduced Eric Hau to the standing room-only crowd. Hau himself was a renewable energy expert at the German Institute for Solar Energy (ISET) and was the EEC's technical point person for the WEGA program. At the Madrid meeting, Hau described the WEGA I results and then held a lengthy discussion on the importance of the continued development of megawatt-scale wind turbines, which he viewed as the next generation of wind power technology.[3] Palz and Hau believed that the challenges of WEGA I should be set aside and that WEGA II should continue to focus on megawatt-scale machines.

However, many in the crowd weren't convinced and a "heated" discussion followed, which spilled out into the hallway.[4] Dr. Jos Beurskens—who was the head of the Renewable Energy Unit and the Wind Energy Unit of the Netherlands (ECN)—later described the discussion as "fierce."[5] Part of the problem in Madrid was that Dr. David Milborrow—an engineer at the Central Electricity Generating Board (CEGB) in the United Kingdom—had just issued a report indicating that the optimal size of a wind turbine was 1 MW. Milborrow's thesis was that, although electrical output increases with the square of the rotor diameter, the machine weight and cost increases with the cube of that diameter. To avoid excessive weight, there needed to be a limit to the size and cost of wind turbines.[6] With Milborrow's report in hand, a faction of the Madrid group believed that WEGA was headed in the wrong direction by building multi-megawatt wind turbines. They believed a more sensible approach was to extend the then-existing 200–400-kW wind turbines to the 750-kW range and then go from there, instead of pushing ahead with wind turbines that were one megawatt and greater in size.

This was another classic debate revolving around the small versus large wind turbine schism that had been part wind power history since American Marcellus Jacobs sent a letter to the United States Congress in 1951 in a successful attempt to scuttle the US FPC's plans to fund multi-megawatt wind turbines in the 1950s. The conflict reared its ahead again when small wind turbine developers mocked NASA's giant wind turbine prototypes in the 1980s. And here it was again in Madrid in 1990. Palz and Hau eventually won the day by indicating that the EEC was already deeply invested in the megawatt-scale program and that failure of this program would likely lead to less funding for all wind energy projects, including the handful of small wind efforts that were valued by the Madrid crowd.

Palz then asked Hau and colleagues Robert Harrison and Herman Snel to examine the question of optimum wind turbine size and report back to the group. Their findings were eventually published in 1993 and later in the book *Large Wind Turbines*.[7] They found that there was no physical limit to the maximum size of wind turbines, but that the flexible elements and materials needed to have a higher strength-to-weight ratio. In other words, megawatt-scale wind turbine could work, but the materials needed to be lightweight, strong, and flexible. The study recommended an optimal range of 2–3 MW based on the known technology at the time. The largest wind turbines today exceed 10 MW in capacity.

The group at the Madrid meeting grudgingly fell in line. Interestingly—as it turns out—this would be one of the last clashes between small and big wind proponents because commercially available wind turbines had already been gradually growing since their introduction anyway. For example, after starting with the 30-kW V-10 machine in 1979, Vestas had introduced the 225-kW V-27 in 1989. Just one year after the Madrid conference, Vestas then introduced their 500-kW V-39 wind turbine; and by the end of the 1990s Vestas was offering the 1.75-kW V-66 wind turbine. The only relevant question in Madrid in 1990 really was whether WEGA II could conduct research on megawatt-scale wind turbine that would prove useful to commercial manufacturers and help accelerate their upsizing efforts.

The WEGA II program went on to build five wind turbines by 1996. They worked in coordination with European wind turbine manufacturers Bonus, Enercon, Nordic, Vestas, and WEG. Three of the five wind turbines led to the improvement of commercial wind turbines, including the projects with Vestas, Enercon, and Bonus. The inclusion of variable-speed configurations and power convertors increased the controllability of wind turbines. This led to the development of digital tools and control strategies that made wind turbines more compatible with grid requirements. Thus, ultimately the EEC's wind research programs were successful because it pooled expertise and resource across European countries that facilitated an intensive exchange of knowledge that led to wind turbine technical advances.[8]

Madrid was an ideal location for the 1990 conference because something special was about to happen in Spain. Up until that time, the race to commercialize wind power had been driven primarily by Denmark, Germany, and the United States. Each of these countries made bold moves and leaps forward across the twentieth century and together these countries had become the three primary markets for the nascent wind power industry. But Spain was now on the move. Through strong government action and industry partnerships, wind power would become a Spanish success story reminiscent of the Spanish economic development miracle that the country experienced between 1955 and 1973.

Spain's wind power journey began after the 1973 global energy crisis. The OPEC oil embargo and tightening global energy supplies led to an increase in the price of oil in Spain from US$1.62/bbl in January 1973, to US$9.31/bbl by the following March.[9] This was problematic for Spain because the country relied on foreign oil for nearly 70 percent of its energy consumption at the time.[10] Although historically heavily reliant on hydropower, between 1955 and 1973, the country had built up its oil-fired generation capacity. In the wake of the crisis, substituting away from oil became a top priority.

In 1975, the Spanish Parliament approved the first National Energy Plan (NEP) designed to promote alternative sources of power.[11] In response to the NEP, the newly democratic government led by Adolfo Suarez Gonzalez announced the creation of the National Institute of Industry's (INI) Solar Program in 1977, the same year that the first democratic elections were held in the country. Thus, Spain embarked upon both a political and an energy transition at the same time. The R&D program was managed by the public utility Gas Y Electricidad S.A. (GESA) in Mallorca. As part of this program, by 1979, GESA had initiated a wind research initiative focused on small wind turbines.

To conduct research and learn about the technology, GESA purchased and installed a 22-kW Danish Windmatic wind turbine in Mallorca in 1982. This led to the subsequent development of a 22-kW wind turbine with a 12-m diameter, which was installed at Palma de Mallorca's Sant Joan de Deu power plant in 1983.[12] In 1984, five versions of the Spanish wind turbine were installed at Spain's first wind farm, Parc Eòlic Pilot de l'Empordà in Garriguella.

The private company Ecotècnia was closely watching GESA's efforts. Ecotècnia was one of the most significant wind power players to emerge from Spain's wind power miracle. Barcelona-based Ecotècnia was founded in 1981 by a group of university professors and graduate studies who were interested in wind power. Ecotècnia's founders had been in contact with government researchers and discussed the possibility of acquiring government funding to build a wind turbine. They ultimately secured funding and built a 15-kW, three-bladed, fixed pitch, stall-controlled wind turbine with a 12-m rotor diameter. The wind turbine was fixed atop a 10-m lattice tower and began generating electricity in 1984.

The Ecotècnia wind turbine worked well and the group later built an upsized 30-kW version and sold twenty-nine of them between 1985 and 1988.[13] The turbines were based on a robust design customized for the winds of mountainous Spain. Ecotècnia expanded aggressively within Spain after Director Antoni Martinez observed wind market activities within Denmark and California, and decided that Spain could follow their lead.

The company also developed increasingly large turbines throughout the 1980s and 1990s. Their 150-kW model was released in 1989 and this led to the development of the 7.5-MW wind farm in Tarifa. In 1997, the company supplied wind turbines for a 15 MW project in Malprica de Bergantiños on the country's northwestern seaboard. The project was made up of 67 Ecotècnia 225-kW wind turbines. The Ministry of Energy pitched in to help finance the project by providing a grant of ESP 600 million. Ecotècnia continued to improve its wind turbines, releasing a 500-kW model in 1995 and a 750-kW model in 1998.[14] By the end of 2000, Ecotècnia had installed more than 280 MW of wind power capacity, with individual wind turbines ranging from 150 to 750 kW in size.

Ecotècnia then expanded outside of Spain by opening an affiliate in France and installing wind turbines in India and Cuba. The long-term objective of Ecotècnia France was to enable Ecotècnia to provide France with enough wind turbines to match the company's market share in Spain. Ecotècnia stated its plans to manufacture wind turbine in France as well. Ecotècnia's move into France caught the eye of France's multinational engineering company Alstom. Alstom eventually purchased Ecotècnia in 2007. Originally named—Alsthom—the company was founded in 1928 from the merger of the French interests of the Thomson-Houston Electric Company, which was then part of GE, and Société Alsacienne de Constructions Mécaniques. Over the course of the twentieth century, Alstom expanded into rail and marine vehicle manufacturing. Alstom sold much of its power business to international competitors GE, Siemens, and MAN Group, and by 2003 was facing a financial crisis. The French government stepped in took a 21 percent stake in the business and provided the company with €5.7 billion as part of a rescue plan.

By 2007, Alstom was looking for acquisitions to revitalize its power business. In March, Alstom acquired a manufacturer of gas turbine components from Calpine Corporation and—in June—the company acquired Ecotècnia in order to establish a footprint in the emerging wind power sector. Ecotècnia was renamed Alstom Ecotècnia, and later renamed Alstom Wind in 2010. Alstom itself was later acquired by GE in 2016 and Alstom's wind power business was integrated into GE Renewable Energy. Thus, the Spanish government's efforts to jumpstart Spain's wind power capability ended up bearing significant fruit.

In addition to Ecotècnia, the Spanish government's wind R&D effort led to the creation of another wind turbine manufacturer named Made. GESA collaborated with Made between 1986 and 1989. As a result of this collaboration, Made commercialized its first wind turbine in 1989 and began operating as a branch of Endesa, Spain's largest electric utility. The company operated until 2003, when it was sold as part of Endesa's privatization effort.[15]

Corporación Energía Hidroeléctrica de Navarra (EHN) is another Spanish wind turbine manufacturer that grew up with government support starting in the 1980s. EHN was founded jointly by the government of the Navarra province and the Spanish public utility Iberdrola. EHN was focused on developing wind projects in Navarra, and—by 1994—it opened its first wind farm with six 500-kW wind turbines.[16] EHN diversified outside of Navarra, and—by 2003—it had installed 561 MW of wind power in Europe and North America. Spanish infrastructure giant Acciona took an interest in EHN and purchased 50 percent of the company in 2003. Acciona Windpower then developed its own prototype wind turbine in 2000 and integrated EHN's technology. Two years later, Acciona purchased the remaining 50 percent making EHN wholly owned by Acciona.[17] That same year, Acciona Windpower released its AW1500 wind turbine, a 1.5-MW model with a 70-m rotor diameter.[18]

15.2   Wind Policy Push

Ecotècnia and Made's success was enabled by an evolving series of proactive Spanish energy policies. The Law of Conservation of Energy 88/1980 was passed in 1980 in the wake of the second international oil crisis. The law outlined established objectives for improving energy efficiency and promoting renewable energy. The law provided wind generators with the right to interconnect to Spain's transmission and distribution network and obligated utilities to purchase the power. The law also encouraged electric utilities to diversify away from oil. This led to a dramatic shift away from oil-fired generation toward coal and nuclear power. Spain built 4.5 GW of nuclear power between 1980 and 1986 and over 5 GW of coal-fired capacity. As a result, the share of electric-sector oil consumption dropped from 68 percent in 1973 to 7 percent by 1984.[19]

In 1986, Spain's first Renewable Energy Plan was released. The plan sought to derive 3 percent of energy consumption from renewable energy by 1992. It was followed by an additional plan in 1989. In 1991, the Spanish government released a new National Energy Plan, which set a goal of raising renewable energy's contribution to the Spanish energy mix to 10 percent by 2000. Royal Decree 2366/1994 of 1994 introduced special pricing for wind power by tariff bands as well as a method for estimating the payment for the produced electricity. The Electric Power Act of 54/1997 passed in 1997 was focused on electric sector liberalization, but it also provided a "Special Scheme" for wind power generators with premium pricing. The Act also increased the national renewable energy target to 12 percent by 2010. In 1998, another Royal Decree 2818/1998 established updated tariffs and provided additional details on the

Special Scheme.[20] And in 1999, the Spanish government set a goal of 9 GW of installed wind power capacity by 2011.

In addition to national policies, several of Spain's autonomous regional governments implemented local content requirements to attract wind turbine manufacturers to their regions to boost industrial development and economic growth. The provinces of Navarra, Galicia, Castile, and Leon and Valencia insisted on local assembly and manufacture of wind turbines and components before granting development concessions.[21] This represented a particularly aggressive use of local support policies that would have a significant impact on the evolution of wind power in the country. In this manner, government policies in Spain created an environment that encouraged Spanish companies to enter the wind power market.

15.3 Denmark's Design

Spanish technology manufacturer Grupo Auxiliar Metalurgico SA (Gamesa) was intrigued by wind power. Founded in 1976 by Tornusa—a private investment company—the company initially focused on the construction and sale of industrial machinery and equipment in the automotive sector, and worked on the development of new technologies for emerging businesses such as robotics, microelectronics, the environment, and composite materials. Gamesa's other business lines include aircraft systems manufacturing and the design and manufacturing of parts using composite materials. Gamesa's niches were the regional jet segment, projects for major aircraft makers, and the helicopter segment.[22]

Always on the lookout for new opportunities, the company had its eyes on the emerging wind power market in Europe and Spain's newly initiated supported policies. But without any previous experience designing or manufacturing wind turbines, it would be difficult to gain a foothold in the wind power space. Anticipating a wind energy boom in Spain, the Gamesa board reached out to Denmark's Vestas to discuss the possibility of a joint venture. Vestas had recovered from the California wind market collapse and by 1994 its technology was at the forefront of the global wind power market. The new Vestas V44 wind turbine had a rated capacity of 600 kW and the company recently released digital control systems that minimized stress on turbine blades and provided grid-friendly power to the transmission system.

Vestas was also looking to become "the largest modern energy company" and was working to supply wind turbines to countries around the world from Great Britain to India to New Zealand.[23] However, with the local content requirements of Spanish provinces, Vestas was unable to enter the Spanish wind market as a foreign manufacturer. Gamesa called at just the right time for Vestas.

A strategic partner within Spain would provide Vestas with a valuable foothold in the nascent Spanish wind power market.

In 1994, Gamesa and Vestas formed Gamesa Eólica based in Pamplona. Vestas's stake was 40 percent and they would provide Gamesa with exclusive rights to manufacture, assemble, and sell Vestas technology in Spain.[24] In exchange, Gamesa would limit its wind power development efforts to the Iberian Peninsula, North Africa, and Latin America. With this agreement in place—in 1995—Gamesa Eólica installed its first wind turbines in the hills of El Perdón, in Spain. The El Perdón wind farm would be powered by fifteen of Gamesa's new 500-kW G39 and twenty-five 600-kW G42 wind turbines. In that year, 45-MW of new wind capacity was installed in the country increasing total installed capacity to 119 MW.[25] Two years later, over 200 MW of new wind power capacity was installed in Spain and total installed capacity approached 500 MW. Gamesa Eólica accounted for 70 percent of the wind turbine market by 1997, with Ecotècnia and Made accounting for most of the remaining balance. Globally, Gamesa and Vestas combined accounted for one-third of the wind power market.[26]

Gamesa Eólica was doing its part to help Spain meets it renewable energy goals. The Gamesa subsidiary was also providing over half of the annual revenue for its parent company Gamesa. Wind power had suddenly become big business, not just in Spain, but across the globe. Gamesa Eólica was well-positioned to explore additional opportunities, and it was starting to feel confined by the Spanish market. It was ready to conquer the world.

In 1998, Gamesa dipped its toe in international waters by managing one of its wind turbines in the Guerrero Negro wind farm, in Baja California, Mexico. In 2000, Gamesa expanded its international footprint, engaging in projects in countries such as Portugal, Italy, France, Greece, Germany, Ireland, the United Kingdom, the United States, China, and Mexico.[27] The initial prohibition from participating in wind projects outside of Spain in deference to Vestas's international ambitions was becoming too much to bear.

Gamesa was also evolving its wind turbine technology beyond the Vestas technology blueprint. Gamesa Eólica's own R&D department was continually developing the technology it had inherited from Vestas. In 1997—for instance—Gamesa introduced the Ingecon variable speed system to its turbines. The system was applied to Gamesa Eólica's biggest seller—the 660-kW G47 turbine—which was based on the Vestas V47. Gamesa Eólica launched its own 1.65-MW G66 unit in 2000. In 2002, Gamesa released its two-megawatt G80 wind turbine, which was based off the Vestas V80 machine.

By 2000, Gamesa had wind farm operations in Mexico, Argentina, Brazil, Portugal, Morocco, Egypt, Italy, and Greece. In 2002, Gamesa announced that it had gained grid connection rights for over 500 MW in Greece and Italy, areas outside the agreement. In the same year, it confirmed further connections in Portugal as well as plans to develop in France, the Dominican Republic,

Mexico, and Brazil. In all, Gamesa planned to install 877 MW in new capacity around the world by 2004.[28]

It had become clear that Gamesa and Vestas were more competitors than partners—and that Gamesa wanted the freedom to sell wind turbines and develop wind farms across the globe. With few good options, Vestas agreed to sell its share of Gamesa for a "fair and reasonable price," in the words of Vestas' managing director, Johannes Poulsen.[29] Vestas got the short end of the stick. Gamesa gained the right to sell anywhere in the world and retained the ability to use Vestas's wind turbines through 2003, including the flagship 2-MW V80 wind turbine, while it phased in its own proprietary technologies. In exchange for access to the Spanish market, it had unwittingly provided the technological knowhow to a formidable global competitor.

Gamesa took the separation quite well. Within a week the company announced an agreement with Germany's Repower to sell its 1.5-MW wind turbine. Gamesa would now prime the German technology pump in the service of its global ambitions. Vestas executives admitted to feeling a "certain wonderment" upon hearing about Gamesa's new partnership with Repower so quickly after their own separation.[30] The separation came back to haunt Vestas. Competition between Gamesa and Vestas was intense throughout the first decade of the twenty-first century. In 2016, Siemens Wind Power— which was created when the German industrial giant Siemens purchased Denmark's Bonus Energy in 2004—joined forces with Gamesa to create Siemens Gamesa Renewable Energy to rival Vestas as the world's largest wind turbine manufacturer.

Gamesa wasn't the only Spanish company to catch Spain's wind power wave and make a splash on the global scene. The Spanish public multinational electric utility—Iberdrola—kept watch as wind power grew in the 1990s. In 2001, they made their move into the emerging wind power market by creating the subsidiary Iberdrola Renovables. Iberdrola Renovables approach was not to manufacturer wind turbines, but to build, own, and operate wind farms across the globe. At the time, Iberdrola's entry into wind power mirrored the movement by other utilities that had expanded to become key wind power investors. In Europe, Portugal's EDP, Italy's Enel, France's EDF, GDF Suez, and Germany's E.ON all moved into wind power. In the United States, utility Florida Power & Light became a large wind farm developer and owner. Iberenova's efforts were successful and remarkably consistent. In less than a decade, Iberenova had become the world's largest wind farm operator and owner.[31]

Iberdrola's wind energy ambitions extended beyond the efforts of Iberdrola Renovables. Iberdrola's subsidiaries—ScottishPower and Avangrid Renewables— also became deeply involved in wind power. ScottishPower's generation portfolio includes nearly 500 MW of wind turbines at the Whitelee and Blacklaw wind farms in Scotland. ScottishPower also owns PPM Energy, an American renewable energy and gas storage developer. PPM Energy is part of ScottishPower-owned

Avangrid Renewables, which has more than 6 GW of owned and controlled wind and solar generation in the United States.[32]

Spain's wind power success doesn't stop there. Spanish company—the Acciona Group—also became a global wind power leader. Acciona itself was established as an international holding company for companies operating in various fields in the industry and the services sectors. The group has divisions that focus on infrastructure, energy, water, and other businesses. Its origins can be traced back to MZOV, a Spanish firm founded in 1862. Acciona's subsidiary Acciona Windpower credits its wind power success to its beginnings in the region of Navarra in 1994. "We were pioneers with a plan of implementing wind power in Navarra when wind wasn't yet looked on as an important economic sector," according to the company's director of marketing, Jose Arrieta.[33] The company made a strategic decision to begin manufacturing wind turbines and created a prototype 1.3-MW wind turbine in 1998. The first prototype was installed in 2000 in the Jaizkibel mountain range. In 2001 and 2002, the machine was improved through two more prototypes located at Aibar and Peña Blanca in Navarra, close to the company headquarters.[34]

The company operated wind turbine manufacturing plants in Barásoain and Castellón, and—in 2007—it opened a new production facility in the United States. In 2008, it released the 3-MW AW3000 machine with a 100-m rotor diameter. Still, finding success in European markets outside of Spain has proved challenging for Acciona Windpower. With stiff competition from Denmark's Vestas, and Germany's Siemens and Enercon, Acciona Windpower was challenged to make inroads into the rest of Europe.

Danish wind turbine manufacturer Nordex faced a quandary too. Founded in 1985 in Give, Denmark by Carsten Pedersen, Nordex had grown steadily over thirty years, but remained focused on Europe with 80 percent of its revenue coming from the continent by 2016. Nordex executed a successful takeover of Acciona Windpower in 2016. The combined company—now called the Nordex Group—has a strong presence across the globe and boasts over 21 GW of installed wind power capacity. The company now has manufacturing facilities in Germany, Spain, Brazil, the United States, and India. The two companies combined generated sales of €2.4 billion in 2015 and have nearly 5,000 employees. The Nordex Group is led by Jose Luis Blanco—former CEO of Acciona Windpower—who has promised to defend and expand the company's market position as "one of the major manufacturers worldwide."[35]

Even without its subsidiary—Acciona Windpower—the Acciona Group continues to play an important role in wind power. By 2017, Acciona Energy had become one of the largest wind power developers in the world with more than 7.6 GW of its own capacity installed across 222 wind farms in fourteen countries.[36] Acciona's strength is its ability to develop large wind projects in the Americas, Africa, and Asia. These non-European wind power markets represented more than 70 percent of the company's business by 2017.

Thus, from just seven megawatts of installed capacity in 1990, Spain's installed wind power capacity grew to over one gigawatt by the end of the decade.[37] Spain's wind power miracle continued in the first decade of the twenty-first century when installed wind power capacity rose to 21 GW by 2010.[38] Spain's wind power accomplishments of the 1980s and 1990s would prove to be durable. By 2017, Spain was the fifth largest market for wind power in the world.[39] Furthermore, two of the top six wind turbine manufacturers in the world in 2017 have Spanish roots. In addition, a couple of the largest wind farm owner and operators in the world are Spanish public multinational electric utility Iberdrola and international developer Acciona Energy.

Thus, Spain was able to permanently vanguard of wind power in the twenty-first century. Like the Spanish conquistadors centuries earlier, Gamesa, Acciona, and Iberdrola expanded their reach beyond the Iberian Peninsula to the Americas, Oceania, Africa, and Asia and they eventually conquered wind power markets around the world, creating newfound wealth for Spain in the process.

However, Spain's wind power rise starting in the 1990s wasn't the only thing happening in the world of wind power. Denmark and Germany continued their wind power journey as well. After falling behind the United States in the 1980s, Europe would surpass America by the end of the 1990s. Europe's share of global wind capacity jumped from 25 percent in 1990 to 75 percent by 2000.[40] The full story of Europe's wind power revival in the 1990s starts in Denmark, as Danish wind turbine manufacturers struggled to rebuild from the wreckage of the 1986 California Wind Rush crash.

Notes

1 Palz, 1978.
2 Palz, 2013, 38.
3 Gipe, 1995, 96.
4 Ibid.
5 Beurskens, 2013, 22.
6 Palz, 2013, 33.
7 Harrison, Hau, and Snel, 2000.
8 Palz, 2013, 35.
9 Cubillo, 2011, 15.
10 Ibid., 18.
11 Ibid., 17–8.
12 Boix, 2013, 474.
13 Ibid.
14 Ibid., 478.

15 Ibid., 485.
16 Ibid., 486.
17 WindPower Monthly, 2005.
18 Acciona Windpower, 2017.
19 Cubillo, 2011, 20.
20 IRENA, 2013, 119.
21 Lewis, 2007, 9.
22 Siemens Gamesa Renewable Energy, 2017.
23 Vestas, 2017.
24 WindPower Monthly, 2000.
25 IEA, 2001, 151.
26 WindPower Monthly, 2002.
27 Siemens Gamesa Renewable Energy, 2017.
28 WindPower Monthly, 2002.
29 Ibid.
30 Ibid.
31 MarketWatch London Bureau, 2007.
32 Avangrid Renewables, 2017.
33 Graber, 2005, S4.
34 Acciona Windpower, 2017.
35 Nordex, 2017.
36 Acciona Windpower, 2017.
37 IEA, 2001, 152.
38 IRENA, 2013, 118.
39 GWEC, 2017, 16.
40 Hatziargyriou and Zervos, 2001, 1766.

Bibliography

Acciona Windpower. 2017. *History*. Accessed December 30, 2017. http://www.acciona-windpower.com/about-us/history.

Avangrid Renewables. 2017. *Wind Business Overview*. Accessed December 30, 2017. http://www.avangridrenewables.us/power-wind.html.

Beurskens, Jos. "Forty Years of Wind Energy Development." In *Wind Power for the World, The Rise of Modern Wind Energy*, edited by Preben Maegaard, Anna Krenz, and Wolfgang Palz, 13–32. Singapore: Pan Stanford, 2013.

Boix, Josep Puig i. "Wind Energy in Spain." In *Wind Power for the World, International Reviews and Developments*, edited by Preben Maegaard, Anna Krenz, and Wolfgang Palz, 473–92. Singapore: Pan Stanford, 2013.

Cubillo, Diego Ibeas. *Review of the History of the Electricity Supply in Spain from the Beginning up to Now*. Madrid, Spain: Universidad Carlos III de Madrid, 2011.

Gipe, Paul. *Wind Energy Comes of Age*. New York: John Wiley, 1995.

Graber, Cynthia. "Wind Power in Spain." *MIT Technology Review* (2005): S1–8.

GWEC. *Global Wind Report 2016*. Brussels, Belgium: Global Wind Energy Council, 2017.

Harrison, Robert, Erich Hau, and Herman Snel. *Large Wind Turbines: Design and Economics*. New York: John Wiley, 2000.

Hatziargyriou, Nikos, and Arthuros Zervos. "Wind Power Development in Europe." *Proceedings of the IEEE* 89, no. 12 (2001): 1765–82.

IEA. *Wind Energy Annual Report 2000*. Golden, CO: International Energy Agency (IEA), 2001.

IRENA. *30 Years of Policies for Wind Energy: Lessons from 12 Wind Energy Markets*. Abu Dhabi, United Arab Emirates: International Renewable Energy Agency, 2013.

Lewis, Joanna I. *A Comparison of Wind Power Industry Development Strategies in Spain, India, and China*. San Francisco, CA: Center for Resource Solutions, Supported by the Energy Foundation, China Sustainable Energy Program, 2007.

MarketWatch London Bureau. 2007. "Iberdrola to Float World's Largest Wind-power Company: Report." *MarketWatch*, 9 9. Accessed December 29, 2017. https://www.marketwatch.com/story/iberdrola-to-float-worlds-largest-wind-power-company-report.

Nordex. *Nordex Annual Report 2016*. Annual Report. Hamburg, Germany: Nordex, 2017.

Palz, Wolfgang. "Wind Power Development in the European Union." In *Wind Power for the World, Interntional Reviews and Developments*, edited by Preben Maegaard, Anna Krenz, and Wolfgang Palz, 23–38. Singapore: Pan Stanford, 2013.

———. *Solar Electricity: An Economic Approach to Solar Energy*. Paris, France: UNESCO, 1978.

Siemens Gamesa Renewable Energy. 2017. *History*. Accessed December 30, 2017. http://www.siemensgamesa.com/en/about-us/history.

Vestas. 2017. *History*. Accessed November 17, 2017. https://www.vestas.com/en/about/profile#!history.

WindPower Monthly. 2000. "From Out of Nowhere to the Top of Spain." *WindPower Monthly*, December 1. Accessed December 29, 2017. https://www.windpowermonthly.com/article/952237/nowhere-top-spain.

———. 2002. "Two Wind Giants Go Head to Head – Vestas and Gamesa Split." *WindPower Monthly*, January 1. Accessed December 30, 2017. https://www.windpowermonthly.com/article/950913/two-wind-giants-go-head-head---vestas-gamesa-split.

———. 2005. "Acciona Becomes 100% Owner of Spanish Developer EHN." *WindPower Monthly*, March 1. Accessed December 30, 2017. https://www.windpowermonthly.com/article/961659/acciona-becomes-100-owner-spanish-developer-ehn.

16

Europe Sails Ahead

I worked out the energy that was available in the wind, but you couldn't make money with it.

—Aloys Wobben, Enercon founder (Lawton 1998)

16.1   Wind Power Reincarnated

The abrupt end of the California Wind Rush sent ripples across the Atlantic. Without demand for new turbines from California, Danish wind turbine manufacturers faced an existential crisis. Vestas itself had to be reborn. The company was scheduled to deliver 1,200 wind turbines to US wind developer Zond in December 1985. But the shipment was delayed because the shipping company went bankrupt. When they finally did arrive, Zond wasn't in a position to purchase them anymore. Vestas found itself with a large stockpile of wind turbines that it couldn't sell. Vestas suspended payments to its creditors at the end of 1986. In 1987, Vestas was resurrected as a smaller company—Vestas Wind Systems A/S—that was focused exclusively on wind power.

At the same time, the domestic market for wind power had stalled. The Danish government—which had initially provided capital grants of up to 30 percent for wind turbine exporters—progressively reduced the incentive to 20 percent, and then 10 percent. The subsidy was repealed in 1988. The only ray of hope during this period was the 1985 agreement between the Ministry of Energy and the nation's electric utilities to install 100 MW of wind power between 1986 and 1990. As part of this agreement, utilities would be required to interconnect wind farms and purchase the electrical output. It was a start.

As a result of the sluggish domestic market, Danish wind turbine manufacturers—led by Vestas—were forced to seek international destinations for their products during this time. Vestas focused on India as a preferred destination for its wind turbines. In 1987, the company was selected as the wind turbine provider

The Wind Power Story: A Century of Innovation that Reshaped the Global Energy Landscape,
First Edition. Brandon N. Owens.
© 2019 by The Institute of Electrical and Electronics Engineers, Inc.
Published 2019 by John Wiley & Sons, Inc.

for six wind energy projects in India sponsored by the Danish International Development Agency (DANIDA), a Danish state-financed aid agency. Overall, however, these were very lean years. The excitement around the reinvention of wind power in the 1970s and early 1980s had given way to the raw reality of inconsistent government policies and low oil prices.

However, the 1986 Chernobyl nuclear accident underscored the world's global energy challenges and highlighted the need for less risky energy options. It also increased public questioning about the role of nuclear energy led to alternative energy solutions research and development.[1] It is in this context that the Secretary-General of the United Nations—Javier Pérez de Cuéllar—asked the Prime Minister of Norway—Gro Harlem Brundtland—to create an organization independent of the UN to focus on environmental challenges and solutions. This new organization was the Brundtland Commission—or more formally—the World Commission on Environment and Development. The Brundtland Commission was first headed by Gro Harlem Brundtland as Chairman and Mansour Khalid as Vice-Chairman. The organization aimed to create a united international community with shared sustainability goals by identifying sustainability problems worldwide, raising awareness about them, and suggesting the implementation of specific solutions.

In 1987, the Brundtland Commission published the first volume of *Our Common Future*, known thereafter as the Brundtland Report. The report provided a clear definition of "sustainable development" as "development that meets the needs of the present without compromising the ability of future generations to meet their own needs."[2] It also called for carbon dioxide emission reduction targets. The report had a strong influence on governments across the globe, particularly in Europe. In December 1990, Germany passed the Electricity Feed Act, which required utilities to purchase electricity from small renewable energy generators and obligated them to pay favorable prices for the power. The contract price was set by German federal law. This was the first implementation of a national feed-in tariff (FIT).

In Spain, the government released a new National Energy Plan in 1991 that set a goal of raising renewable energy's contribution to the Spanish energy mix to 10 percent by 2000. Royal Decree 2366/1994 of 1994 introduced special pricing for wind power by tariff bands as well as a method for estimating the payment for the produced electricity. The Netherlands also moved to implement the recommendations of the Brundtland report. In the Third Energy Memorandum of 1995, the Dutch government set a goal of stabilizing carbon dioxide emissions at their 1990 level by 2020. To help reach the target, the government estimated that 750 MW of wind power would be required by 2020.[3] In the United Kingdom, the government's wind energy research program—which was initiated in 1979—transitioned to a commercial deployment program known as the Non-Fossil Fuel Obligation (NFFO). The first NFFO order—NFFO-1 starting in 1990—called for 28 MW of new wind power.

By 1995, over 250 MW of new wind power capacity was installed in the United Kingdom because of the NFFO.[4]

The bottom line is that European wind power policies enacted at the end of the 1980s and early 1990s were effective in reigniting wind power manufacturing and project development in Europe. In fact, Europe's wind power push in the 1990s moved the region to the vanguard of global wind power development. The initial policies of this era were followed up with a second round that deepened support for wind. In Denmark, the government issued its fourth energy plan in 1996 that proposed a renewable energy consumption goal of 14 percent by 2005. An even higher target of 20 percent by 2003 was introduced in 1999. Copenhagen then elected to end its fixed feed-in tariffs for wind and simply required utilities to meet the wind energy targets proposed by the national energy plan. By 2003, all wind turbines were connected to the grid under the new renewable portfolio standard. The financial remuneration provided to wind farm owners was made up of the market price plus a premium.[5]

In Spain, the Electric Power Act of 54/1997 passed in 1997 was focused on electric sector liberalization, but it also provided a "Special Scheme" for wind power generators with premium pricing. The Act also increased the national renewable energy target to 12 percent by 2010. In 1998, another Royal Decree 2818/1998 established updated tariffs and provided additional details on the Special Scheme.[6] In Germany, the 1990 electricity feed-in law was incorporated into the Act on the Reform of the Energy Sector. In addition, the seminal Renewable Energy Sources Act (Das Erneuerbare-Energien-Gesetz or EEG) came into force in 2000. The Act—which provided the main stimulus for the national wind market in Germany—established a feed-in tariff for each kilowatt-hour produced, and it awarded priority connection to the grid for power generation based on renewable energy. The Act implemented a two-component tariff—which was designed for wind energy—with an initial fixed tariff for a period of five years, and a second period of fifteen years with a tariff level modulated by the local wind conditions. An important element of the Act was the obligation for power utility companies to purchase renewable energy at set tariffs over a period of twenty years.[7]

While these country-level policies were important in establishing a foundation of wind power policy support in Europe, the seminal moment occurred in 2001 when the European Union (EU) passed its Directive on Electricity Production from Renewable Energy Sources 2001/77/EC. The RES Directive took effect in October 2001 and set indicative renewable energy targets for its fifteen member states. The directive was nonbinding, but it provided a roadmap that individual countries could follow to enable the EU to achieve its 21 percent renewable electricity target by 2010.

The European policy push was a godsend for wind turbine manufacturers in Denmark, Spain, and Germany, who had been struggling through the 1980s. In the decade of the 1990s, these wind turbine manufacturers were able to stabilize their footing as the demand for wind turbines picked up in Europe and began to expand selectively in other parts of the world like India, where installed wind power capacity reached 1,100 MW by the end of the 1990s.[8]

Europe's wind power policies provided an environment that enabled the company to advance wind turbine technology. After the frenzy of the California Wind Rush from 1981 to 1985—when wind turbines came in all shapes in sizes—in the 1990s, commercial wind turbines had begun to settle on certain design standards, but the technology didn't completely converge. Three-bladed, upwind, stall-controlled wind turbines with gearboxes and induction generators modeled after the "Danish Concept" were the most common configuration. However, although stall regulation was the preferred approach, pitch-controlled machines continued to make inroads. As digital control systems became more sophisticated, pitch-controlled wind turbines began yielding better energy output and they experienced lower rotors loads, which ensured a longer life for the whole wind turbine. Danish manufacturers Bonus and NEG Micon continued to stay faithful to stall regulation, while Vestas gradually adopted pitch regulation.

The upwind configuration was now favored because it resulted in lower blade fatigue because the blades do not have to pass through the tower's wind wake in the upwind design. Remember all of the effort that Ulrich Hütter, NASA, and the German government had to put into creating wind turbines with teetering hubs to absorb the vibrations caused by the downwind design? This is avoided with an upwind design, which is why the wind industry settled upon this approach in the 1990s.

By the 1990s, three-blade machines were preferred because they provided the best balance among blade stiffness for tower clearance, aerodynamic efficiency, and tower shadow noise. Rotor noise and esthetic considerations also strongly supported the choice of three blades rather than two or one.[9] However, several manufacturers continued to offer two-blade designs in order to keep the rotor weight down, including Dutch companies Lagerwey and NedWind, and the Swedish manufacturer Zephyr Energy, which was started in 1988 in Falkenberg by Leif Svensson.

Overall, wind power innovation progressed steadily throughout the 1990s and by the end of the decade wind turbine manufacturers could adopt a mix and match approach to designing their wind turbines. In fact, during this period, wind manufacturers largely became wind turbine assemblers rather than manufacturers, selecting off-the-shelf combination of components tailored to meet their needs.[10]

However, the most visible sign of wind power innovation in the 1990s was the increase in wind turbine size. Building large wind turbines always made sense in theory, which is why government-led American and German wind power research focused on creating megawatt-scale wind turbines in the 1970s and 1980s. And this is why the European Economic Community focused on large wind turbines as part of the WEGA program that started in the 1980s. However, as a practical matter—up until this time—wind turbine manufacturers couldn't construct wind turbines that could withstand the high structural loads that larger wind turbines experienced relative to their small-scale brethren.

But as the 1990s progressed, Vestas and others were able to break through the megawatt-scale size barrier. This breakthrough occurred because of advances in the strength, flexibility, and weight of wind turbine components. Blades made from steel and aluminum were gradually discarded in favor of blades fabricated from composite materials, first fiberglass and later carbon fiber. For example, in 1990, Vestas introduced a new blade design that weighed 1,100 kilograms (kg), down from 3,800 kg in the previous design. Enercon started handmaking its blades of fiberglass composite materials. In addition, advances in the understanding and quantification of structural loads led to international design standards for wind turbines.[11]

In 1989, the electric capacity of Vestas's flagship wind turbine—the V27—was 225 kW. This was up substantially from the original 30-kW V10 wind turbine that Vestas released a decade earlier. Still, the 1989 model had yet to break the megawatt barrier. But by 1997, Vestas had introduced the V66 turbine. At the time, it was the world's largest commercial wind turbine. Its capacity of 1.65 MW was fifty-five times greater than the V10. Vestas prospered during this period. By 1998—with its turbines representing 22 percent of the world market—the company carried out a stock listing on the Copenhagen Stock Exchange. The share offer was eight times over-subscribed.[12]

Figure 16.1 After twenty years of gradual scale-up—in 1999—Vestas broke the 2-MW barrier with its V80 wind turbine. The V80 would become a mainstay of Vestas's product offerings for over a decade.

16.2   The Wind Power Wizard

In 1987, the government-led 3-MW Growian wind turbine project in Germany successfully employed a variable-speed wind turbine design. During the mid-1980s, a consortium of two Dutch aerospace companies designed a series of variable-speed wind turbines with financial support from the Dutch government. Three two-bladed, upwind variable-speed wind turbines were built: two in the Netherlands and one on Curaçao Island. The Dutch machines used power conditioning to transform the variable frequency AC output to fixed frequency for the European grid.[13]

The German wind turbine manufacturer Enercon too was at the leading edge of variable-speed design when it introduced its 55-kW E15 variable-speed wind turbine in 1985. They followed up with the 80-kW E17 in 1987 and the 300-kW E33 in 1989. Variable-speed wind turbines were also being developed simultaneously in the United States by a consortium composed of the Electric Power Research Institute (EPRI), California-based utility Pacific Gas & Electric (PG&E), and U.S. Windpower, which had by then changed its name to Kenetech in a bit of strategic rebranding after the California wind bust. This effort would eventually lead to the development of Kenetech's KVS-33 wind turbine.

This was a positive development for wind power because as it turned out, the conditioned power output from variable-speed machines was of higher quality than the unconditioned output from constant-speed machines. In addition, improvements in power electronics also enabled wind turbine manufactures to address the power quality issues that had vexed the wind industry from the beginning.* These problems were significant enough to stop many electric utilities from interconnecting wind turbines into their networks. The challenges were acute for small power networks or "weak grids," where a single wind turbine-induced power disturbance could bring down the entire network.

More sophisticated control systems were developed that enabled wind turbines to extract increasing amount of energy from the wind and provide electrical output that was acceptable for the grid. These advances not only facilitated wind turbine scale-up, but also improved performance and drove down costs. For example, Enercon developed a highly flexible grid-management

---

* One of the fundamental challenges for wind power had always been providing electric output that was suitable for the utility transmission systems. Output quality challenges had vexed wind turbine designers from the very beginning. Denmark's Poul la Cour himself considered the problem and build a system of pulley belts that he called "Kratostate" in the 1890s to keep the rotational speed of his wind turbine constant so that it produced electrical output with a fixed frequency and voltage (Nissen 2009).

system which adapted the output to the needs of the grid. By electronically monitoring and adjusting voltage, frequency, output, and the power factor, Enercon's pulse-width modulated inverters were able to support weak grids.[14] Vestas introduced a propriety configuration for its constant-speed wind turbines in 1994 to ensure that even electrical output was provided to the grid. It was the first time that a fully integrated solution to this challenge was offered.[15]

The net result of all of these innovations was that by the end of the 1990s, wind power technology had grown from a small-scale technology with questionable reliability to a utility-scale solution with increasingly sophisticated power electronics to ensure smooth output and grid compatibility. Although most wind turbines were still prone to failure at least once a year, overall reliability levels approached 98 percent by the turn of the twenty-first century.[16] Capacity factors, which represent the average annual utilization of wind turbines, where between 25 and 30 percent by 2000.[17] A twofold improvement from the utilization levels of the wind turbines installed during the California Wind Rush. Although all of the bugs weren't worked out, after a century of progress, wind turbine technology was at long last positioned to plug into bulk power networks around the world. The heavy lifting to get wind turbine technology size and reliability to an acceptable level was done in Europe in the 1990s.

Much of that heavy lifting was accomplished thanks to German wind innovator Aloys Wobben. When Wobben started his company, Enercon, in the backyard of his house in 1984, his goal was to build the most innovative wind turbines in the world. He achieved that goal and he never stopped innovating.[18] So when the time came to develop Enercon's first 500-kW wind turbine, it needed to be something special. The company had been manufacturing its flagship 330-kW E33 wind turbine since 1988 and had done very well.

Wind turbine unreliability and component failure was still an issue. Those components that were most likely to fail were blades, generators, and gearboxes. Wobben and his team wanted to develop a wind turbine that tackled these problems head on. To improve the reliability of their blades, Enercon redesigned their blade tips to reduce turbulence, noise, and increase power output. They also devleoped redesigned blades using a large flange root and a double row bolt connection that provided additional strength by creating even load distribution.

However, improving the reliability of the generator and the gearbox would require an even more creative solution. The sheer number of moving parts in both the generator and the gearbox created mechanical stress on these components. Removing moving parts or slowing them down would reduce failure rates. As it turns out, this very problem had been examined and an innovative solution had already been proposed in Germany decades ago. In the 1930s,

German engineer Hermann Honnef believed that wind power could become a viable replacement for coal-fired electricity and he developed an ambitious plan that outlined the possibilities. He published his plans in multiple books and articles and provided a roadmap for the development of large wind turbines in Germany.[19] He proposed the first gearless—or "direct drive"—wind turbine. He also proposed a power conversion approach to produce AC output at a fixed frequency, and he suggested erecting turbines at heights that would allow them to reap the benefits of higher wind speeds at greater altitudes.

Honnef's proposed wind turbine design used two hubs that rotated in opposite directions. The hubs served as the primary components of the turbine's ring generator. One hub was to be fitted with a copper coil and the other would be fitted with magnets. When the wind blew, the hubs would contrarotate, and the copper coil would move across the magnetic field to generate electricity. By integrating the electric generator directly into the rotating hubs in this manner, Honnef was able to eliminate the gearbox entirely.

Honnef's direct drive wind turbine concept remained dormant for fifty years before Gerd Otto—a researcher based in the former German Democratic Republic—created new interest in the design and applied for a patent on the technology in 1986.[20] Further, in 1990, Professor Friedrich Klinger and his wind energy research team at the Saarland University of Applied Sciences modernized Honnef's concept and visited wind turbine manufacturers in 1990 to discuss the advantages of the direct drive system. These visits ultimately led to conceptual designs and prototypes by Goldwind, Lagerwey, Zephyros, MTorres, Leitwind, Hyundai, IMPSA, and Siemens. Wobben and the Enercon team entertained Klinger in the spring of 1990.[21]

Inspired by Honnef's proposals—and emboldened by Klinger's updated designs—Wobben and his team set about to create the world's first direct drive wind turbine in 1990. By 1992, they had successfully developed a wind turbine where the rotor and the generator were directly connected to form one consolidated unit. Unlike geared systems with many moving parts, Enercon's drive system required only two slow-moving rolling-element bearings. And unlike conventional fast-running generators, Enercon's low-speed multipole generator was subjected to little mechanical wear. The reduced number of rotating parts in Enercon's direct drive system increased the wind turbine's lifespan and reduced the machine's operating expenses.[22]

The design employed a variable-speed approach that used a multipole synchronous generator to produce electric output which was converted first to DC and then to fixed-frequency AC. This required power conversion equipment, which added to the weight and cost of the system. So, Enercon's implementation of direct drive increased the reliability and reduced the maintenance costs of the wind turbine by eliminating the gearbox, but it also weighed more and required higher capital costs.

Figure 16.2 In 1993, Enercon released its E40 gearless wind turbine. It was the world's first gearless, variable-speed wind turbine. It was the culmination of a vision first articulated by German's Hermann Honnef in the 1930s.

Enercon released its gearless 500-kW E40 wind turbine in 1993. It was commercially successful and over 1,400 were sold by 1998.[23] In practice, Enercon's E40 demonstrated higher reliability and 30–40 percent lower operating costs than conventional wind turbines.[24] Enercon's direct drive design had delivered as promised and the design approach began to spread to other manufacturers.

Klinger went on to found the Forschungsgruppe Windenergie (FGW) at the University of Applied Sciences in Saarbrücken. FGW continued research on direct drive concepts. In 1997, FGW developed the direct drive prototype GenesYs 600. Vensys was founded in 2000 to commercialize GenesYs 600. By 2003, Vensys had installed a 1.2-MW prototype and licensed the technology to China's Xinjiang Goldwind Science and Technology. China's Goldwind later made Vensys's direct drive model the centerpiece of its wind turbine technology platform and would go on to become the largest wind turbine manufacturer in the world.[25]

To summarize, the E40 combined two important innovations: direct drive and variable-speed. Variable-speed wind turbines were fully embraced and would quickly supplant constant-speed wind turbines. Between 1995 and 2004, the market share of variable-speed wind turbines grew from 14 to 73 percent.[26]

However, direct drive wind turbines didn't become the single technology of choice as expected. There were some drawbacks. For one thing, they were heavier and had a higher initial cost relative to geared systems. In addition, conventional geared designs themselves continued to demonstrate improvements that made them increasingly attractive over time. Direct drive designs still compete with geared systems today. There has proven to be no single winner in this technology competition.

Direct drive wind turbine began to achieve greater acceptance around 2010 when improvements—including the use of permanent magnets and new generator arrangements—led to lighter, more affordable direct drive models. In addition, interest in permanent magnet generators accelerated after magnet prices fell in 2012.[27] By 2013, direct drive turbines accounted for 26 percent of global turbine sales.[28] Current trends indicate that direct drive turbines will account for an increasing share over time as wind turbines continue to increase in size and as offshore installations grow.

16.3   The Wind Virus

Enercon wasn't the only wind turbine manufacturer in Europe working on developing a direct drive wind turbine. As noted, Klinger's 1990 roadshow encouraged several manufacturers to study the feasibility of the direct drive approach. For example, Dutch wind turbine manufacturer Lagerwey took direct drive very seriously and converted its commercial wind turbine line from geared to direct drive in 1995. Lagerwey released its LW50 model, a 750-kW three-bladed gearless wind turbine with a synchronous generator in 1995. Enercon's Wobben was so threatened by the move that he contacted all of Enercon's component suppliers and promised to terminate his contracts with them if they agreed to also provide components to Lagerwey.[29] There's no record to indicate whether or not Wobben followed through on his threat, and—if so—whether Lagerwey's business was adversely impacted. But Lagerwey was eventually forced to file for bankruptcy in 2003. However, Lagerwey founder, Henk Lagerweij—along with Aart van de Pol, André Pubanz, and Albert Waaijenberg—resurrected Lagerwey in 2006 when they created Lagerwey Wind.

Lagerweij's long wind power journey was interesting. It began in 1972, when an eighteen-year-old Lagerweij built a homemade wind turbine from two wooden boards connected to an electric generator and mounted it on a fallen tree trunk in the forest near Kootwijk in The Netherlands. Lagerweij would later recall "that was the moment I was caught by the wind virus."[30] In 1979, he developed his first 35-kW wind turbine the LW10—along with his brother-in-law Gijs van de Loenhorst—and he founded Lagerwey. His timing was right because the world was in the midst of the second global energy crisis and oil prices were elevated, which made wind power economically attractive in some cases.

Lagerweij was committed to wind turbine innovation from the start. He was one of the first to commercialize the variable-speed concept.[31] In addition, he was at the forefront of developing the benefits of flexible rotor structural design elements to decrease the loading. Unsurprisingly, many successful types of turbines were designed and manufactured. By 1985, Lagerwey had introduced an 80-kW wind turbine and sold over sixty machines. However, when oil prices fell starting in 1985, the demand for wind turbines declined in the Netherlands and Lagerwey was forced to file for bankruptcy.

The 1986 Chernobyl nuclear accident raised awareness of the need for alternative energy across Europe and Lagerweij found it easy to raise capital to restart Lagerwey. He continued to innovate, releasing a 250-kW wind turbine in 1992, setting his sights on direct drive designs. The 1995 release of Lagerwey's 750-kW LW50 model represented an evolutionary step-change after focusing on geared machines the pervious fifteen years. The reincarnated Lagerwey Wind is still focused on direct drive technology today and recently installed two 4.5-MW L136 direct drive prototypes.

Lagerweij's wind power accomplishments haven't gone unnoticed in his home country. In April 2018, he was knighted in the Order of the Netherlands Lion—a recognition of his years of pioneering work in wind energy. The Order of the Netherlands Lion is a Dutch order of chivalry founded by King William I of the Netherlands in 1815.[32] And—in an interesting twist—direct drive competitor Enercon acquired Lagerwey Wind in 2018. In hindsight, Enercon might have saved some cash if they had simply made the purchase in 1995 when it was clear that Lagerweij was an innovator in the direct drive wind power technology arena.

By the turn of the twenty-first century, wind turbines had been scaled up to fit with the bulk power system and many of the rough edges had been smoothed out. However, for wind power to thrive, one more hurdle needed to be cleared: wind turbines needed to produce power that was economically competitive relative to other electricity generation options. This had been one of the factors that had been holding wind power back for decades. In 2000, after a decade of progress in Europe, it was time to ask the question again: Was wind economically competitive yet?

A look at wind power cost and performance during this time indicates that considerable progress had been made. In 1980, wind turbine costs per kilowatt were around US$3,500 in today's dollars. Costs fell to less than US$2,000/kW by 2000 because of wind turbine up-sizing and technology advances that lowered component costs.[33] Annual production from wind turbines rose too. From a capacity factor of around 15 percent for the best wind sites in 1980, by 2000, wind turbines at the most favorable sites could achieve capacity factors up to 30 percent.[34] Taken together, this meant that the cost of electricity production from wind turbines decreased from approximately US$300/MWh to around US$80/MWh.[‡] That is an astounding 75 percent drop in the cost of electricity from wind power between 1980 and 2000.

However—as usual—progress had also been made in improving the cost and performance of conventional electricity generation technologies, namely gas- and coal-fired generators. In 2000, a state-of-the-art natural gas combined-cycle power plant cost US$1,000/kW and had an electrical conversion efficiency approaching 50 percent.[35] This meant that half of the energy contained in the natural gas was converted to electricity. With natural gas prices in the range of US$3 per million Btu in Europe at the time, this meant that the cost of electricity production from a new natural gas combined-cycle power plant was about US$40/MWh.[§] Admittedly, this was the best-case scenario for gas-fired generation and gas supplies were not available everywhere, even in Europe and North America, still though, even with all of the wind turbine advances that occurred in the 1990s, the best conventional technology was clearly less costly than new wind turbines.

The upshot is that while wind power had made substantial economic progress, it still was not able to compete without policy support by the turn of the twenty-first century. This wasn't necessary problematic because Europe had demonstrated how effective policies can be in pulling wind power into the market in the 1990s. But it did mean that wind power was unlikely to thrive in countries that didn't have strong and stable wind power support programs. Europe had demonstrated just how to do it, but would any other countries follow their lead?

Thus, due to the combination of support policies and wind turbine technology advances, the baton of global wind power leadership was passed from the United States to Europe during the 1990s. In 1990, there was approximately 300 MW of wind power capacity installed across Europe compared to roughly 1,500 MW in the United States. America accounted for over 80 percent of global capacity in 1990. By the end of the decade, global wind power capacity reached 13,500 MW. Europe alone accounted for

---

‡ The cost of production from a wind turbine with a US$3,500/kW capital cost and a 15 percent average annual capacity factor is US$3,500 × 10 percent capital carrying charge ÷ 15 percent × 8,760 hours/year = US$266/MWh. Adding 20 percent for O&M yields US$320/MWh. Likewise, the cost of production from a wind turbine with a US$1,800/kW capital cost and a 30 percent average annual capacity factor is US$1,800 × 10 percent capital carrying charge ÷ 30 percent × 8,760 hours/year = US$68/MWh. Adding 20 percent for O&M yields US$82/MWh.

§ The cost of production from a natural gas combined-cycle plant with a US$1,000/kW capital cost and a 75 percent average annual capacity factor with a 7,000 Btu/kWh heat rate at a US$3/mmBtu fuel price is US$1,000 × 10 percent capital carrying charge = US$15/MWh, adding 20 percent for O&M yields $18/MWh. Fuel costs are US$3/mmBtu × 7,000 Btu/kWh = US$21/MWh. Combining capital, O&M, fuel costs yields about US$40/MWh.

nine gigawatts or two-thirds of global capacity. Without the benefit of a decade of support polices like Europe's, America's share of global wind power capacity fell below 20 percent.[36]

A concerted effort would be needed to reignite American wind power efforts and enable the United States to get back in the wind power game. Fittingly—across the Atlantic—Kenetech was plotting its return with an innovative wind turbine of its own, and a legal strategy that—if successful—would lock Enercon out of the United States market. The next part of the wind power story takes place back in America, as Kenetech and Zond struggle to survive after the 1986 California wind market bust.

Notes

1 IRENA, 2013, 26.
2 World Commission on Environment and Development, 1987, 27.
3 IEA, 2001, 133.
4 Ibid., 171–3.
5 IRENA, 2013, 60.
6 Ibid., 119.
7 Ibid., 69.
8 IEA, 2001, 30.
9 See Tangler (2000) for a technical discussion of rotor and blade design considerations.
10 WindPower Monthly, 1997.
11 Dehlsen, 2003.
12 Backwell, 2018, 13–4.
13 Gipe, 1995, 216–7.
14 Lawton, 1998.
15 Vestas, 2017.
16 Echavarria et al., 2008.
17 Lantz, Wiser, and Hand, 2012, 10.
18 SuccessStory.Com, 2008.
19 For example, see Honnef (1939).
20 Klinger, 2013, 459.
21 Ibid., 451.
22 Enercon, 2015.
23 Schröm, 1999.
24 Lawton, 1998.
25 Klinger, 2013, 453–4.
26 Varrone, *The Impact of Variable Speed Concepts on Wind Power*, 2013b.
27 Mäkinen, 2014.
28 offshoreWIND.biz, 2014.

29 Kruse and Owens, 2018.
30 Lagerwey, 2015.
31 Gipe, 1995, 216–7.
32 Lagerwey, 2018.
33 Lantz, Wiser, and Hand, 2012, 5.
34 Ibid., 10.
35 Author's estimates.
36 European and global capacity numbers for 1990 and 1999 from IEA (2001). U.S. wind power capacity figures from 1990 taken from CEC (1991).

Bibliography

Backwell, Ben. *Wind Power: The Struggle for Control of a New Global Industry.* New York: Routledge, 2018.

Blaabjerg, F., Z. Chen, R. Teodorescu, and F. Iov. "Power Electronics in Wind Turbine Systems." CES/IEEE 5th International Power Electronics and Motion Control Conference: Conference Proceedings, Shanghai, China. IEEE Power Electronics Society, 2006.

Carlin, P.W., A.S. Laxson, and E.B. Muljadi. *The History and State of the Art of Variable-Speed Wind Turbine Technology.* Technical Report. Golden, CO: National Renewable Energy Laboratory, 2001.

CEC. *Results from the Wind Project Performance Reporting System: 1986 Annual Report.* San Francisco, CA: California Energy Commission (CEC), 1987.

———. *Results from the Wind Project Performance Reporting System: 1987 Annual Report.* San Francisco, CA: California Energy Commission, 1988.

———. *Results from the Wind Project Performance Reporting System: 1990 Annual Report.* San Francisco, CA: California Energy Commission, 1991.

Christensen, Benny. "History of Danish wind power." In *The Rise of Modern Wind Energy: Wind Power for the World*, edited by Preben Maegaard, Anna Krenz, and Wolfgang Palz, 642. Singapore: Pan Stanford Publishing, 2013.

Dehlsen, James, interview by U.S. Department of Energy. 2003. *From Zond to Enron Wind to GE Wind* WINDExchange.

Echavarria, E., B Hahn, G. J. W. van Bussel, and T. Tomiyana. "Reliability of Wind Turbine Technology Through Time." *Journal of Solar Energy Engineering* 130 (2008): 1–8.

Enercon. *Enercon Wind Energy Converters: Technology & Service.* Aurich, Germany: Enercon, 2015.

Gipe, Paul. *Wind Energy Comes of Age.* New York: John Wiley, 1995.

Honnef, Hermann. "High Altitude Wind Power Plants." *Elektrotechnik und Maschinenbau* 57 (41–42) (1939): 501–6.

IEA. *Wind Energy Annual Report 2000.* Golden, CO: International Energy Agency (IEA), 2001.

IRENA. *30 Years of Policies for Wind Energy: Lessons from 12 Wind Energy Markets.* Abu Dhabi, United Arab Emirates: International Renewable Energy Agency, 2013.

Klinger, Friedrich. "Direct drive wind turbines." In *The Rise of Modern Wind Energy: Wind Power for the World*, edited by Preben Maegaard, Anna Krenz, and Wolfgang Palz, 445–62. Singapore: Pan Stanford, 2013.

Kruse, Andy, interview by Brandon Owens. 2018. "Former VP of Sales & Marketing, Southwest Windpower," May 23.

Lagerwey. 2015. *Lagerwey History.* Accessed May 12, 2018. https://www.lagerwey.com/about-lagerwey/history.

———. 2018. "Wind Pioneer Henk Lagerweij Appointed Knight in the Order of the Netherlends Lion." April 26. Accessed May 23, 2018. https://www.lagerwey.com/blog/2018/04/26/wind-pioneer-henk-lagerweij-appointed-knight-in-the-order-of-the-netherlands-lion.

Lantz, Eric, Ryan Wiser, and Maureen Hand. *IEA Wind Task 26: The Past and Future Cost of Wind Energy.* Technical Report. Golden, CO: National Renewable Energy Laboratory, 2012.

Lawton, Michael. 1998. "Power from the Sky." *Evolution: Business and Technology from SKF.* May 15. Accessed January 9, 2018. http://evolution.skf.com/power-from-the-sky.

Mäkinen, Jukka-Pekka. 2014. PMG vs. DFIG – the Big Generator Technology Debate. Technology Paper, The Switch. https://www.offshorewind.biz/wp-content/uploads/2014/03/Technology_Point-_PMG_DFIG_06032014.pdf.

Milborrow, David. 2017. "E&T." *Recent Developments in Wind Turbine Design.* September 26. Accessed January 9, 2018. https://energyhub.theiet.org/users/66328-david-milborrow/posts/20629-recent-developments-in-wind-turbine-design.

Nissen, Povl-Otto. "The Scientist, Inventor and Teacher Poul la Cour." In *Wind Power-the Danish Way*, by The Poul la Cour Foundation, 6–11. Vejen, Denmark: The Poul la Cour Foundation, 2009. Accessed August 22, 2017. www.poullacour.dk.

OffshoreWIND. 2014. *Wind Turbine Gearbox and Direct-Drive Systems, 2014 Update – Global Market Size, Gearbox Refurbishments, Competitive Landscape and Key Country Analysis to 2020.* September 14. Accessed January 10, 2018. https://www.offshorewind.biz/2014/09/19/report-on-wind-turbine-gearbox-and-direct-drive-systems-out-now.

Reagan, Ronald. "Message to the Congress Transmitting Proposed Legislation." In *Weekly Compilation of Presidential Documents* 21, 707–8. Washington, DC: Office of the Federal Register, National Archives and Records Administration, 1985.

Schröm, Olive. 1999. "Betrayal Among Friends." *Die Zeit.*

SuccessStory.Com. 2008. *Aloys Wobben Success Story.* Accessed December 18, 2017. https://successstory.com/people/aloys-wobben.

Tangler, James L. *The Evolution of Rotor and Blade Design*. Conference Paper. Golden, CO: National Renewable Energy Laboratory, 2000.

Varrone, Chris. "The Selfish Invention." *Renewable Energy Focus*, 14, no. 3 (2013a): 34–7.

———. 2013b. "The Impact of Variable Speed Concepts on Wind Power." *Renewable Energy Focus*. Accessed January 12, 2018. http://www. renewableenergyfocus.com/view/32093/the-impact-of-variable-speed-concepts-on-wind-power.

Vestas. 2017. *History*. Accessed November 17, 2017. https://www.vestas.com/en/about/profile#!history.

Wikipedia. 2017. *Brundtland Commission*. December 14. Accessed 1 8, 2018. https://en.wikipedia.org/wiki/Brundtland_Commission.

WindPower Monthly. 1997. "Still a Paradise of Possibilities." *WindPower Monthly*, October 1.

World Commission on Environment and Development. *Our Common Future*. Oxford, UK: Oxford University Press, 1987.

17

Reigniting American Wind Power

> *As a nation, if we treated wind energy with the priority it deserves – based on, among other reasons, national security considerations, best energy value, environmental urgency, and jobs creation, almost any of which essentially justify elevating the priority for wind – I'm quite confident we could get the job done.*
>
> —James Dehlsen, founder of Zond Energy Systems (2003)

17.1   Picking up the Pieces

Wind power's rapid journey from a promising energy alternative to the star of California's 1980s energy revolution ended abruptly. The clock struck midnight with the passage of the Tax Reform Act of 1986. The Act terminated the federal investment tax credit for wind power. Further, attempts to extend that state's wind energy tax credit past their scheduled termination at the end of 1986 also failed.

Both the California wind developers and the Danish wind turbine manufacturers went into survival mode. U.S. Windpower took a two-pronged approach. They attempted to leverage their capabilities as a wind turbine manufacturer and began to market their turbines internationally. In 1988, they changed their name to Kenetech Corporation in order to facilitate the transformation from a US wind developer to an international wind turbine manufacturer. At the same time, they continued to develop domestic wind farms where they could complete economically. This meant that they needed to initiate a relentless effort to cut costs and improve wind turbine productivity, however possible. They hired electric utility executive—Dale Osborn—to help facilitate the transition. They approached the 1990s with an uncertain future.

Zond continued its focus on wind farm development even after the tax credits expired. It was a challenging time and the company was forced to work

The Wind Power Story: A Century of Innovation that Reshaped the Global Energy Landscape, First Edition. Brandon N. Owens.
© 2019 by The Institute of Electrical and Electronics Engineers, Inc.
Published 2019 by John Wiley & Sons, Inc.

out arrangements with some 300 banks and suppliers to extend credit, while trying to stabilize and survive. Zond went into full survival mode, liquidating equipment and laying off a large part of their staff. For the next several years, the only progress Zond made was by reengineering virtually every aspect of the wind projects they had built, with the goal of increasing revenue through greater productivity. They succeeded in increasing output by 22 percent from 1986 to 1989.[1] However, Zond founder—James Dehlsen—knew that the company needed another showcase project to demonstrate that wind power was alive and well.

Dehlsen choose the town of Gorman, an isolated community on the boundary between Los Angeles and Kern countries. Unfortunately for Zond, the project attracted a lot of negative attention because of its potential to kill condors, an endangered species that was going to be released from the LA Zoo. Opponents were able to successfully convince the LA County Board of Supervisors to deny an operations permit for the wind farm.

After the defeat of the Gorman project, Zond found itself in a do or die situation. Either it completed another big project, or it would cease to exist. Zond was dealing with the very real prospect of not being able to pay its debts.[2] There was one final possibility—Zond had identified a site in Tehachapi known as Sky River. Sky River was away from raptor migration zones, so there was unlikely to be any environmental opposition like the Gorman project faced. It also had very good wind resources and it was large enough to accommodate a wind farm of at least 80 MW. California standard contract for nonutility generators—the ISO4—was still available, so if a project could be built at Sky River, then the utility—Southern California Edison (SCE)—would be required to buy the output.

There were a couple of things standing in the way of Sky River. First, although it was close to a transmission line, SCE was already using the full transmission line capacity. There was no room for the wind power to flow across it. Second, Zond didn't have the money to invest the required US$160 million in the project. Under normal circumstances, these obstacles would have been enough to sink the project. But Zond was on its last breath, so the circumstances were not normal.

Zond would solve both of these problems using other people's money. First, Zond partnered with another developer—SeaWest—in a joint venture. Together Zond and SeaWest paid for a seventy-five million 220-kV transmission line to connect the Sky River site with the uncongested portion of SCE's grid. Second, Zond secured a US$157 million in investment from the American utility Florida Power & Light (FP&L), which began a long and successful run as a wind power investor with the Sky River project.[3] At the time, it was the largest wind facility financing in the world. Zond's ability to secure funding from FP&L is a credit to Dehlsen's business acumen because wind power technology was still emerging—and therefore risky—and because there were no tax incentives available.

Thus, Zond was able to successfully build the 77-MW Sky River wind farm in 1991 and the project enabled Zond to survive. It also launched FP&L into the wind business. Although PURPA prohibited American utilities from owning wind farms, there was no prohibition against providing equity or debt to wind farm developers like Zond or Kenetech. FP&L would eventually become the country's leading utility wind power investor and start a new trend in the process. The same thing happened in Europe as Spain's Iberdrola, Portugal's EDP, Italy's Enel, France's EDF, GDF Suez, and Germany's E.ON, all moved into wind power investing in the 1990s.

With Sky River off the ground, Dehlsen and his team shifted their focus to building their own wind turbine. Up until this point, Zond had purchased virtually all the wind turbines for its projects from Vestas. However, Dehlsen felt that Zond needed to be vertically integrated to survive without tax credits. If Zond was able to manufacture its own wind turbines, it could cut out Vestas's profit margin and make wind power more economically competitive. Ironically, in order to build a wind turbine that would enable Zond to survive without government support, Zond turned to the government for help. In 1993, Zond was awarded grants for turbine development from the National Renewable Energy Laboratory (NREL), in Golden, Colorado.

A year earlier, NREL had joined forces with EPRI to launch the Utility Wind Turbine Verification (TVP) Program. The goal of the program was to help electric utilities gain field experience with wind turbines by testing precommercial turbines at various sites. TVP was seeking a wind turbine manufacturer that needed to test its wind turbine prototypes. The grant provided Zond with funding for the development of its first wind turbine—the 550-kW Z40. After two years of development, the first version of the Z40 was completed in 1995. It was a 500-kW constant-speed wind turbine with a 40-m rotor diameter. A total of twenty-three Z40 wind turbines were installed in two separate locations between 1996 and 1998. These projects enabled Zond to fund the development of the turbine and test the machines in an active farm. The TVP host utilities—Central & Southwest (CSW) in Texas and Green Mountain Power (GMP) in Vermont— were generally pleased with the Z40 turbines because they didn't experience any catastrophic failures. Thus, Zond had successfully managed the transition from wind farm developer to a fully integrated wind power company with its own commercial wind turbine offering. By the time the Z40 was fully tested through the TVP program, Zond was already working on its next wind turbine, the 750-kW Z46.

At the same time, another wind turbine manufacturer—Germany's Tacke Windtechnik—deployed its wind turbines through the TVP program. Tacke started in 1985 when wind turbine designer Günter Wagner partnered with Franz Tacke to develop wind turbines. Together, they designed and built a series of wind turbines that were eventually installed in the Tehachapi area in California. By the end of the 1980s, Tacke had two working models: a 45-kW and a 150-kW version. By the mid-1990s, Tacke was working on a 600-kW version of its wind turbine, which it was able to complete through the TVP program. By 1996, Tacke had sold several hundred 600-kW wind turbines and had become the second largest German wind turbine manufacturer.

After the California Wind Rush ended, the CEC has reported that the federal and California state treasuries lost approximately US$630 million and US$770 million, respectively, that would have been paid as taxes but for the implementation of the energy tax credit programs.[4] This is the amount that was collected in avoided taxes by wind farm investors. Whether or not tax payers paid a fair price for this wind power or not is unknowable. However, it is apparent that the credit created inefficiencies and waste by incentivizing the development of wind farms littered with turbines that were unable to produce much electricity. However, this issue was spotted early on and remedies were proposed. In fact—in 1985—in the U.S. House of Representatives a bill was proposed that changed the investment credit to a "production credit," that provided a tax credit for electricity generation instead of wind farm investment.[5] The bill also proposed tying the credit value to the price of oil and providing an orderly credit phase out if oil prices increased.

This bill didn't become law, but by 1992, Iowa Senator Chuck Grassley was looking for ways to make Iowa's Renewable Portfolio Standard (RPS) less costly for the state's electric utilities. After combing through the literature on the California tax credit experience of the 1980s, they revived the idea of the production tax credit (PTC) from the 1985 House of Representatives Bill. Grassley's staffers included a provision for a ten-year 1.5-¢/kWh production tax credit for wind power and other renewable energy sources in the Energy Policy Act (EPACT) that was moving through the US Congress at the time.

Grassley's PTC fit well within EPACT's stated aim, which was to reduce US dependence on petroleum and improve air quality by addressing all aspects of energy supply and demand, including alternative fuels, renewable energy, and energy efficiency. When the bill was passed by Congress in the fall of 1992, the PTC provision remained intact. President George H.W. Bush signed the bill into law in October and granted wind power projects a valuable new tool in their arsenal. The reason that the PTC was so effective was because it provided a very large tax benefit for investors. The PTC itself provided investors with a large enough tax benefit that it enabled them to reduce the price of wind power by 2.5 ¢/kWh.* When accelerated depreciation was coupled with the new PTC, wind farm investors could achieve up to 50 percent of their return from these federal tax benefits.

---

* Appendix B of Owens (2002) explains that the ability of the PTC to reduce the price of electricity from a wind farm is equal to the face value of the PTC divided by (1 − marginal income tax rate). This is known as the "grossed up" value of the tax credit. If the face value of the credit is 1.5 ¢/kWh (US$15/MWh) and the marginal income tax rate is 39 percent, then the ability of the PTC to reduce the price of wind power is US$15/MWh ÷ (1−39 percent) = US$24.60/MWh.

Although the PTC went a long way toward covering the still present cost gap between wind power and conventional generation options, it was not enough to stimulate wind power development on its own. Successful deregulation efforts starting with the 1978 Natural Gas Policy Act in 1978, resulted in extremely low prices in the United States by the 1990s, with prices falling below US$2.50/mmBtu by 1998. This meant that electricity generation from natural gas-fired combined cycle power plants was around US$35/MWh versus about US$80/MWh from wind power without tax incentives.[†] Adding the benefits of the PTC reduced the cost of wind to US$55/MWh. This was close, but it still wasn't enough to make wind power economically competitive with conventional generation sources. Wind needed additional policies to support its development. Iowa would provide the answer.

17.2 Iowa Innovates

In 1983—at the height of the California Wind Rush—Iowa state legislators began looking for ways to support alternative energy technologies. They came up with a simple solution: just require the state's electricity utilities to install renewable energy projects or purchase renewable energy. No need for state tax incentives or fancy utility contracts that specified the purchase price. In 1983, Iowa became the first state in the United States to adopt a renewable portfolio standard (RPS) by enacting the Alternative Energy Production law, although it wasn't referred to as an RPS at the time. The Iowa law required its investor-owned utilities to own or to contract 1.5 percent of their power from renewable energy sources.

Iowa's utilities fought the RPS for a decade in court but ended up settling in 1997 and agreeing to add more renewable energy to their generation portfolios. As a result, the three largest utilities signed agreements to purchase the output of a 112-MW wind farm in Buena Vista County, a 76-MW facility near Storm Lake in northwestern Iowa, and a 42-MW wind farm in Cerro Gordo County near Clear Lake. Zond won the Buena Vista and Storm Lake projects and installed their new Z46 wind turbine at these sites. Kenetech and FPL partnered on the Clear Lake project.

---

† The cost of production from a wind turbine with a US$1,800/kW capital cost and a 30 percent average annual capacity factor is US$1,800 × 10 percent capital carrying charge ÷ 30 percent × 8,760 hours/year = US$68/MWh. Adding 20 percent for O&M yields US$82/MWh. The cost of production from a natural gas combined cycle plant with a capital cost of US$1,000/kW and a 75 percent average annual capacity factor is US$1,000 × 10 percent capital carrying charge ÷ 75 percent × 8,760 hours/year = US$15/MWh. If the plant has a conversion efficiency of 45 percent, then the fuel cost of generation is US$2.50/mmBtu × 7,500 Btu/kWh = US$19/MWh. Adding the two together, yield a total cost of production from the gas plant of about US$34/MWh.

The Iowa law was enacted with little fanfare and the national discussion still focused on the California Wind Rush and its aftermath. Still, in the wake of the California wind market bust, and the subsequent years in the late 1980s and early 1990s in which the wind power industry was on life support, some states began to look for ways to increase their renewable energy industries. In Minnesota, the 1994 Prairie Island Law required Northern States Power Co. (NSP) to build or buy 400 MW of wind power by 2002 in return for being allowed to store nuclear waste on the site of the utility's Prairie Island nuclear power plant. NSP's first 25-MW wind facility was completed near Lake Benton in southwestern Minnesota in 1994. Kenetech won the bid for the 25-MW wind farm and installed seventy-three wind turbines there in 1994. A subsequent 107-MW wind farm was required to meet the state requirement. Zond won the bid to build the second wind farm. This facility—located near Lake Benton on the Buffalo Ridge—consisted of 143 750-kW Zond Z46 wind turbines.

Zond's Lake Benton project caused a ruckus in the wind industry. Zond had beaten rival Kenetech by promising to deliver wind power at US$35/MWh. Even considering the impact of the PTC, it was far below what anyone believed could be delivered. Zond would take a loss on the project just to get its new, yet to be constructed, Z46 wind turbine deployed in a utility-scale, commercial setting. Kenetech protested mightily, partially because the land for the project was owned by Kenetech and would have to be condemned by the state of Minnesota for Zond's use.[6] Kenetech's objections didn't matter in the end because FERC ended up invalidating the whole auction because of claims of false bidding by all of the participants.[7]

After the California wind market bust, Kenetech was focused on making wind power economically competitive. With the PTC in place, if they could further reduce costs and increase wind farm performance, they might be able to push down the cost of wind power and make it competitive with coal- and gas-fired generation. Even in the low natural gas price environment, Kenetech was fully committed to making wind power cost competitive. According to Kenetech General Counsel James Eisen in 1994: "there are a lot of Republican-minded people in management who think it's important that a private company succeed on its own."[8]

To build the most economically competitive wind turbine in the world, Kenetech would have to pull a rabbit out of a hat. Not a problem. There was one unexploited option for reducing the cost of wind power that Kenetech had not yet explored: variable-speed operation. Variable-speed wind turbines provide a 10 percent performance boost over industry-standard fixed-speed machines. Also importantly, the peak mechanical stresses caused by wind gusts can be reduced by limiting the torque acted upon the wind turbine by the generator and allowing the wind turbine to speed up in response to wind gusts. Variable-speed wind turbines were already commercially available in Europe. However, the technology had not yet been introduced in the United States for large wind turbines.[‡] If Kenetech was going to achieve its objective of making wind power cost competitive, it needed to develop its own variable-speed machine.

Kenetech's desire to develop a variable-speed wind turbine was compounded by interest in the technology by the Electric Power Research Institute (EPRI). EPRI organized an advanced-turbine feasibility study on the topic in 1998. Kenetech co-sponsored the study and three utilities—PG&E, Virginia Power, and the Bonneville Power Administration—participated. The basic question was whether modern power electronics could be incorporated into a variable-speed wind turbine to produce a machine that generates electricity for 5¢/kWh. EPRI concluded that such as goal was achievable and proceeded to organize a consortium for a five-year program headed by Kenetech that would lead to the development of a variable-speed wind turbine that could generate electricity for 5¢/kWh. PG&E and Virginia Power joined the group to provide utility perspectives on the machine.

The consortium developed technical specifications for the proposed wind turbine. It would be a 300-kW three-bladed, upwind, variable-speed machine with a 33-m rotor diameter. The 16-m blades would be made of fiberglass composite materials. It would have three times the power rating of Kenetech's workhorse 56–100 wind turbine. The variable-speed concept would be based around the Darlington transistor, essentially a high-powered, high-frequency on–off switch. The transistor-based converters would rectify the generator's wild AC and then convert it back into relatively harmonic-free, constant-frequency AC for relay to a step-up transformer and supply to the utility grid. The electronic controller would also enable the turbine to consume reactive power from the utility system or to supply it to the system, depending on the need, or to operate at unity power factor.[9] The wind turbine would become known as the KVS-33 and it would be America's first grid-friendly wind turbine.

Kenetech moved forward, sinking US$70 million in the development of KVS-33. The other consortium partners pitched in another US$5 million. Kenetech borrowed heavily to finance development of the KVS-33. By 1993, the company owed US$230 million in debt. In late 1993, Kenetech closed a successful initial public stock offering that raised US$96 million to repay debt and proceed with the development and marketing of the KVS-33. In February 1994, Kenetech launched a marketing campaign for the KVS-33 aimed at utilities. By the summer of 1994, Kenetech had installed its prototype KVS-33 machines at its wind farms in Buffalo Ridge, Minnesota and Palm Springs, California. Kenetech was confident about its new KVS-33 wind turbine. Kenetech's president—Dale Osborn—stated that the KVS33 would be the "best wind turbine in the world".[10]

---

‡ The Bergey Windpower Company—headquartered in Norman, Oklahoma—introduced its first variable-speed, utility-connected wind turbine in 1980. By 1983, Bergey was offering a 10-kW version of its machine.

Figure 17.1 Kenetech's KVS-33 wind turbine was a 300-kW variable-speed machine. After sinking millions in its development, Kenetech's fortunes sank when initial versions of the wind turbine proved to be unreliable.

The cracks started to show in the summer of 1994. Henry Hermann—an analyst in Dallas who covered the industry for W. R. Lazard, Laidlaw & Mead— warned investors about the problems with the KVS-33 after visiting Buffalo Ridge in August. "They invested tremendous amounts in a machine that did not work very well," Mr. Hermann later told the *New York Times*. According to Hermann, when he visited Buffalo Ridge, sixteen of the seventy-three KVS-33 wind turbines were not operating on a windy day. All the machines were making loud noises that suggested to Herman that meaningful mechanical issues existed, and several turbine blades had broken off of the wind turbine and had visible cracks at the blade joints.[11] *Windpower Monthly*—a trade publication—reported in September 1994 that there were cracked blades and damaged generators at wind farms in Buffalo Ridge and Palm Springs.[12]

The truth was, Kenetech's KVS-33 was experiencing blade problems, and the additional cost and efficiency penalty associated with the power electronics wiped out the benefits of variable-speed operation in the initial versions. Much more time would be needed to fully test and fix the machine. But it was time that Kenetech didn't have. They had already begun selling the machine. They told investors that they expected 1,000 KVS-33 wind turbines to be sold in 1995 alone. So Kenetech pressed forward anyway. By 1995, ninety KVS-33 wind turbines were installed in Spain. The blades broke off several of the turbines causing the project to be immediately shut down.[13]

Kenetech executives assured investors that the problems with the KVS-33 could be fixed and they remained publicly optimistic. However, the technical problems of the KVS-33—along with poor earnings from 1993—created a financial storm by the end of 1994 that the company would not recover from. Once investors started to believe that the KVS-33 was flawed—whether it was fatally flawed or not—they lost confidence in Kenetech and started to look closely at the company's financials. With the company's large debt load, it was clear that if it could not meet its financial commitments then the company wouldn't be financially solvent.

From a peak stock price of US$29.50 in March 1994, the stock price plunged to US$3.25 in August 1995 after the company reported a quarterly loss and investors lost confidence.[14] In September 1995, Gold, Bennett & Cera of San Francisco, a law firm that specializes in class-action shareholder suits, filed a complaint against Kenetech contending that its top officers had given analysts "misinformation, causing them to mislead investors."[15] The suit asserted that some top executives and directors had falsely claimed Kenetech's turbines could compete with coal by producing electricity for 5 ¢/kWh. It also asserted that they continued to push the product even after flaws had been identified.

By March 1996, Kenetech stock was worth less than a dollar and the company reported an annual loss of US$250 million for 1995. Kenetech's CEO at the time—Gerald Alderson—hired a turn-around specialist to save Kenetech from possible bankruptcy, but to no avail. In the Spring of 1996, the company filed for bankruptcy. What started in 1978, when U.S. Windpower founder Russell Wolfe met with William Heronemus in his office at the University of Massachusetts, ended nearly twenty years later. Kenetech's demise shook the global wind power industry. The promising—but complicated—KVS-33, had become an albatross that ultimately led to Kenetech's demise. Kenetech's implosion made Enercon's E40 look all that more impressive.

17.3 The Selfish Invention

As late as 1994—while the KVS-33 was still under development—Kenetech apparently still had some big unanswered questions about variable-speed wind turbine technology. That's when Bob Jans—director of Kenetech's Netherlands office—Kenetech employee Ruth Heffernan, and German wind industry consultant Ubbo de Witt visited one of Enercon's E40 wind turbines in Oldenburg, Germany to see how Enercon had successfully implemented its variable-speed approach. The three carried out an inspection of the machine in March 1994. They took pictures and examined the machine's control system. The problem was, they inspected the Enercon machine without permission from the project owner or Enercon itself.[16]

When Enercon caught wind of the breach two years later, it filed a formal complaint and the Oldenburg police launched an investigation. A 1998 report commissioned by the European Parliament asserted that America's National Security Agency (NSA)—under a program named "Echelon"—had the capacity to intercept and analyze all electronic communications in Europe. The German news magazine *Der Spiegel* published an article in 1999 which claimed that NSA tapped Enercon's telephone lines under the Echelon program.[17] According to the accusations, NSA could have been listening in on Enercon's calls for years and passing the intelligence to Kenetech.

Whether or not NSA committed industrial espionage on behalf of Kenetech is unknown, but something interesting did come from the event. Kenetech had filed a patent for its variable-speed wind turbine back in 1992. US Patent number 5,083,039 was granted in 1993. Several different models of Enercon's variable-speed wind turbines that fit this description had already been commercially available since 1985. Enercon should have filed its own US patent for its variable-speed machine well before 1992, but it did not. Whether or not Kenetech received information from NSA or other sources that influenced the patent language or provided information that encouraged Kenetech to apply for a patent on technology that Enercon had already developed is unknown.

As soon as the patent was granted, Kenetech put it to good use. The company filed a complaint with the US International Trade Commission (ITC) in 1995 to prohibit Enercon from importing its variable-speed wind turbines for a wind farm project that it was developing at Big Spring in Texas with its partner the New World Power Corporation. ITC ruled in favor of Kenetech and issued an order prohibiting Enercon from importing its turbines into the country. After Kenetech filed for bankruptcy in 1996, Zond swooped in and purchased the patent from Kenetech. Enercon appealed the ITC ruling and moved to prohibit Zond from stepping in to take Kenetech's place as the new patent owner, but Enercon failed on both counts. ITC's ruling was reaffirmed and the original Kenetech patent was used to lock Enercon out of the US market until it lapsed in March 2010. This constrained Enercon's wind turbine activities and forced it to focus primarily on the European market.

As it turned out, the variable-speed wind turbine patent outlived both Kenetech and Zond. Wind industry consultant Chris Varrone has described the variable-speed wind turbine as the "selfish invention" because it appears to have a life of its own regardless of the success or failure of its original inventors.[18] Variable-speed wind turbines would prosper irrespective of the condition of Kenetech, Zond, or Enercon. Between 1995 and 2004, the market share of variable-speed wind turbines grew from 14 to 73 percent.[19]

Zond's turbine manufacturing volume climbed sharply because of the Storm Lake and Lake Benton projects. However, despite Dehlsen's best efforts to raise capital, Zond was still cash-starved. There were no good capital solutions for the emerging wind industry, and things tightened up significantly after Kenetech's bankruptcy. Capital was increasingly difficult to come by, just as Zond's needs were ramping up. Dehlsen concluded that Zond needed a deep-pocket partner if it was going to avoid bankruptcy too.

In the spring of 1997, Dehlsen arranged for Houston-based natural gas company Enron to purchase Zond for US$100 million. Dehlsen described it as a "partial sale," because Enron took control in stages.[20] Zond's CEO Ken Karas took over as the CEO of Enron Wind Corporation (EWC). At the time, Enron was a respected diversified energy company with a broad presence in both domestic and international markets. It had US$19 billion in assets, and operated one of the largest natural gas transmission systems in the world. It was also the largest purchaser and marketer of natural gas and the largest nonregulated marketer of electricity in North America. Enron was exactly what Zond needed.

At Zond's urging, Enron Wind purchased Germany's Tacke Windtechnik in 1997. Although it was the world's fifth largest wind turbine manufacturer at the time, Tacke itself had fallen into bankruptcy. The combination of Tacke and Zond would form the basis of EWC moving forward. In addition, Finn Hansen—the director of Zond's manufacturing operations and the former Vestas CEO—was put in charge of EWC's wind turbine manufacturing operations. At long last, an American wind turbine manufacturer was well-positioned to compete with the Europeans.

By acquiring Tacke, EWC inherited its TZ 750-kW wind turbine which would be the basis for EWC's wind turbine platform moving forward. EWC developed a larger TZ 1.5-MW wind turbine in 2000 and tested it in Tehachapi in March. The TZ 1.5-MW wind turbine was a variable-speed machine that combined the best components from Tacke and Zond. With Kenetech's original variable-speed patent in hand, EWC was free to develop and deploy its variable-speed technology throughout the United States. By combining Tacke's technology expertise with Zond's field experience, EWC believed it now had a successful technology platform that could compete internationally.

EWC continued to grow. Sales rose from US$50 million in 1997 to US$800 million in 2001.[21] Then the unexpected happened: EWC's parent company—Enron—filed for bankruptcy in December 2001. A subsequent investigation by the United States Securities and Exchange Commission and the Department of Justice discovered that Enron had inflated its earnings by hiding debts and losses in subsidiary partnerships. Because of the Enron bankruptcy, the assets of EWC became available for purchase in 2002.

EWC's assets were put up for auction in April 2002. The live auction pitted two industrial giants—GE and Caterpillar, Inc.—against each other. During the auction in Manhattan, GE beat out Caterpillar for the right to buy the assets of EWC. But in doing so it bid up the purchase price by more than US$100 million, a sum that had staggered observers of the wind industry. GE won the auction with a bid of US$358 million.[22] However, by December of 2002, GE was back in court trying to get a US$160 million discount on the purchase price based on their updated assessment of the value of EWC.[23]

Now, at this point, GE had a long history with wind power. The hydropower division had participated in the development of the Smith–Putnam wind turbine in 1941. GE was one of the primary contractors for NASA's federal wind energy program in the 1970s and 1980s. However, GE watched on the sidelines as wind power grew in the 1990s. After Enron's collapse, GE executives re-evaluated the wind energy space. Steve Zwolinski—President and CEO of GE's Hydropower division—led the evaluation and convinced other executives that wind power was worth the risk. Zwolinski served as the first president of GE Wind Energy.

After the purchase of EWC, GE went to work applying its lean production processes and famed management approach to wind power. By the end of 2003, GE had increased its wind turbine sales to US$1.2 billion.[24] GE Wind Energy had become the second largest wind turbine manufacturer in the world—behind only Vestas. Given is size, global scope, and relationships with traditional electric utilities, GE was well-positioned to grow its wind business. Furthermore, GE was able to develop a robust wind turbine technology supply chain using some existing suppliers from its other power businesses.

However, the transition from EWC to GE Wind Energy was not entirely smooth. The reliability of the TZ 1.5-MW technology was not close to GE's expectations. GE launched a comprehensive effort to examine and improve every major subsystem in the turbine and initiated an effort to develop larger 2- and 3-MW wind turbines with improved power electronics and simplified generators. GE's effort to re-engineer the wind turbine technology platform proved to be more involved than either Zwolinski or GE CEO Jeff Immelt expected when they acquired EWC in 2002. According to Immelt "[Enron's wind business] was a very broken model that took us three years to fix."[25]

GE invested heavily to improve the 1.5-MW wind turbine. Its investment would pay off because GE's 1.5-MW wind turbine would soon account for half of all wind turbines in the United States. GE made wind power technology part of its global industrial system that included design, manufacturing, delivery, and operations. Wind power became part of the global industrial system which enabled it to thrive in the twenty-first century. Under GE's watch, wind power's maturation was accelerated and the technology was finally ready to complete on firm footing within the global energy system. GE Wind Energy also put the United States back atop the emerging global wind power business. By 2003, the top five global wind turbine manufacturers were: Vestas, GE Wind, Enercon, Gamesa, and NEG Micon.[26]

Thus, at the dawn of the twenty-first century, not only had wind power been reignited in the United States through a combination of federal and state-level policies, America finally had a global wind turbine manufacturer that could keep pace with its European counterparts. At this point, the primary question for GE was whether the US market was big enough to support its ambitions, particularly considering the uncertainty caused by periodic lapses in the federal PTC. And if GE was going to seek its wind power fortunes beyond American shores, breaking into the European market was going to be challenging given the intensity of competition among European wind turbine manufacturers.

Fortunately for GE—and the entire wind industry—other parts of the world beyond the United States and Europe were beginning to awaken to wind power. Wind power was already awakening in Asia, a region that would ultimately becoming the largest wind power market in the world. The next chapter of the wind power story begins in India, where a textile business owner was in the process of creating a new wind turbine manufacturer that would soon rival the best of the West.

Notes

1 Dehlsen, 2003.
2 Asmus, 2001, 185.
3 Ibid., 186.
4 Starrs, 1988, 140.
5 Ibid., 143.
6 Asmus, 2001, 175.
7 Berger, 1997, 169.
8 Serchuk, 1995, 278.
9 Moore, June 1990, 23.
10 Berger, 1997, 164.
11 Asmus, 2001, 176.

12 Salpukas, 1995.
13 Ibid.
14 Ibid.
15 Ibid.
16 Windpower Monthly, 1999.
17 Ibid.
18 See Varrone, *The Selfish Invention*, 2013a.
19 Varrone, *The Impact of Variable Speed Concepts on Wind Power*, 2013b.
20 Dehlsen, 2003.
21 Bloomberg News, 2002.
22 Ibid.
23 WindPower Monthly, 2002.
24 Fairley, 2005.
25 Backwell, 2018, 27.
26 BTM, 2004.

Bibliography

Asmus, Peter. *Reaping the Wind*. Washington, DC: Island Press, 2001.

Backwell, Ben. *Wind Power: The Struggle for Control of a New Global Industry*. New York: Routledge, 2018.

Berger, John J. *Charging Ahead*. New York: Henry Holt and Company, 1997.

Bloomberg News. 2002. "G.E. to Buy Enron Wind-Turbine Assets." *The New York Times*, April 12.

BTM. *International Wind Energy Development – World Market Update*. Copenhagan, Denmark: BTEM Consult ApS, 2004.

CEC. *Results from the Wind Project Performance Reporting System: 1987 Annual Report*. San Francisco, CA: California Energy Commission, 1988.

Dehlsen, James, interview by U.S. Department of Energy. 2003. *From Zond to Enron Wind to GE Wind*. WINDExchange.

Fairley, Peter. 2005. "The Greening of GE." *IEEE Spectrum*, July 1. https://spectrum.ieee.org/energy/environment/the-greening-of-ge.

Moore, Taylor. "Excellent Forecast for Wind." *EPRI Journal* June (1990):14–25.

Owens, Brandon. *An Ecomomic Valuation of a Geothermal Production Tax Credit*. Technical Report. Golden, CO: National Renewable Energy Laboratory (NREL), 2002.

Salpukas, Agis. 1995. "A Giant of Wind Power Stumbles Badly." *New York Times*, December 27.

Serchuk, Adam Harris. "Federal Giants and Wind Energy Entrepreneurs: Utility-Scale Windpower in America, 1970–1990." PhD thesis, Virginia Polytechnic Institute and State University, 1995.

Starrs, Thomas A. "Legislative Incentives and Energy Technologies: Government's Role in the Development of the California Wind Energy Industry." *Ecology Law Quarterly* 15, no. 103 (1988): 103–58.

Varrone, Chris. "The Selfish Invention." *Renewable Energy Focus* 14 (2013a): 34–7.

Varrone, Chris. 2013b. "The Impact of Variable Speed Concepts on Wind Power." *Renewable Energy Focus*. Accessed January 12, 2018. http://www.renewableenergyfocus.com/view/32093/the-impact-of-variable-speed-concepts-on-wind-power.

Windpower Monthly. 1999. "Trans Atlantic Espionage Claimed by Germa Wind Company." *Windpower Monthly*.

———. 2002. "Corporate Dance Starts Over Sales Price – GE Demands 50% Refund from Enron for Wind Company." *WindPower Monthly*, December 1.

18

India's Wind Power Path

Asia is the market and the location where production is inexpensive. The technological expertise and specialist knowledge comes from the industrialized world. The aim is to combine them.
     —Tulsi Tanti, founder Suzlon Energy (Der Spiegel Online 2011)

18.1   Coping with the Crisis

The history of wind power spans the twentieth century. Activities were dominated by Denmark, Germany, the United States, and—by the end of the twentieth century—Spain. However, surprisingly—by 2017—two Asian countries—China and India—were among the top five in both installed wind power capacity and wind power capacity additions globally. And today, with 36 GW already installed, India is committed to installing another 24 GW by 2022.

To understand Asia's unexpected wind power rise, it is useful to start with an examination of India's wind power history. Like Spain, India's wind power journey began in the wake of the 1973 global energy crisis. The rise in oil prices hit the country particularly hard and led the government to focus on developing alternative sources of energy supply. These efforts would eventually lead to the creation of a both a domestic wind power market and wind turbine manufacturing industry in India.

India's path to wind power success began in January 1974 as the government tried to cope with the rising cost of oil imports that placed the country in a bleak position as the nation was beset by inflation, political dissension, lagging growth, a spiraling population, and unchecked poverty. The cost of the oil-derivative fuel kerosene—the country's primary cooking fuel—had climbed by 50 percent in two months. Gasoline prices has risen 80 percent.[1] Food production—a key to India's stability—was expected to decline by at least

The Wind Power Story: A Century of Innovation that Reshaped the Global Energy Landscape, First Edition. Brandon N. Owens.
Published 2019 by John Wiley & Sons, Inc.

3 million tons during the spring harvest because of the rising oil price and a shortage of petroleum-based fertilizers.

The oil crisis impacted India negatively in three ways: first the jump in oil prices dramatically increased the country's import bill even with conservation efforts; second, the decline in oil availability impacted agricultural and industrial production; third, higher cost oil increased the cost of fertilizers and reduced supply, which drove up the cost of food. Eighty-five percent of India's oil imports were used for agriculture and industrial production.[2] The oil crisis represented an existential threat to the country's economic, political, and social stability.

It didn't have to be this way because India had other options. The country's coal reserves were estimated to be as high as 200 billion tons, yet production was just 75 million tons per year and dropping. The government and the private sector were just not organized or focused enough to increase domestic coal production to offset oil imports. Increasing coal use wasn't the only option. Coal already accounted for 60 percent of primary energy consumption and there were unexploited domestic sources such natural gas.[3] In addition, there was the possibility of developing renewable energy technologies—such as wind and solar power—which the Americans and Europeans had already started examining in the wake of the oil crisis.

To explore new options for domestic energy security, the Commission for Additional Sources of Energy (CASE) under the Department of Science & Technology was formed. The Commission was charged with creating programs for the development of new renewable energy sources, and coordinating renewable energy R&D activities. Like government-led renewable energy activities in the United States and Europe, the programs that were launched after the crisis and were intended to lead to quick results ended up taking years to gain momentum. The Commission's activities were rolled into the newly formed Department of Non-Conventional Energy Sources (DNES) under the Ministry of Energy in 1982. DNES was eventually given Ministry status and renamed the Ministry of New and Renewable Energy (MNRE) in 2006. DNES commissioned the Indian Institute for Tropical Meteorology (IITM) to undertake the country's first wind resource assessment, which it did in 1983.[4]

On the basis of the IITM resource assessment, DNES purchased a 40-kW Dutch turbine from Polenko B.V.—a manufacturer from the Netherlands—and had it installed in Gujarat as part of a joint venture between the Gujarat Energy Development Agency, JK Synthetics, and the Gujarat Electricity Board.[5] The Polenko wind turbine in Gujarat was the first successful demonstration of grid-connected wind power in India.

In 1986, DNES followed up on the Polenko project by offering grants for wind farms across India to demonstrate wind power technology. In response, Vestas, Micon, and Windmatic were involved in the development of a series of small wind farms that demonstrated the technical and economic viability of wind turbines in India starting in 1987. The demonstration projects were installed in the Indian states of Gujarat, Maharashtra, Orissa, and Tamil Nadu. For many Indians, these wind farms represented their first exposure to wind power. Unlike America, Europe, or even Central Asian countries such as Iran—that have a history of using windmills—India had no such tradition.[6] This was something entirely new. The encouraging results from these projects created enthusiasm for wind power across India.

In 1988, 20-MW wind farm projects were installed in Gujarat and Tamil Nadu. These projects were funded by the Danish International Development Agency (DANIDA). DANIDA is the Ministry of Foreign Affairs of Denmark, which provides humanitarian aid and development assistance to other countries, with a focus on developing countries. Given Denmark's position as an international wind turbine manufacturer at the time, DANIDA hoped its aid to India would result in new projects for Denmark's wind turbine manufacturers. Vestas was chosen as the wind turbine manufacturer for these projects.[7] These utility-scale wind farms were successful and further increased Indian excitement around wind power. The 20-MW farm at Lamba in Gujarat—near the birthplace of Mahatma Gandhi—ignited the public's imagination.[8] It was the success of these demonstration projects that helped provide real data on the economic feasibility of wind energy power in India.[9]

At the same time, government policymakers were getting more serious about wind power. The Seventh Five-Year Plan (1985–1990) contained several wind power incentives. The most important of which was the 100 percent accelerated depreciation tax benefit. This benefit provided wind power investors with the ability to write off 100 percent of their capital investment in the first year of the project. This incentive was later reduced to 80 percent, and then eventually phased out in 2012. It was the primary driver for India's wind power development in the 1990s and the first decade of the twenty-first century. In addition, the Seventh Five-Year Plan provided wind farm owners with a five-year income and sales tax exemption. It also required each state's Electricity Board to purchase wind power and develop special tariffs.

In addition, a dedicated Indian Renewable Energy Development Agency (IREDA) was set up under DNES to provide loans for energy efficiency and renewable energy projects throughout the country. Starting from a US$4 million base in 1987, IREDA sanctioned renewable energy projects worth approximately US$3.6 billion through 2011.Commercial wind power capacity grew from 2.2 to 37 MW between 1985 and 1990.[10] The first wind power projects were developed in Tamil Nadu, Karnataka, and Gujarat.

In 1991, additional reforms were enacted to support wind power development. As part of a broader economic liberalization plan, the Indian electricity market was opened to both domestic and foreign private investors. Import duties and taxes were reduced for all Independent Power Producers (IPPs) and secured rates of return were offered to foreign investors through guaranteed tariffs that were based on project costs. In addition—in 1992—DNES was elevated to a full-fledge Ministry, now called the Ministry of Non-conventional Energy Sources (MNES). MNES subsequently issued guidelines to advance power generation from renewable sources. As part of the Eighth Five-Year Plan (1992–1997), MNES set a target of 500 MW of new wind power. In 1993, MNES set the wind power tariff price at INR2.25/kWh or 11.25¢/kWh. Wind power now had the policy incentives and government support needed to enable it to grow.

Although India was now a hospitable environment for the development of wind power, the country still did not have any domestic wind turbine manufacturing capability. Danish and German manufacturers had been eyeing India since the successful DANIDA projects in the late 1980s. By the mid-1990s, these firms were dreaming of making money in India and eager to cash in on India's wind power potential.

But India is a unique environment and developing wind power projects in the country is not simply a matter of foreign manufacturers selling wind turbines to domestic wind farm developers. As anyone who has tried knows, local knowledge and presence is required to get business done in India. Certain aspects of India's business environment have long suffered from legal and physical infrastructure challenges. There are also issues with the protection on Intellectual Property (IP) rights. Danish and German wind turbine manufacturers were well aware of this, but rushed into India in the mid-1990s to sign licensing agreements and set up local manufacturing facilities nonetheless.

Domestic manufacturing was thus set in motion by a series of licensing agreements between 1995 and 2005 with Danish (Vestas, Micon) and German (Enercon, Nordex, DeWind, Südwind) firms. Vestas established a joint venture with local partner, RRB. NEG Micon set up NEPC Micon with the Indian NEPC group, and Enercon India (EIL) was founded in 1994 by Enercon and the Mehra Group. The two entities held 51 and 49 percent of the joint venture, respectively. The joint venture entered into an exclusive license agreement with Enercon for the full use of its equipment and technology. Enercon was provided with nineteen foreign patents in India. Enercon India set up a production facility and began manufacturing Enercon's well-regarded E40 direct drive wind turbine.

Yogesh Mehra was the Managing Director who ran the local Enercon operation. Enercon became ubiquitous in India alongside Vestas. EIL would eventually become the second largest wind turbine manufacturer in India. In the meantime, Enercon had lost its patent battle with Zond in America. The variable-speed technology patent rights were now owned by GE Wind Energy and Enercon was effectively locked out of the United States wind market until 2010. Against this backdrop, Enercon's growing presence in India was an important part of its international business strategy.

The EIL licensing agreement expired in 2004 and Enercon and Mehra spent several years attempting to renegotiate a contract. Mehra controlled a minority stake in the company, but had managed to exclude Enercon company representatives from the board.[11] By the time that negotiations started, Enercon believed that the Mehra family was not being transparent about its operations. Enercon also wanted to continue to adhere to its more conservative approach of sustained, moderate growth. For his part, Mehra believed that the parent company was starving EIL of critical component supplies and wanted to take back full control.[12] Mehra was also planning more rapid growth than Enercon envisioned and was eyeing a stock market listing for EIL.[13]

In 2008, while they debated company strategy with Mehra in Mumbai, two Enercon executives were summoned to the local police station. They were questioned for five hours and authorities told them that they were suspected of conspiring against their Indian partners.[14] From that point forward, EIL executives progressively limited their contact with the German executives. Enercon had been effectively banished from India. Mr. Mehra filed to revoke Enercon's technology patents in India. The Indian justice system eventually ruled against Enercon. According to Enercon's lawyer, the court decision was based on the premise that India's economic development needs have priority over any foreign company's patent rights and cautioned that any new technology patent in India could be nullified in the future.[15] Because of the ruling, EIL—which Mehra renamed Wind World—was free to manufacture and install Enercon wind turbines not only in India, but across the globe. As of 2016, Wind World was still the second largest wind turbine manufacturer in India.[16]

EIL is a cautionary tale for Western firms. The large and fast-growing markets of Asia represent a unique challenge for European and American wind turbine manufacturers. Because of IP protection challenges and an ardent desire on the part of both India and China to develop their own domestic technology champions, Western wind turbine manufacturers have found that there are real risks to committing to these markets. However, give the sheer size and expected growth rates of India and China, international wind turbine manufacturers also couldn't afford to ignore them.

Much to its credit, Enercon has maintained its position as a top global wind turbine manufacturer despite having a limited presence in the US and Indian markets. This can't last forever though. Enercon—in particular—needs an expanded international footprint to grow. Enercon re-entered India in 2017, and is now set to offer its 3.5-MW machine to Indian customers.[17] Enercon plans to participate in competitive bids and expects new projects to be commissioned by 2019.

Vestas too separated from its Indian partner RRB in 2004. Vestas wanted to focus exclusively on providing and installing wind turbines in India, not servicing them. However, success in India has historically meant technology providers must also provide turnkey project development and maintenance services.[18]

This not only means taking on the risk of building site infrastructure and maintaining the turbines, it also means acquiring land and all of the official and unofficial aspects of that endeavor in India.

However, despite this disconnect, the company ultimately found that to have continued success in India, it needed to maintain a strong commitment to the country. Vestas has a manufacturing facility for wind turbine hubs at Oragadam, one for blades in Ahmedabad, and a repair center in Coimbatore. In addition, to support wind turbine sales in India, the company opened a technology center in Chennai to help monitor wind speeds across India and create optimal wind turbine solutions for any site in the country. Today, nearly 10 percent of Vestas's workforce is in India and the company plans to increase its headcount in India by 50 percent soon.[19]

18.2 A Star Is Born

Tulsi Tanti reached a breaking point in 1994. It was hard enough keeping the twenty-person family textile business in Surat running given intense international competition and the challenges associated with finding reliable workers. But the frequent electricity outages and skyrocketing energy prices made the job near impossible. There had to be a better way. Tanti was aware of the Indian government's efforts to promote wind energy, but neither he, nor his three brothers, knew anything about it. Still—if left unchecked—his problem with electricity reliability from the local utility would put him out of business. He had to act quickly.

As an experiment, Tanti contacted German wind turbine manufacturer Südwind and ordered two wind turbines. Südwind was founded by students at the Technical University of Berlin in the early 1980s. They had been commercially successful in the German market throughout the 1980s. When Tanti called in 1994, they were producing high-quality wind turbines for their local market.

Tanti's original plan was to take his factory off the grid. If the utility couldn't supply him with power, he would generate his own. It sounded so easy. It turns out that Tanti had to order the turbines from Südwind, work with another local company for installation, and contact yet another company to come out and maintain the machines. By the time they were up and running, Tanti was at his wits' end. Maybe unreliable power from the utility company wasn't so bad after all?

Still, the Südwind wind turbines did work well and served an important need. Given India's power problems, they could spread like wildfire across the whole country. But wind power development in India would have to be much easier. What India needed was an integrated wind power company—customers needed turn-key solutions, not a lot of run around. If he could solve that problem by providing wind turbine installation and maintenance, then it might be worth going into the wind energy business and leaving textiles behind. Tanti and his three brothers sold the family textile business to raise US$600,000 in seed money for their new wind power venture in India. In April 1995, they founded Suzlon Energy in Pune, near Bombay.[20]

However, there was one problem with Tanti's plan: Suzlon didn't know anything about wind turbines and had no technology to offer. They would have to rely upon Südwind to provide the technology and the expertise. Suzlon's first project was a 3.5-MW ten-wind turbine project for Indian Petrochemicals in Dhank, Gujarat.[21] Then, in 1998—much to the surprise of Tanti—Südwind filed for bankruptcy. The company had been sitting on a high debt load and their bank suspended its credit after they got behind in their payments.[22] Although this was unfortunate for Südwind, it was a stroke of luck for Tanti. Unlike, Enercon India, which would later have to go through a painful divorce to acquire its German wind turbine technology, Südwind's technology fell into Suzlon's lap.

Suzlon acquired the rights to Südwind's technology, its research laboratory, and hired its former employees. Suzlon's approach was to tap the technology expertise from Europe and install it in India. Tanti knew that the best technology expertise was in Europe and the United States, while the biggest markets resided in the developing economies like India and China. According to Tanti "the technological expertise and specialist knowledge comes from the industrialized world. The aim is to combine them."[23] By 2003, the company had research and manufacturing plants in the Netherlands, India, and Germany.

Suzlon grew throughout the late 1990s and the first half of the 2000s. Suzlon's growth was also driven by its growing European technology portfolio, which enabled it to control an increasing share of its component supply chain. By the mid-2000s—because of strong global wind power growth—wind turbine components were increasingly globally constrained. For example, bearing manufacturers had been increasingly unable to keep up with growing demand from wind turbine manufacturers, creating a bottleneck. Other components such as gearboxes, casting, and forgings were also in short supply.[24] Suzlon's acquisitions of AE-Rotor (Netherlands), Einundzwanzigste Vittorio Verwaltungs (Germany), and Hansen Transmissions (Belgium) helped the company minimize the impact of component constraints.

Suzlon's rise was also fueled by the Indian government's continued support for wind power, which had resulted in a 100-fold increase in installed capacity between 1994 when capacity stood at 41.30 MW, and 2005 when capacity had grown to 4.43 GW.[25] The biggest policy driver was the Electricity Act of 2003. The Act required State Electricity Regulatory Commissions (SERC) to take steps to advance renewable energy. SERCs become responsible for interconnecting wind farms to the grid, purchasing the wind-generated electricity, providing preferential compensation for wind farm owners through feed-in tariffs, and committing to generating fixed quotas of wind power through mandated Renewable Purchase Specifications (RPS).[26] The 80 percent accelerated depreciation and ten-year holiday tax incentives remained in place.

MNRE issued guidelines to all state governments to create an attractive environment for the export, purchase, wheeling, and banking of electricity generated by wind power projects. Ten Indian states implemented quotas that set specific targets for renewable energy generation and introduced preferential tariffs. In addition, several states added their own financial incentives for wind power projects including energy buy-back—known as net metering in Europe and the United States—preferential grid connections, and transportation charge and electricity tax exemptions.[27]

The wind-resource rich states of Gujarat, Karnataka, Tamil Nadu, Andhra Pradesh, Maharashtra, and Madhya Pradesh were those that implemented the most aggressive renewable energy policies. The Indian government would later institute a "Green Energy Corridor" program to facilitate the transfer of power from states with high levels of renewable energy installations to other parts of the country. Together, these incentives were effective in driving up India's installed wind power capacity. Between 1996 and 2005, Suzlon's installed wind power capacity in India grew from 270 kW to 1 GW.[28] By 2005, 44 percent of India's installed capacity was manufactured by Suzlon. Sixty percent of all new capacity additions also came from Suzlon. In 2005, Suzlon went public on the Bombay Stock Exchange. Shares were oversubscribed by forty-three times within ten minutes and the company quickly grew to a market capitalization of US$8 billion. Suzlon had become the tenth largest wind power company in the world.[29]

However, India was not large enough for Tanti's ambitions. He viewed Suzlon as an international wind player and he set his designs on global expansion. To climb to such heights, Tanti again drew upon European expertise. This time he targeted the business know-how of Per Pedersen. Pederson was the CFO of NEG Micon, one of the original Danish wind power manufacturers. He was used to high growth and high pressure environments and had overseen NEG Micon's global expansion. When NEG Micon merged with Vestas in 2004, Tanti asked him to join Suzlon to lead its global expansion. Pedersen joined Suzlon in June 2004 and he was put in charge of international sales and marketing based in Denmark.

Tanti then cast his eyes upon Germany's Repower in 2007. Repower's roots could be traced back to Jacobs Energie of Heide in Germany. In 1988, Jacobs purchased the license to a small wind turbine design produced by German industrial giant MAN with funding from the German government. Jacobs increased the wind turbine capacity to 33 kW and continued to build successfully larger wind turbines, eventually reaching 1.5 MW in 1998. Repower was formed in 2001 when Jacobs merged with Husumer Schiffswerft (HSW), BWU, and pro+pro energy systems. Repower began producing its 1.5 ProTec MD wind turbine in 2001.

By the time they fell into Tanti's sights, Repower was Germany's third largest wind turbine manufacturer. It had a strong presence in Europe and was expanding into China and Australia. Just as important, Repower had developed massive 5- and 6-MW wind turbines for expansion in the emerging offshore wind power market segment. By combining Suzlon and Repower, Tanti saw an opportunity to become a truly global wind power company. The combined company would have a footprint in both the developed and emerging worlds.

At this point, the global wind power industry had been on a tear. Rapid technology improvements and expanding support policies in countries around the world propelled the industry on first in the first decade of the twenty-first century. From just 3.5 GW in 1994, global installed capacity had grown to 94 GW by the end of 2007 with 100 GW in sight for 2008. The industry was growing at a rate of 30 percent per year.[30] If Tanti wanted to purchase Repower in order to create a leading globally integrated wind power company, he would have stiff competition. Wind power wasn't a secret anymore; in fact, French nuclear company Areva already had designs for Repower.

Areva started its life as France's version of Westinghouse. Founded in 1958 by Westinghouse and a coalition of French industrial companies to license Westinghouse's Pressurized Water Reactor technology, by the early 2000s the company was looking to diversify beyond nuclear power. In 2006, Areva created its Renewable Energies Business Group and took a 30 percent stake in Repower. By 2007, Areva was interested in full ownership and submitted a bid of €105 per share in January 2007. Suzlon countered with a €126 per share bid in February and the takeover battle was on. To strengthen its position, Suzlon partnered with the Portuguese steel company Martifer, which itself had a 23 percent stake in Repower. By May 2007, the strength of Suzlon had convinced Areva not to increase its bid. Instead, Suzlon and Areva reached an agreement that would enable Suzlon to pay €1.35 billion for Repower while keeping Areva as its key transmission and distribution equipment supplier. Suzlon had won, but it paid a hefty price. Pedersen was placed at the helm of Suzlon's Repower to lead its continued international expansion.

By 2008, the combined Suzlon and Repower had 8 percent of the global wind power market and annual revenues of US$1.7 billion. From its inception through 2007, Suzlon had sold a cumulative US$4.1 billion worth of wind turbines, US$500 million in India, and US$3.6 billion around the world. While accounting for close to half of India's installed capacity, it was now expanding into China, the United States, and Australia.[31] Life was good for Tanti, Suzlon, and India's wind power industry.

However, after the global financial crisis struck in 2008, the global wind power industry slowed down. Suzlon had purchased Repower with borrowed money. The company got behind on its obligations and failed to repay US$209 million of its debt in 2012. After eighteen months of negotiations, Suzlon was able to restructure its debt and remain solvent. Suzlon renamed Repower

"Senvion" in 2014 and then began looking for suitors to purchase the company. Reducing debt had become more important than hanging on one of its crown jewels. Plus—by this time—Suzlon had already fully integrated Suzlon's off-shore wind technology into its portfolio. In January 2015, Suzlon sold Senvion to the American private equity firm Centerbridge Partners for €1 billion, less than it had paid in 2007.[32]

In the end, Tanti's contribution to the global wind power industry wasn't just related to creating a new wind turbine manufacturer in India. Rather, his primary contribution was recognizing that the wind power component supply chain could be relocated to low-cost countries to reduce wind turbine costs and improve their international competitiveness. Keep in mind that wind turbines are simply the sum of their components and that these components are manufactured separately and only later assembled into a complete wind turbine. Tanti globalized the wind power component supply chain by purchasing European component suppliers and then opening manufacturing facilities in India. Low cost component suppliers would quickly emerge in the rest of Asia and later in South America. By 2015, over half of all wind turbine components would be manufactured in South and East Asia.

18.3 Gamesa Gets It

Spanish wind turbine manufacturer Gamesa was ready to enter the Indian wind power market. They understood the importance of what Suzlon had accomplished by opening the door to low cost component manufacturing in India and they had their own plans to open production facilities in India. They also sympathized with Wind World's need to free itself from the shackles of its technology agreement with Enercon. They had done the same thing when they exited their arrangement with Vestas in 2002, even though Gamesa's original wind turbine technology came from Vestas. And now—free of that arrangement—they were ready to move without restriction across the globe in search of wind power riches.

Ironically—in many ways—Vestas opened the door for Gamesa in India when they separated from RRB in 2004 to focus exclusively on providing and installing wind turbines in India, not servicing them. The managing director of Vestas's Indian subsidiary—Ramesh Kymal—had tried, without success, to convince his European bosses to reconsider. He knew that success in India meant providing installation and maintenance solutions, increasing investment in India, and using local knowhow to acquire land. Kymal wanted to be part of a wind company that could provide a total solution by buying the land, getting permits, identifying the wind site, and setting up the wind turbines for its customers.

In 2010, Gamesa set up shop in India, and put Ramesh Kymal in charge of building Gamesa India from scratch and setting up local manufacturing operations. Gamesa India opened new factories in Chennai and Baroda in the state of Gujarat. Within a year, the company was the third largest wind turbine supplier in India—sitting behind domestic suppliers Suzlon and Wind World. Gamesa CEO—Jorge Calvet—went even further, shutting down wind R&D operations in Spain and shifting them to low-cost destinations like India.[33]

Calvet had clearly learned the lesson that Suzlon's Tanti taught the wind industry. He had also been burned by the 2008 global financial crisis which strongly impacted the European wind market, and he therefore understood the importance of diversification. Gamesa had gone from selling over one-third of its wind turbines to its home market in Spain in 2009, to full global diversification without any home market sales by 2011. India was an important part of Gamesa's diversification plans. Gamesa's "India strategy" would become a key part of their globalization playbook moving forward. When German industrial giant Siemens purchased Gamesa in 2016 to form Siemens Gamesa Renewable Energy, they wisely kept Kymal at the helm in India.

18.4 The Path Forward

Suzlon's rocket ride and India's wind power market growth paint a sunny picture. However, it's not all clear skies for India. Indeed, it's useful to examine wind power's journey in India against the backdrop of India's broader energy situation. India faces problems meeting the energy needs of its population in a way that doesn't undermine future growth, adversely impact human health, and harm the environment. A lot of ground needs to be covered to bring sustainable energy to its 1.3 billion people while creating opportunities for increased productivity and economic growth. Three-quarters of Indian energy demand is currently met by fossil fuels.[34] Coal is the largest share of the power sector, accounting for 70 percent of generation.

Against this backdrop, wind power is seen as an important part of the energy solution for India. However, India's energy situation creates headwinds for wind power that limits the rate of growth. The biggest need is for enormous amounts of low-cost, reliable power, and that need is often met with coal-fired capacity. That's why coal is the fuel of choice in India. Remember, this is a country with a modest per capita income of US$1,670 and per capita energy use of about one-tenth of US per capita energy use.[35] Simply getting energy to communities, businesses, and factories that need it is the most important objective.

Wind power faces additional headwinds in India. First, the existing T&D network is not adequate to absorb all the additional wind power and wheel it to demand centers around the country. That's why the country implemented a "Green Energy Corridor" initiative to facilitate the development of high-voltage transmission lines, substations, and associated equipment to improve connectivity between India's western and southern regional power grids. Second, although India's wind power support policies look great on paper, it's the local distribution companies that are obligated to interconnect wind farms and purchase wind power at above market costs. Many local distribution companies are already under financial duress and have become further financially strained by wind subsidies. As a result, states have been unable to comply with RPS goals. In 2017, to reduce the level of the wind power subsidy, the Indian government scaled back the accelerated depreciation incentive to 40 percent and eliminated the generation-based incentive that was added on top of the tariff in 2009.

In 2017—to make wind more competitive—India followed an international trend by conducting a wind auction to procure wind power at the lowest cost. Brazil started the trend when it began using an auction-based procurement approach as part of its electric sector reform in 2004. By 2009, they held their first wind auction and discovered that the received bids were much lower than they were previously paying for wind power. India followed Brazil's lead in February 2017 by holding a wind auction of its own. The bids for one gigawatt of new wind power came in at INR3.45/kWh (US5.2¢/kWh), a lower price than any of the feed-in tariffs offered in any states. A second auction resulted in even lower prices at INR2.64 (US4¢/kWh).

In fact, the bids were so low that nearly all the states stopped purchasing power using their feed-in tariff pricing. Wind turbine manufactures in India have begun to cry foul, warning that such low prices are unsustainable for the wind industry in India. Siemens Gamesa's Kymal called the auction prices an "aberration" and believes the auction prices will rise when they move to lower wind resource states.[36] India policy makers are now discovering that there is a delicate balance between reducing the cost of wind power policies and driving wind turbine manufacturers out of business.

Nonetheless, the Indian national government continues to set strong goals for wind power. In February 2015, India committed to installing 60 GW of wind power by 2022. As of the end of 2018, India had a total of 36 GW of installed capacity. That means the country will have to add an average of 6 GW per year through 2022 to hit the target. That's not completely out of the question, but the highest amount of annual capacity additions ever added was 5 GW in 2017.[37] However, India has already defied the odds by quickly becoming the fourth largest wind power market in the world and creating a homegrown wind power champion in Suzlon.

If India's wind power rise over the last two decades appeared sudden and dramatic, then the wind power world was in for an even greater shock. Just as India was establishing its foothold as a major player on the global wind power scene, China was preparing to completely transform the global wind power industry with its own playbook.

Notes

1 Weinraub, 1974.
2 Ibid.
3 Deb and Appleby, 1995, figure 9.
4 Mani and Mooley, 1983.
5 Hossain, 2013, 256.
6 Ibid., 256–7.
7 Vestas, 2017.
8 Hossain, 2013, 258.
9 IRENA, 2013, 87.
10 Ibid.
11 Backwell, 2018, 72.
12 Ramesh, 2017.
13 Ewing and Vikas, 2011.
14 Ibid.
15 Backwell, 2018, 72.
16 Ibid.
17 Ramesh, 2017.
18 Backwell, 2018, 68.
19 Simhan, 2018.
20 Schiessl, 2008.
21 Suzlon, 2017.
22 Wind Energy Research Group, Münster University, 1996.
23 Der Spiegel Online, 2011.
24 Barton-Sweeney et al., 2008, 6.
25 GWEC, 2017, 49.
26 IRENA, 2013, 90.
27 GWEC, 2009, 12.
28 Suzlon, 2017.
29 Barton-Sweeney et al., 2008, 4.
30 GWEC, 2017, 18.
31 Barton-Sweeney et al., 2008, 9.
32 Sanjai, 2015.
33 Mishra, 2011.

34 IEA, 2017, 704.
35 As of 2016, according to World Bank (n.d.).
36 Tendulkar, 2018.
37 GWEC, 2018.

Bibliography

Backwell, Ben. *Wind Power: The Struggle for Control of a New Global Industry.* New York: Routledge, 2018.

Barton-Sweeney, Alexandra, Joan Elias, Constance Baley, and Douglas Rae. *Suzlon: An Indian Wind Company Goes Global.* Business Case Study. New Haven, CT: Yale School of Management, 2008.

Deb, Kaushik, and Paul Appleby. *"India's Primary Energy Revolution: Past Trends and Future Prospects."* Conference Paper. New Delhi, India: National Council of Applied Economic Research, 1995.

Der Spiegel Online. 2011. "Wind Turbine Tycoon Tulsi Tanti: 'We Should Not Fear China, But Learn from Them Instead.'" *Der Spiegel*, March 8. Accessed Janurary 18, 2018. http://www.spiegel.de/international/world/wind-turbine-tycoon-tulsi-tanti-we-should-not-fear-china-but-learn-from-them-instead-a-749647.html.

Ewing, Jack, and Bajaj Vikas. 2011. "German Energy Company Hits Headwinds in India." *The New York Times*, March 23.

GWEC. *Indian Wind Energy Outlook 2009.* Brussels, Belgium: Global Wind Energy Council, 2009.

———. *Global Wind Report 2016.* Brussels, Belgium: Global Wind Energy Council, 2017.

———. *Global Wind Statistics 2017.* Brussels, Belgium: Global Wind Energy Council, 2018.

Hossain, Jami. "Wind Power Development in India." In *Wind Power for the World: International Reviews and Development*, edited by Preben Maegaard, Anna Krenz, and Wolfgang Palz, 255–65. Singapore: Pan Stanford Publishing, 2013.

IEA. *World Energy Outlook 2017.* Paris, France: International Energy Agency, 2017.

IRENA. *30 Years of Policies for Wind Energy: Lessons from 12 Wind Energy Markets.* Abu Dhabi, United Arab Emirates: International Renewable Energy Agency, 2013.

Mani, A., and D. A. Mooley. *Wind Energy Data for India.* New Dehli, India: Allied Publishers, 1983.

Mishra, Ashish K. 2011. "Second Wind for Gamesa in India." *Forbes India*, September 23.

Ramesh, M. 2017. "German Wind-Turbine Major Enercon Plans Second Coming into India." *The Hindu BusinessLine*, May 28.

reNews. 2018. "Vestas Serves Seconds in India." *reNews*, February 1.

Sanjai, P. R. 2015. "Why Did Suzlon Sell German Subsidiary Senvion." *LiveMint.*

Schiessl, Michaela. 2008. "The Rise of Indian Wind Power: Tulsi Tanti's Success Story." *Der Spiegel*, June 8. Accessed Janurary 19, 2018. http://www.spiegel.de/international/business/tulsi-tanti-s-success-story-the-rise-of-indian-wind-power-a-559370.html.

Simhan, Raja. 2018. "Vestas' New Design Centre in Chennai to Scope Out Global Wind Mill Sites." *The Hindu BusinessLine*, April 19. Accessed May 24, 2018. https://www.thehindubusinessline.com/todays-paper/tp-others/tp-variety/article23609505.ece.

Suzlon. 2010. *Annual Report 2009–10.* Annual Report. Pune, India: Suzlon.

Suzlon. 2017. *History.* Accessed Janurary 19, 2018. http://www.suzlon.com/about/history.

Tendulkar, Suuhas. 2018. "The INR 2.64/kWh Tariff is an Aberration." *WindPower Monthly.*

Vestas. 2017. *History.* Accessed November 17, 2017. https://www.vestas.com/en/about/profile#!history.

Weinraub, Bernard. 1974. "India, Slow to Grasp Oil Crisis, Now Fears Severe Economic Loss." *The New York Times*, Janurary 20.

Wind Energy Research Group, Münster University. 1996. *News.* Accessed Janurary 19, 2018. http://www.iwr.de/wind/neu/new_96_2.html.

World Bank. n.d. *National Accounts Data and OECD National Accounts Data Files.* Washington, DC. http://data.worldbank.org.

19

China's Wind Power Surge

> *Wind is our wealth given by nature. We should cherish it and make good use of it.*
>
> —Dr. He Dexin, Chairman of the Chinese Wind Energy Association (2014)

19.1 The Birth of Goldwind

In April 1986, three Vestas 55-kW wind turbines rolled off the truck in the city of Rongcheng in Shandong Province. Vestas was thrilled with the sale. The company was coping a crisis due to the end of the California Wind Rush and had just been jilted by American wind power developer Zond, who failed to pay for 1,200 turbines that Vestas had already delivered in December 1985.[1] Vestas hoped China's purchase represented the opening of a new market for their wind turbines. The Vestas turbines installed at the Shandong Rongchang Wind Farm marked the first wind farm put into operation in China. Later the same year, the Belgian government provided the Chinese with a grant to purchase four 200-kW Windmasters from the Netherlands.[2]

In October 1989, the Danish International Development Agency (DANIDA) provided grants to support the purchase and installation of thirteen Bonus 150-kW wind turbines in the city of Dabancheng in China's Xinjiang Autonomous Region. Dabancheng is a natural wind tunnel in China's Xinjiang province making it one of the best sites for wind power in the world. The wind runs through the narrow corridor between the Junggar basin and the Taklamakan desert. This is where Marco Polo wrote of hearing the voice of a genie calling from the whirlwinds.[3] Chinese entrepreneur and businessman—Wu Gang—grew up in the area. He remembers braving the blistering cold as part of a team that did the wind measurements to help get the Dabancheng site ready. This was young Mr. Wu's introduction to wind power. But was there a future in China for this new technology?

The Wind Power Story: A Century of Innovation that Reshaped the Global Energy Landscape, First Edition. Brandon N. Owens.
© 2019 by The Institute of Electrical and Electronics Engineers, Inc.
Published 2019 by John Wiley & Sons, Inc.

These installations were part of an effort to experiment with wind power technologies in China. This effort was a consequence of Deng Xiaoping's December 1978 announcement that declared China's intention to open the door to foreign businesses that wanted to be set up in China. The "Open Door" policy encouraged foreign trade and investment. After the adoption of the policy in 1978, China's economic growth accelerated and the Chinese energy sector underwent remarkable expansion and growth.[4] China's wind power experiments in Shandong and Xinjiang were both an attempt to invite foreign wind power businesses into China, and to find new ways to help meet the country's growing energy demand. If the Chinese economy continued to develop as expected, then the Chinese government was going to have to figure out new ways to meet rising electricity demand.

The country had been trying to create its own wind power turbines starting in 1980. Starting at a small scale, Gu Weidong began research into remote wind power in 1980.[5] The National State Committee of Science and Technology then funded R&D that focused on the development of a 100-kW wind turbine in the 1980s. By the early 1990s, they had produced 200- and 250-kW versions. However, like many of the government-led R&D efforts in Europe and America, these research projects didn't lead to commercial wind turbines. Instead, China continued to import commercial wind turbines from Europe to support its wind power ambitions throughout the 1990s.

By 1996, China had imported and installed 600 MW of European wind turbines. China's effort to develop wind power using imported European wind turbines was buttressed by the German government's "Gold Plan," which it launched in 1995 to support wind power development efforts in emerging economies. The Gold Plan provided the Chinese government with grants to cover two-thirds of the costs of installing wind turbines.

Instead of continuing to conduct its own research to develop homegrown wind turbines, the Chinese government shifted gears and focused on supporting wind power development in other ways. In 1994, Chinese policy makers decided to require electric utilities to facilitate the interconnection of wind turbines at the nearest grid point and compelled them to purchase the output from small wind farms. However, these national directives were not fully embraced by Chinese utilities and wind power grew only slowly. In 1995, the State Development Planning Commission, General Office of the State Science and Technology Commission, and the State Economic and Trade Commission made a "Notification on Launching New Energy and Renewable Development Outline" that included specific wind power targets for the country. This was the first time that wind power targets were explicitly considered.[6] The stated target was one gigawatt by 2001.[7] This led to the purchase and installation 500 MW of wind capacity. Danish manufacturer Nordex's 1.3-MW S62 wind turbine was the turbine of choice.[8]

Wu Gang's first experience with the Dabancheng Wind Farm in 1989 whet his appetite for wind energy. He watched as the Chinese government progressively increased its support for wind power throughout the 1990s. He earned an engineering degree from the Dalian Institute of Technology and then joined Xinjiang Wind Energy Co. in December 1987. He rose to deputy general manager by 1995. By 1998, he concluded that there was a future for wind power in China and he founded China's Xinjiang Goldwind Science and Technology to manufacture wind turbines and develop wind farms in China. Mr. Wu believed that to be successful in China, he'd have to offer a fully integrated service—from turbine manufacturing all the way to wind farm operations and servicing. Wind power wasn't going to grow in China unless customers could get the entire package. From the outset, Goldwind presented its capabilities as a "one stop" shop.

Mr. Wu started from scratch. Like Tulsi Tanti—the founder of India's Suzlon in 1995—Mr. Wu would need to strike a deal with a European wind turbine manufacturer to acquire wind turbine knowhow and technology. After searching for the right solution, Wu Gang eventually concluded that Germany's direct drive wind turbine technology was the right solution. He figured out what Germany manufacturers Enercon and Vensys, and Dutch manufacturer Lagerwey already knew—the direct drive approach was an innovative way to increase wind turbine reliability, and he felt that high reliability would be the key to wind power's success in China.

Vensys was founded by Professor Friedrich Klinger in 1997 as a spin off from the Forschungsgruppe Windenergie (FGW) at the University of Applied Sciences in Saarbrücken, Germany. In 1997, FGW developed the direct drive prototype Genesys 600. Klinger founded Vensys in 2000 to commercialize Genesys 600. By 2003, Vensys had developed and installed a 1.2-MW direct drive prototype. Wu Gang approached Klinger in 2003 to strike a licensing deal. He succeeded, and Goldwind became Vensys's first partner. Goldwind was granted a license to produce and sell Vensys wind turbines. By 2007, Goldwind was busy erecting Vensys 1.5-MW wind turbines in China. In 2008, Goldwind acquired Vensys outright.[9]

Goldwind's acquisition of Vensys eliminated the possibility of any future IP conflict like that experienced by Enercon India. Goldwind's implementation of direct drive technology was different from Enercon's because it included a permanent magnet synchronous generator (PMSG). In a PMSG, the excitation field is provided by a permanent magnet instead of a coil. The magnets for excitation consume no extra electrical power, and the absence of a mechanical commutator and brushes or slip rings means low mechanical friction losses. Basically, this means that the generator is more efficient, easier to maintain, and more compact. In a PMSG, the number of magnetic poles and the rotational speed of the generator determine the frequency of the generator output, but the output must still pass through a power converter before it is exported to the grid. During this period, as wind turbines became larger and more advanced, wind turbine manufacturers began looking to PMSG designs to enhance reliability and serviceability, reduce weight, and comply with grid codes.[10] Goldwind led the way with its 1.5-MW permanent magnet direct drive (PMDD) wind turbine.

However, there are several drawbacks to PMDD designs that Goldwind had to contend with. First, manufacturing PMDD wind turbines is challenging because the tiny air gap of only a few millimeters between the rotor and the stator requires straight tolerances and maintaining such standards when matching large components is a serious production challenge. Also, PMSG require expensive rare earth metal magnets. The volatility of rare earth metal prices has discouraged several international wind turbine manufacturers from adopting PMDD designs, and has encouraged others to develop rare earth metal substitutes. Another interesting twist is that as wind turbine technologies have matured, gearboxes—which had historically experienced high failure rates—have improved and now only account for 4–8 percent of all wind turbine failures. Meanwhile, increasingly complex power electronics are accounting for upwards of 30 percent of all wind turbine failures.[11] From a wind turbine reliability perspective, the idea of swapping out a gearbox for a larger power conversion system isn't as attractive as it was when direct drive wind turbine designs were first introduced in the 1990s.

Nonetheless—after identifying the PMDD technology as the core of its wind turbine design—Goldwind has successfully developed a full line of PMDD wind turbines that have demonstrated reliability in the field. As the wind industry moves forward, larger offshore wind turbines PMDD designs are likely to become increasingly attractive. Mr. Wu had found a commercially successful and differentiated technology solution. With Goldwind at the forefront, China's installed wind power capacity jumped from 568 MW in 2003 to 1.25 GW by the end of 2005.[12]

Goldwind's wind power development efforts were aided by the Chinese government. Under its Tenth Five-Year Plan (2001–2005), the Chinese government introduced the concept of a "Mandatory Market Share" of renewable energy in the national electricity supply; and, as part of a broader electricity market reform, the government also introduced market-based mechanisms such as concession tendering for wind projects in 2002.[13] This program promoted domestic projects through competitive bidding, requiring wind turbines to be manufactured with 70 percent domestically produced content.

The concession projects also helped to determine the right tariff levels for wind power in China. During this period, the tariffs were either determined under these concessions or through government-approved tariffs. The "concession policy" accelerated wind power development, but a standard wind tariff was not yet in place across China. The wind power market continued to grow, reaching 2.6 GW by the end of 2006.[14]

Despite these programs, Chinese wind power installations still lagged behind Europe, the United States, and even its Asian neighbor India, which boasted 6.27 GW of installed capacity by the end of 2006.[15] The challenge was that regional electric utilities in China continued to favor conventional generation sources in spite of the Central Government's desire to advance wind power. There was no national legal framework that required grid companies to favor wind power over other sources of generation. The national Renewable Energy Law—which was passed in 2005 and enacted in 2006—did just that. Although it did not include tariffs or specific targets for wind power, it provided the legal framework to advance renewable power and it provided a framework for follow-up regulations.

In 2007, the Medium and Long-term Development Plan for Renewable Energy laid out the Chinese government's commitment to wind power and put forward national targets and policy measures for implementation. It included a mandatory market share of 1 percent of nonhydro renewable energy in the country's total electricity mix by 2010 and 3 percent by 2020. To further encourage the wind industry, the government issued Notice 1204, which required all wind turbines installed in China to have 70 percent of their content manufactured in China.[16] Trade lawyers would argue that China's local content requirement is a violation of the rules of the World Trade Organization (WTO), which China joined in 2001. However, instead of fighting the requirements, international wind turbine manufacturers flocked to China to open local production facilities. The US government would later press China about Notice 1204 resulting in China's revocation of the notice in 2009. By that time, however, domestic and international wind turbines manufacturers were already producing over 70 percent of their wind turbine components in China for those turbines that were intended to be installed in China.

These policies opened the floodgates for Chinese wind power development and domestic wind turbine manufacturing. By the end of 2007, there were forty domestic wind turbine manufacturers. By the end of 2008, the number had risen to seventy.[17] International wind turbine manufacturers flooded the Chinese market too. Spain's Gamesa, Denmark's Nordex and Vestas, America's GE Wind Energy, and India's Suzlon, all rushed to form joint ventures with Chinese manufacturers to produce local content.[18] China's wind power capacity surged to 12 GW by 2008.[19]

19.2 The Rise of Sinovel

The creation of state-owned wind turbine manufacturer—Sinovel—was the result of government efforts to reinvigorate heavy industry in northern China, which had fallen on tough times by the turn of the twenty-first century. Sinovel had started as a subsidiary of the state-owned industrial group Dalian Huarui

Heavy Industry Group (DHHI). DHHI itself began as the Dalian Iron Works in 1914. By the twenty-first century, DHHI had grown into a sprawling industrial empire that included port machinery, coke oven machinery, and metallurgic machinery.[20] These markets were stagnant in 2004 when National Development and Reform Commission (NDRC) vice-chairman Zhang Guobao reached out to DHHI to find ways to give it a boost. Han Junliang was tapped to become the president and chairman of a new wind power company with headquarters in Beijing that would be formed out of a DHHI electrical equipment company in 2004.

Sinovel secured a license to manufacture 1.5-MW wind turbines from German wind manufacturer Fuhrländer of Liebenscheid, Germany. Founded in 1960, Fuhrländer itself was a family-owned steel forging business that was transformed into a wind turbine manufacturer in the 1980s. They had built a good reputation as small but solid wind turbine manufacturer and service organization. Fuhrländer would later run into trouble with "postponements of projects by customers" that had led to "unpredictable delays in project payments" and would be forced to file for bankruptcy in 2012.[21]

Mr. Han leveraged his connections to secure investors and ramp up Sinovel's production capability. For example, a subsidiary of New Horizon Capital, co-founded by Wen Yunsong—the son of Chinese premier Wen Jiabao—bought a 12 percent stake in the company. By 2006, the factory in Dalian could produce 1,500 1.5-MW wind turbines per year. In September of 2006, the first Sinovel 1.5-MW wind turbine was put into operation. In April 2007, thirteen Sinovel wind turbines were preselected for a trial run at China's first localized megawatt-scale wind farm, the Huaneng Weihai Phase I Project. Sinovel relied upon Chinese government tenders to fill its order book from here on out. The Chinese government didn't explicitly state that it wouldn't award contracts to foreign wind turbines, but no foreign firms won significant tenders from these orders.[22]

In 2008, Sinovel was awarded a 1.8-GW order for the first phase of the Gansu Wind Farm. Gansu was part of the government's "Wind Base" program. The National Energy Administration (NEA) selected seven locations throughout the country with high wind resources. Their plan was to install 138 GW of wind power at these seven locations by 2020. They would attempt to achieve this by conducting competitive bids, selecting domestic wind turbine manufacturers, providing government-backed financing, and then requiring regional electric utilities to interconnect the wind turbines and purchase the electricity.

China's Wind Base program heavily favored domestic project developers and wind turbine manufacturers. All of the new wind power capacity that was built as a result of the Wind Base program would be financed by five national power companies—China Power Investment Group, Guodian, Datant, Huaneng, and Huadian. The wind turbines were supplied primarily by the "Big 5" wind turbine manufacturers in China at the time: Sinovel, Goldwind, XEMC, Shanghai Electric

Group, and Ming Yang. The Gansu Wind Farm reported the fasted growth, with over 5 GW installed by 2010. Altogether, nearly 23 GW of new wind power capacity was installed at the Wind Bases by 2010.[23]

China's wind power industry was further supported by the introduction of a feed-in tariff in 2009. The twenty-year tariffs ranged from RMB0.51/kWh (8 ¢/kWh) to RMB0.61 (9 ¢/kWh), depending upon the wind farm's underlying wind resources. The feed-in tariff was the final piece of the policy puzzle, providing investors with long-term financial stability and making wind farms financially bankable. Electric utilities were required to purchase all the electricity generated from wind farms at the feed-in tariff rate but were able to apply for subsidy reimbursement from a new Renewable Energy Fund. The Fund itself was financed from a nationwide surcharge on the price of electricity that started at RMB.002/kWh (0.03 ¢/kWh) and was later raised to RMB.004/kWh (0.06 ¢/kWh).[24]

Government policies supporting both on- and off-shore wind power led China's domestic wind turbine manufacturers to rapidly increase the size of their wind turbines. There was simply no way to achieve the targets defined in the Wind Base program or the offshore goals without installing very large wind turbines. Sinovel—which had just started producing its 1.5-MW wind turbine—almost immediately began to design and development work for a 3-MW wind turbine at its facility in Yancheng.

By 2010, Sinovel had become the world's second largest wind turbine manufacturer, behind only Vestas. With nearly 4.4 GW of installed capacity and 11.1 percent of the global market, it was poised to become number one. Sinovel's initial public offering that year was the most expensive ever on the Shanghai stock exchange, the shares were listed at RMB90 (US$13) each. Further, four of the top ten wind turbine manufacturers in the world were Chinese in 2010—Sinovel, Goldwind, Dongfang Electric, and United Power.[25]

Just ten years earlier, the country had just 346 MW of installed wind power capacity, and its only wind turbine manufacturer—Goldwind—was still sitting on the launch pad without any wind turbine technology. In a decade, Chinese wind power growth exploded. Installed capacity ballooned to nearly 45 GW. Not only was that the highest total in the world, it was 5 GW greater than the installed capacity in the United States, a country that had been developing wind turbine technology on and off for over 100 years. China's rapid rise shifted the gravitational center of the wind power world in less time than it often takes to build a transmission line in the West.

In the fourth place globally in 2010, Goldwind adopted a more measured pace compared to Sinovel. Goldwind would eventually take the top spot in 2015, but its path to the top would be marked by more steady progress and measured persistence compared to its domestic sibling Sinovel. While Sinovel remained focused on meeting domestic demands with the help of government policies that favored Chinese manufacturers, Goldwind began to look for opportunities internationally. Wu Gang understood that if the Chinese market ever became oversaturated, then Goldwind would need other outlets for his wind turbines. Goldwind was the first Chinese wind turbine manufacturer to sell wind turbines aboard. Goldwind exported its 750-kW GW50 wind turbines to Cuba in 2008. In 2009, Goldwind sold three turbines in the United States.[26] By 2010, Goldwind had established a wholly owned subsidiary in Chicago to make inroads in the American market. Goldwind's American subsidiary, along with its production base in Germany, and Goldwind's acquisition of Vensys, gave the company a foothold in the three largest wind turbine markets in the world.

As it began its global expansion, Goldwind was able to leverage its low-cost supply chain to offer wind turbines at a lower price than its rivals in America, Europe, and India. By 2010, because of the growth of domestic wind turbine manufacturers and the influx of European and American wind turbine component suppliers, 70–95 percent of wind turbine assembly could be met with Chinese components.[27] Like India's Suzlon, Chinese wind turbine manufacturers were able to leverage their low-cost supply chains to increase their international competitiveness. This occurred at a time when international wind turbine prices had fallen below US$1,000/kW, a price point that was squeezing out much of the profits out for American and European manufacturers. However, China's advantage began to erode somewhat as international rivals themselves began to source many of their wind turbine components from China-based component manufacturers. In addition, Chinese wind turbine manufacturers like Goldwind, Sinovel, and Dongfang remained dependent upon American and European control systems and power converters from manufacturers like AMSC Windtec, Mita-Teknik, and ABB.[28]

In addition to the relationships with Führlander, Vensys, and AMSC Windtec, Chinese manufacturers established license linkages with many European wind turbine designers such as Aerodyn, DeWind, Norwind, and Repower. These relationships led to the effective co-design of the next generation of wind turbines from Chinese manufacturers that went into serial production starting in 2010.[29] In many cases, Chinese, European, American, and Indian wind turbine manufacturers more closely resembled collaborators—than competitors—all driven by the same goal of expanding the global wind power market.

Between 2010 and 2017, Goldwind gradually expanded its footprint outside of China. The company created seven subsidiaries: Goldwind Europe, Goldwind USA, Goldwind Australia, Goldwind MENA, Goldwind South America, Goldwind Africa, and Goldwind Asia. By the end of 2017, Goldwind's installed wind power capacity outside of China reached 2 GW.[30] Not bad, but Goldwind's installed capacity in China alone itself was already 36 GW by that time, so the company's international installations still remained a small part of its overall operations.

Mr. Wu acknowledged the gap between Goldwind's efforts on the international stage and the role played by established players such as Vestas when he said "Take Vestas. Their products are sold in more than thirty countries. Ours are sold in only seventeen countries. This is a gap. As a Chinese company, we lag far behind our foreign competitors in internationalism." True, but given China's closing speed and the quality of Goldwind's technology, it wouldn't be long before Goldwind was able to meet Vestas eye-to-eye on the international stage.

By 2010, China's wind power surge had resulted in a record-breaking expansion of installed wind power capacity. But with massive amounts of wind power being added to the Chinese electric power grid, there had to be some limits to growth. And with the global outbreak of the Great Recession in 2008, storm clouds were looming on the horizon. As it turned out, for China's wind power industry, change was just around the corner.

19.3 The Year of Adjustment

The surge in installed wind power capacity that occurred in the wake of the 2005 Renewable Energy Law and its follow-up regulations led to a globally unprecedented increase in installed wind power capacity. But it also started to create real problems for the Chinese electricity grid. The year 2010 represented the apex of the first phase of Chinese wind power growth. Wind power capacity additions peaked at 18.9 GW in 2010 and didn't get back to that level until 2014. In the interim, Chinese wind turbine manufacturers, the grid operator, electric utilities, government regulators and policy makers grappled with the impacts of the wind power surge.

Several troubling issues had become apparent by 2010 and they started to have an impact on Chinese wind power in 2011. The first problem was that the windiest regions in China were far away from China's population centers. While China's best wind sites were in the north, northeastern, and northwestern areas, most of its population was situated along the eastern and southern coastal regions. As a result, most of the wind power capacity had been installed in the "Three Northern Areas," which includes the Inner Mongolia, Gansu, and Xinjiang. This is far away from the population and electricity demand centers in the eastern and southern provinces such as Jiangsu and Guangdong. In fact, the Chinese government's Wind Bases program was focused primarily on adding gigawatts of wind power capacity to Gansu, Inner Mongolia, and Xinjiang. This is problematic because the Three North Area provinces had weak grids and limited transmission capacity for moving power to the demand centers in the east and south.

The transmission systems in the Three Northern Areas were unable to absorb the amount of wind power that was being added to the system. As a result, many wind farms were either left disconnected from the grid, or they were connected but the wind power output was curtailed instead of being fed into the system. In effect—because of transmission limitations—a significant fraction of the installed wind power capacity was under utilized. In East Inner Mongolia in 2011, 23 percent of all wind power output was curtailed. Nationally, more than 10 billion kWh of wind power was not generated because of curtailments in 2011. In 2012, this figure doubled to 20 billion kWh and the national curtailment rate reached 17 percent.

In response to these challenges, the NEA instituted a policy that required all wind power projects to undergo approval before starting construction. The goal was to coordinate wind power development with available power transmission capacity and grid interconnection capability. NEA's policy slowed down wind power development in China and led to a bottleneck in the development process as the agency worked to ensure that the grid could absorb any new wind power that entered the system. As a result, national wind power installations dropped by one-third between 2010 and 2012.

To tackle the bottlenecks in the transmission system, the NEA, State Grid, and Southern Grid began planning the construction of twelve new long-distance transmission lines. Several of the lines were designed to connect the Inner Mongolia and Hebei provinces, where vast wind development was taking place with load centers in densely populated areas. By 2014, several high voltage lines were under construction, such as the Xinjiang Hami-Zhengzhou 800-kV line. However, it took some time to build transmission capacity, even in China. The challenge of wind power curtailment in China is unlikely to disappear anytime soon. Even as recently as 2015, 50 billion kWh of wind power was curtailed in China, and provinces such as Gansu and Xinjiang still had curtailment rates as high as 40 percent.

Problems started to appear on the financial side of things as well. Capital for new wind power projects became increasingly difficult to come by. Wind power projects had previously been sheltered from capital constraints because the primary project developers were China's state-owned utilities with access to government-backed money for infrastructure projects. However—in 2012— this situation changed. Utilities were suffering losses from their coal-fired power plants and were under pressure to improve their balance sheets. In addition, China implemented tighter monetary policy and some banks lowered the credit lines of wind power and it therefore became more difficult for wind farm developers to obtain financial support through bank loans.[31]

Wind power in China became a victim of its own success. The feed-in tariffs— which were introduced in 2009—were funded through the Renewable Energy Fund. The government raised funds for the Renewable Energy Fund through a surcharge paid by all electricity consumers. By 2011, wind power had grown so quickly that the government began delaying payments. Only half of the wind power projects received feed-in tariff premiums in 2012.

China's wind power industry entered a period of adjustment starting in 2011. The challenges were clear, but the solutions were not. Exactly how would the country address problems with wind turbine quality, curtailment of wind energy due to an overabundance of wind power in the wrong locations, and the growing financial pressures brought on by the country's wind power surge? And could China recover and again lead the world in annual wind power additions? The next few years would be pivotal for the future of wind energy in China.

To recover from the slowdown in China's wind power surge, the first thing that needed to be done was increase the renewable energy surcharge to pay for the quantities of wind power that were already on the grid. So—in 2013—the government increased the surcharge to RMB.015/kWh (0.25 ¢/kWh) to correct this problem.[32] Despite continuous improvements in wind power technology, wind power was still more expensive than conventional generation options such as coal-fired electricity, particularly considering high wind power curtailment levels. As such, all of China's electricity consumers were going to have to pay to keep this wind surge going.

China's wind power policies continued to evolve. Later, NEA began promoting the idea of a "Green Certificate" program in place of the feed-in tariffs. The idea was to create a secondary market for the sale and purchase of green certificates that represent zero-emissions electricity generation. Every wind farm owner would be able to sell a green certificate to a willing buyer for each kilowatt-hour of wind power generated. The revenue stream from green certificate sales would create additional project revenue that would offset the loss of the feed-in tariff. The challenge was that there was not strong demand for green certificates in China without an underlying national or provincial-level RPS that required utilities to purchase a specified amount of wind power.

The headwinds faced by the Chinese wind power industry after 2010—and Goldwind's slower than expected globalization efforts—were evidence of an additional underlying challenge for China's wind turbine manufacturers: quality. China had witnessed—by far—the highest number of trip incidents, faults, and accidents caused by failures and malfunctions of wind turbines, including massive power outages and even some fatalities. And the problems grew worse as the country added more wind power to the grid. Furthermore, the quality problems were not limited to one or two wind turbine components, rather, equipment faults were spread across all parts including the pitch system, frequency conversion system, electrical system, control system, gearbox, generator, and yaw systems. Yes, China can add large amounts of wind power to their transmission system, but whether the wind turbines functioned properly and delivered electricity to consumers was another question altogether.

Relative to their international counterparts, Chinese wind turbines were delivering low quantities of electricity. In 2007, the average capacity factor of China's wind power fleet was 11 percent. There was only marginal improvement in 2011 and performance increased to 13 percent. These numbers compared well to the performance of the wind turbine installed in California in the 1980s. For example—in 1986—all Danish wind turbines installed back in California achieved an average capacity factor of 14 percent. The notoriously unreliable US machines were only operating at an average capacity factor of 9 percent.[33] The Chinese had vaulted to the top of the global wind power market by installing wind turbines that had reliability that was on par with thirty-year-old technology.[34]

The biggest issue was that most of the Chinese wind turbines installed through 2011 could not meet the requirements of safe grid integration. One of the problems was that most Chinese wind turbines lacked low-voltage ride through (LVRT) capability. LVRT is the capability of electric generators to stay connected in short periods of lower electric network voltage. It is needed to prevent a short circuit from causing a widespread loss of generation. Without LVRT, wind turbines were unable to stay connected to the grid during low-voltage events. Unplugging from the gird during system faults created a cascading effect and turned minor voltage fluctuations into blackouts.

A second issue was that the reactive power compensation devices in some wind farms could not satisfy grid safety requirements. Most wind farm turbines either were not capable of regulating reactive power dynamically or their capacity and regulation speed were incapable of satisfying grid operation requirements. A third issue was construction faults in the design of some wind farms. For example, some wind farms were designed by referring to the design standard of the conventional distribution network, which resulted in using improper grounding methods. As a result, operators were unable to cut off the selected line when there was single phase fault, which led to accidents and grid disconnection. In other cases, major construction quality deficiencies existed. This occurred because quality, testing, control, and acceptance inspections were not carried out with any degree of rigor or standardization.

As summarized in 2013 by Zeng Ming, Xue Song, Li Ling-yun, Cheng Huan, and Zhang Ge of the School of Economics and Management at North China Electric Power University: "At present, China's wind power system is not sound, and technical standards lag far behind those in other nations. National technical standards have not been set, and technical standards for the centralized control of wind farms—along with standards for system design, integration, and monitoring—are still on the way."[35]

One solution was for the Chinese government to impose interconnection standards that required wind farm owners to adhere to technical guidelines to interconnect their wind turbines. This would force wind turbine manufacturers to produce products that can meet those standards. In August 2011, NEA issued the industry standard on the "Wind Power Interconnection Technical Regulations," (NB/T 31003-2011) which specified the design of technical requirements for large-scale wind power plant connection. This was followed by the "Technical Rule for Connecting Wind Farm to Power System" (GB/T 19963-2011), in December 2011 that established national compulsory interconnection requirements. In addition, China's wind power certification organizations began technical exchanges and cooperation with international wind power research certification institutions such as DNV GL, Garrad Hassan, NREL, Risø, TÜV NORD, and DEWI-OCC.[36]

That was a good start, but standards and certification would do little to address the curtailment of existing wind farms. Too much wind was built in areas with too little demand or transmission capability to move the power to load centers. It would simply take some time to build out the transmission network. In addition, it would take time for China wind players to adapt to the new standards, and the process of improving wind turbine quality was slow. The entire supply chain needed to be improved and much of the technology and know-how simply wasn't available within China at the time. Europe and America had been working on this problem for thirty years—the struggle to make wind turbines reliable and fully grid-friendly was a long, protracted effort that had taken down its fair share of wind turbine manufacturers.

19.4 Sinovel's Stall

One way to accelerate the effort was for Chinese wind turbine manufacturers to partner with American and European firms on the control systems and power electronics required to make wind-generated electricity suitable for the Chinese grid. This is the genesis of Sinovel's partnership with AMSC. Founded in 1987, Boston-based American Superconductor (AMSC) specialized in electronic controls and systems ranging from converter and control cabinets to advanced software systems. AMSC deepened its wind power commitment in 2007 with the purchase of Windtec.[37] Windtec was founded by Gerald Hehenberger—the son of a farmer from Upper Austria—in 1995. Windtec was a wind turbine design and engineering firm that specialized in wind power control systems. Hehenberger held twenty-seven patents worldwide and had built a small, but well-respected business.[38] AMSC's purchase was an ideal fit. Windtec's control system was already using AMSC's flagship product—the PowerModule—to facilitate connection to the grid. Together with Windtec, AMSC was able to offer turnkey wind power control systems and power electronics. AMSC had the exact suite of software needed by Chinese wind turbine manufacturers to make their turbines grid-friendly.

Sinovel had been seeking a solution for the control and interconnection of its wind turbines for several years. In 2007, they decided to partner with AMSC on the design of electrical systems for their new 1.5-MW wind turbine. Sinovel announced a US$70 million order for wind turbine design specifications and electrical systems from AMSC. AMSC would design Sinovel's new 1.5-MW wind turbine, hand the specs over to Sinovel, and then allow Sinovel to license the AMSC software.[39]

The arrangement initially worked well. In 2008 and 2010, AMSC and Sinovel signed two component supply contracts to the value of nearly US$900 million.[40] Sinovel continued to grow, capturing over a quarter of the massive Chinese market for new installations. After winning 1.8 GW of orders in a government tender in 2008, the company followed up with 2 GW in 2009 from the Inner Mongolia and Hebei concessions, and another 1.35 GW in 2010. By the end of 2010, the Sinovel had 14 GW of orders on hand. Company founder and President Mr. Han projected an annual growth rate of 30 percent over the next five years.[41]

As discussed, starting in 2010, the Chinese government made efforts to reign in wind power development to address curtailment and more effectively manage the way new wind farms were interconnected. This meant that Sinovel could no longer expect huge government orders like it had in the past. Sinovel's inventory began to pile up and its cash position worsened. The company posted a US$155 million cash flow deficit in 2010. At that time, Sinovel refused to accept and pay for contracted shipments from AMSC.

Sinovel's January 2011 IPO on the Shanghai Stock Exchange didn't go well. Sinovel stocks sank below the offer price because of fears of a domestic wind power slow down. The company was now in a tailspin that it wouldn't begin to pull out of until 2014. Its eventual turnaround was the result of cost control, the transfer of part of its debts to two other companies, collecting debts from some clients, and debt forgiveness from some of its suppliers. But AMSC was not one of the suppliers that was offered debt forgiveness. By the end of 2014, Sinovel had 2.49-GW in contracted orders awaiting delivery. The company returned to profitability, but its outlook is much more modest. After peaking with 4.4 GW of new capacity in 2010, by 2015, Sinovel's new wind turbine installations had dropped tenfold to 435 MW.[42]

Sinovel's decision to refuse to accept and pay for AMSC equipment left lasting scares for both companies. AMSC brought a suit against Sinovel in both Chinese and American courts. The China court dismissed the case for lack of evidence. An appeal by AMSC in a higher Chinese legal venue was still pending in 2018. However, the American suit was escalated by the Obama Administration in 2013. In 2018, the American court found that Sinovel was guilty of stealing AMSC technology. In the court case, AMSC alleged that Sinovel's offenses went far beyond simply refusing to pay for contracted shipments. Allegedly, Sinovel convinced an AMSC engineer to secretly download the source code for AMSC's LVRT software application and provide a copy to Sinovel in March of 2011. With the software in hand—according to AMSC—Sinovel was then free to install the software on hundreds of new wind turbines and retrofit existing ones.

The case led directly to the cancellation of Sinovel's 1 GW contract with Mainstream Renewable Power in Ireland in 2011. Brazilian wind power developer Desenvix demanded the right to inspect twenty-three wind turbines that it had ordered because they were afraid that they contained stolen AMSC software.[43] Sinovel is now working with Danish control electronics provider Mita-Teknik and third-party testing agencies to certify its new wind turbines to regain international credibility.

AMSC never fully recovered because so much of its revenue was tied to its arrangements with Sinovel. US attorney Scott Blader noted that "Sinovel's illegal actions caused devastating harm to AMSC."[44] From a high stock price of US$430 in late 2010, AMSC shares plummeted to US$35 within a year. The combined value of AMSC's assets fell by more than US$1 billion and almost 700 jobs were lost. As of 2017, the company was still running an operating loss.[45]

Much has been made about China's wind power curtailments and its efforts to expand the transmission system to move wind power from its high wind resource areas to load centers. However, there's more to the solving this challenge than building transmission lines. Fact is, the way that Chinese electric power networks operate needs to be revised. This requires a bit of explaining.

Recall, at the turn of the twentieth century, the growing AC power generation and transmission system facilitated the adoption of the central station model of electricity production and distribution, which favored increasingly large power plants.* Increasingly large steam turbines were the favored technology for the central station model because they were efficient, they could be built at the megawatt-scale, and they could be turned on and off at will—a feature known as "dispatch" in electric industry parlance—to ensure optimal operation of the power system. Wind turbines possessed none of these characteristics and were therefore of little use to electric utilities at the time. The central station model may not have been the only way to produce and deliver electricity to meet the needs of homes and businesses, but it was the dominant approach, and wind power had a limited role to play within this system.

The way to operate a traditional electricity system in order to balance electricity demand and supply is straightforward. Power plants are dispatched in order based on their marginal generation cost until electricity demand is fully

---

* The central station model of power generation and delivery is basically a hub and spoke model where a large power plant is centrally located and transmission lines are built to distribute power to customers near and far. This is opposed to a distributed power model where smaller, distributed power plants are located at or near customers.

satisfied. Marginal generation cost is just the cost of producing each kilowatt-hour of electricity. For some large plants—like coal—the marginal cost is very low because coal is relatively inexpensive. These generators are dispatched first and then the rest of the higher marginal cost generators are dispatched until demand is fully met. The higher marginal cost generators such as gas turbines are generally more flexible than the larger, lower-cost generators that run most of the time.

However, wind power is different. First, there is very little marginal generation cost. After the turbines are paid for and properly serviced, there's very little variable operations costs. When the wind blows, the electricity is basically free. Second, wind turbines can't be dispatched at will. To accommodate wind power, the rest of the generators on the system must be dispatched around the incoming wind-generated electricity. This means that the entire system must have flexible generators and large balancing areas. Furthermore, it helps if the wind farms are geographically disbursed and there is a market mechanism that allows variable resources like wind power to sell excess energy or purchase shortages to accommodate the natural characteristics of wind.

Today large amounts of wind power are being added to electricity grids that are still based upon large and dispatchable power plants across the globe. In North America and Europe, wind power is being added to legacy central station electricity networks. European and American electricity networks are being adjusted to become increasingly flexible. In addition, power market structures are being adapted to better encourage system flexibility. In the emerging economies of China and India, wind power is being added to newer transmission networks that are still based on the central station model because these countries are heavily reliant upon large, central-station coal-fired power plants. Lack of flexibility in these power systems, coupled with lack of competitive electricity market where energy, capacity, and other services can be priced are barriers to higher levels of wind power penetration. The bottom line is that China's power system is going to have to evolve if it is going to continue to absorb and deliver increasing amounts of wind power.

19.5   Recalibration

China's wind power market rebounded starting in 2014 with the addition of nearly 24 GW of new wind power capacity. That represented a near doubling of capacity additions from the previous year and put the country over the 100-GW mark for installed wind power capacity. Although it still accounted for less than 3 percent of total national electricity generation, wind power was on the rise again. The Chinese government continued its efforts to attempt to balance the wind market and create a growth path that was sustainable. Part of this meant introducing new feed-in tariffs for offshore wind projects—an area that the government had previously eyed for future growth—and a reduction in the feed-in tariff for onshore projects to constrain the pace of onshore wind power project development. Onshore feed-in tariffs were set to RMB0.49/kWh (7¢/kWh) to RMB0.61/kWh (9¢/kWh).[46] The expectation was that the government would continue to lower the onshore feed-in tariff in the future, to constrain the ongoing costs of wind power development. And indeed, in 2019, NDRC announced that tariffs paid to onshore wind projects will be cut to RMB0.29 yuan (4¢/kWh) in 2020.

The Chinese government also issued a compulsory certification process requiring all wind turbines and their components to be certified before entering a government tendering process. Furthermore, the NEA introduced a performance data requirement that created an evaluation system for wind turbine quality assessment, including a reporting system for turbine faults and incidents. The State Grid and Southern Grid began installing twelve new transmission lines.

These policies coupled with an anticipated future reduction in the onshore feed-in tariff led to even stronger growth in 2015. Thirty-one gigawatts of new wind power capacity was added in 2015, increasing China's installed wind power capacity to 145 GW.[47] Now, over 3 percent of China's electricity was coming from wind power. Goldwind maintained its pole position as China's top wind turbine manufacturer, followed by Guodian United, Envision, Mingyang, and CSIC, all of which installed over two gigawatts of new capacity in 2015. For its part, Sinovel had fallen from the ranks of the top manufacturers in China, but the company began a pivot to the offshore market by installing 170 MW of new offshore capacity. However, wind curtailment remains a lingering problem in China, with an average natural curtailment rate of 15 percent, and regional rates approaching 40 percent.[48]

On the policy side the concept of "energy transition" made its way in the Thirteenth Five-Year Plan that covers the period from 2016 to 2020. Chinese government officials had become increasingly concerned about air pollution and the need to curb climate impact, and together with traditional supporters of fossil fuels, began promoting the key role that wind power can play in China's clean energy future. The government set a target of peaking greenhouse gas emissions by 2030, and increasingly counted on wind power to play a key role in helping the government to achieve its targets. The Plan set an explicit wind capacity target of 210 GW by 2020. Additional targets were set for the newly added Central Southern region to force the geographic diversification of wind farms.[49]

In addition to setting new onshore wind power targets, the Thirteenth Five-Year Plan also introduced a goal of installing 5 GW of offshore wind power by 2020.[50] The Chinese government had previously laid the groundwork for the movement to offshore wind power. In 2010, the NEA and the State Oceanic Administration jointly published "Interim Measures for the Administration of Development and Construction of Offshore Wind Power." These guidelines helped encourage offshore wind power development by setting out provisions for project approval procedures, as well as criteria for project development and construction. Tender procedures are the the preferred method of selecting the offshore projects, and foreign investors are allowed to hold a minority stake in offshore wind developments.[51]

The attractiveness of offshore wind power from the government's perspective lies in the fact that there are favorable wind resources along the eastern coast of China where the major population centers reside. In November 2011, the government announced the winners of four offshore wind projects along the coast of the Jiangsu province. The winning price ranged from RMB0.62/kWh (9¢/kWh) to RMB0.74/kWh (11¢/kWh). After a slow start, offshore wind power has begun to pick up in China. From 1.5 MW installed in 2007, offshore wind power installations jumped to just under 400 MW in the wake of the new offshore wind feed-in tariff policy. By 2015, over 1 GW of offshore capacity had been installed.[52] In 2016, new installations jump 64 percent to 592 MW, pushing China's offshore wind power installed capacity to 1.6 GW. It's still a small fraction of the onshore total, but momentum is growing, and China's wind power players are learning how to make offshore wind power projects work.

In addition to the offshore push, the Chinese government has more recently began to encourage wind turbine manufactures to seek international markets for their products. In 2015, Chinese President Xi Jinping introduced the "One Belt One Road" initiative, which refers to the Silk Road and Economic Belt and the twenty-first Century Maritime Silk Road. The Belt and Road initiative is geographically structured along the six corridors and the Maritime Silk Road. The government challenged its domestic manufacturers to build products for overseas markets. The government set up a US$40 billion Silk Road Infrastructure fund to support manufacturers entering markets. At the same time, the Asian Infrastructure Investment Bank announced that over US$160 billion would be made available for infrastructure projects.[53] These funds are likely to have a positive impact on China's wind power industry and its efforts to globalize.

In the wake of China's wind power recalibration, the Chinese wind power industry is again growing at a breakneck pace relative to other countries around the world. Wind power additions reached 20 GW in 2017. Although curtailment remains an issue, the recalibrated Chinese wind power industry appears to have adapted to a more sustainable pathway in response to some of the early challenges posed by its wind power surge. China's wind power surge looked eerily like a supercharged version of the California Wind Rush of the 1980s. There, a massive boom was followed by an equally large bust when the government pulled the plug. But times have changed. China isn't pulling the plug, rather, it is adjusting the rules of the game to better accommodate its wind power ambitions.

The Chinese are currently leveraging the lessons learned in Europe and the United States over the past three decades. So, what looks like a wind power "frenzy" from the outside is also a necessary learning process that will enable China to make wind power an important part of its electricity system and an important part of meeting its environmental goals.[54] Given China's proven ability to quickly adapt and successfully execute to win, it's unwise to bet against their wind power success. Expect China to remain a twenty-first century wind power trailblazer in the years ahead.

Notes

1 Vestas, 2017.
2 Dexin, 2014, 99.
3 Shepard and Hornby, 2016.
4 Lam, 2005.
5 Weidong, 2014, 174.
6 Dexin, 2014, 152.
7 IRENA, 2013, 49.
8 Dexin, 2014, 116.
9 Vensys, 2016.
10 Lawson, 2012.
11 Ibid.
12 GWEC, 2017, 37.
13 IRENA, 2013, 49.
14 GWEC, 2017, 37.
15 Ibid., 49.
16 IRENA, 2013, 50.
17 Ibid., 51.
18 Ibid., 50.
19 GWEC, 2017, 37.
20 DHHI, 2016.
21 Knight, 2012.
22 Patton, 2013.
23 IRENA, 2013, 51.
24 Ibid., 52.
25 Lema, Berger, and Schmitz, 2012, 9.
26 Goldwind International, 2017.
27 Lema, Berger, and Schmitz, 2012, 10.
28 Ibid., 16.
29 Ibid., 20.
30 Goldwind International, 2017.
31 Dexin, 2014, 107.

32 GWEC, 2014, 43.
33 Data derived from the summary data table of CEC (1987, 30).
34 See Gao et al. (2016) for a more detailed comparison of wind turbines in China versus the United States.
35 Ming et al., 2013.
36 Dexin, 2014, 137.
37 AMSC, 2013.
38 Visionaer, n.d.
39 WindPower Monthly, 2007.
40 WindPower Monthly, 2008.
41 Patton, 2013.
42 Richard, *Sinovel Found Guilty in AMSC Case*, 2018b.
43 Backwell, 2018, 48.
44 Richard, *Sinovel Found Guilty in AMSC Case*, 2018b.
45 Richard, *AMSC Loss Widens as Inox Deliveries Slow*, 2018a.
46 GWEC, 2014, 39.
47 GWEC, 2017, 37.
48 GWEC, 2014, 38.
49 GWEC, 2016, 38.
50 GWEC, 2017, 37.
51 IRENA, 2013, 53.
52 GWEC, 2016, 32.
53 Wikipedia, 2018.
54 Harrabin, 2016.

Bibliography

AMSC. 2013. *About Us*. Accessed February 10, 2018. http://www.amsc.com/about/index.html.

Backwell, Ben. 2018. *Wind Power: The Struggle for Control of a New Global Industry*. New York: Routledge.

CEC. *Results from the Wind Project Performance Reporting System: 1986 Annual Report*. San Francisco, CA: California Energy Commission, 1987.

Dexin, He. "Wind Energy Development in China." In *Wind Power for the World: International Reviews and Developments*, edited by Preben Maegaard, Anna Krenz, and Wolfgang Palz, 91–171. Singapore: Pan Stanford Publishing, 2014.

DHHI. 2016. *About DHHI*. Accessed February 10, 2018. http://www.dhhi.com.cn/index.php?g=Portal&m=Company&a=index&id=11.

Gao, David Wenzhong, Eduard Muljadi, Tian Tian, Mackay Miller, and Weisheng Wang. *Comparison of Standards and Technical Requirements of GridConnected*

Wind Power Plants in China and the United States. Technical Report NREL/
TP-5D00-64225. Golden, CO: National Renewable Energy Laboratory (NREL),
2016.

Goldwind International. 2017. *Company Profile.* Accessed February 10, 2018.
http://www.goldwindinternational.com/about/company.html.

GWEC. *Global Wind Report: Annual Market Update 2013.* Brussels, Belgium:
Global Wind Energy Council (GWEC), 2014.

———. *Global Wind Report: Annual Market Update 2015.* Brussels, Belgium:
Global Wind Energy Council, 2016.

———. *Global Wind Report 2016.* Brussels, Belgium: Global Wind Energy Council,
2017.

Harrabin, Roger. 2016. "China Embarked on Wind Power Frenzy, Says IEA." *BBC,*
September 20.

IRENA. *30 Years of Policies for Wind Energy: Lessons from 12 Wind Energy Markets.*
Abu Dhabi, United Arab Emirates: International Renewable Energy Agency,
2013.

Knight, Sara. 2012. "Fuhrländer struggles on through insolvency." *WindPower
Monthly,* November 1. https://www.windpowermonthly.com/article/1156685/
fuhrlander-struggles-insolvency.

Lam, Pun-Lee. "Energy in China: Development and Prospects." *China Perspectives*
59 (May–June 2005). Accessed January 8, 2017. http://journals.openedition.
org/chinaperspectives/2783.

Lawson, James. 2012. "Which Wind Turbine Generator Will Win?" *Renewable
Energy World.* http://www.renewableenergyworld.com/articles/print/special-
supplement-wind-technology/volume-2/issue-3/wind-power/which-wind-
turbine-generator-will-win.html.

Lema, Rasmus, Axel Berger, and Hubert Schmitz. *China's Impact on the Global
Wind Power Industry.* Discussion Paper 16/2012. Bonn, Germany: German
Development Institute, 2012.

Ming, Zeng, Xue Song, Li Ling-yun, Cheng Huan, and Zhang Ge. 2013. "Wind
Power Incidents in China: Investigation and Solutions." *Power,* April 1.
Accessed February 10, 2018. http://www.powermag.com/wind-power-
incidents-in-china-investigation-and-solutions/?printmode=1.

Patton, Dominique. 2013. "The Rise and Stall of Sinoval." *RECharge,* Janurary 6.

Richard, Craig. 2018a. "AMSC Loss Widens as Inox Deliveries Slow." *WindPower
Monthly,* February 8. https://www.windpowermonthly.com/article/1456484/
amsc-loss-widens-inox-deliveries-slow.

———. 2018b. "Sinovel Found Guilty in AMSC Case." *WindPower Monthly,*
Janurary 25. Accessed February 10, 2018. https://www.windpowermonthly.
com/article/1455469/sinovel-found-guilty-amsc-case.

Shepard, Christian, and Lucy Hornby. 2016. "Wind Turbine Maker Goldwind
Looks Beyond Xinjiang." *Financial Times,* June 20.

Vensys. 2016. *History.* Accessed February 10, 2018. http://www.vensys.de/energy-en/unternehmen/historie.php.

Vestas. 2017. *History.* Accessed November 17, 2017. https://www.vestas.com/en/about/profile#!history.

Visionaer. n.d. *Inspired by the Wind.* Accessed February 10, 2018. http://www.visionaer.info/56_0_en_a_279_Visionaries-/Inspired-by-the-Wind.html.

Weidong, Gu. "Non-Grid-Connected Wind Power and Offshore "Three Gorges of Wind Power" in China." In *Wind Power for the World: International Reviews and Development,* edited by Preben Maegaard, Anna Krenz, and Wolfgang Palz, 173–98. Singapore: Pan Stanford Publishing.

Wikipedia. 2018. *One Belt One Road Initiative.* February 8. Accessed February 10, 2018. https://en.wikipedia.org/wiki/One_Belt_One_Road_Initiative.

WindPower Monthly. 2007. "China's Sinovel Places Biggest Order Yet for American Wind Turbine Control Electronics." *WindPower Monthly*, September 1. Accessed February 10, 2018. https://www.windpowermonthly.com/article/959341/chinas-sinovel-places-biggest-order-yet-american-wind-turbine-control-electronics.

———. 2008. "Turbine Manufacturer Sinovel Places 10 GW Order." *WindPower Monthly*, July 1. Accessed February 10, 2018. https://www.windpowermonthly.com/article/951884/turbine-manufacturer-sinovel-places-10-gw-order.

20

The Globalization of Wind Power

Wind energy has become a global movement and an important source of energy supply.

—Preben Maegaard (2014)

20.1 Into the Fire

After over a century of interment wind power technology development and policy support, the wind power industry entered the twenty-first century with strong tailwinds. The technology itself had matured to the point where mega-watt-scale turbines had high reliability and wind power was becoming increasingly economic. European policy makers had laid a solid foundation to support wind power growth and wind power was growing rapidly in Spain, Denmark, and Germany. Growing federal- and state-level policy support in the United States had rekindled wind power growth there and positioned it to succeed in the twenty-first century. And, India and China wind markets were showing signs of life as well. It had been a long uphill battle over the course of the twentieth century, but wind power had finally arrived as a legitimate player in the global electricity landscape.

However, it was a bit too early to break out the Champagne. Wind power's very existence as a legitimate power technology meant that it was subject to the same fluctuating fortunes that other electric power technologies had already experienced. Electricity market participants in the twenty-first century are exposed to swings in electricity demand growth, changes in fuel prices and availability, an uneven and uncertain policy landscape, the impacts of electricity restructuring, and intense technology competition to name a few challenges; not to mention the need to contend with global economic cycles. Furthermore, the increasing globalization of the wind power supply

The Wind Power Story: A Century of Innovation that Reshaped the Global Energy Landscape, First Edition. Brandon N. Owens.

chain brought a whole new set of issues such as those related to international policies and risks. In many ways, wind power had moved out of the frying pan and into the fire.

In Europe, wind power experienced a roller coaster ride during the first two decades of the century. By 2000, the EU had installed 12.9 GW of capacity, which accounted for nearly 75 percent of the world total.[1] Beyond the core markets of Spain, Denmark, and Germany, the countries of Italy, Portugal, France, and the Netherlands implemented favorable wind power policies and began to exhibit growth. While these country-level policies were important in establishing a foundation of wind power policy support in Europe, the seminal moment occurred in 2001 when the EU passed its Directive on Electricity Production from Renewable Energy Sources 2001/77/EC. The Directive took effect in October 2001 and set indicative renewable energy targets for its fifteen member states. The directive was nonbinding, but it provided a roadmap that individual countries could follow to enable the EU to achieve its 21 percent renewable electricity target by 2010.

According to the Directive, each member state was obligated to generate a specified portion of its electricity from renewable sources by 2010. In the meantime, the member states were left to their own devices and could determine the appropriate country-specific approach to hit the Directive's target on time. This led to a potpourri of new wind power policies across Europe. Most countries followed Germany's lead and implemented feed-in tariffs. Italy mirrored the UK approach and implemented a quota system. The Netherlands used a combination of feed-in tariffs and tax incentives. Others—like Belgium—coupled their incentive schemes with "green certificate" systems that enabled wind power generators to sell green certificates for every kilowatt-hour of electricity they produced. Ireland and France opted for a pure tendering approach.

By 2005, the variety of policy approaches briefly led the EU to consider issuing a new Directive requiring some sort of "harmonization" of renewable energy policies across countries. This idea didn't advance, but the Commission's analysis did demonstrate that feed-in tariffs had been the most effective policy approach to date. Regardless of which policy approach was most effective, one couldn't argue with the results. Between 2000 and 2005, installed European wind power capacity grew from 17.5 to 40.5 GW. Europe was still the global focal point for wind power, but growing wind power markets in the United States, China, and India began to erode the European lead. Nonetheless—by 2005—69 percent of the world's installed wind power capacity still resided in Europe.[2]

The growth of wind power in Europe—and emerging growth across the globe—led to supply chain shortages. Wind manufacturers and their component suppliers had not anticipated such rapid growth in global wind power. Although many of them had been planning for at least a decade for a wind power surge, when it finally arrived everyone was a bit surprised. There was unprecedented pressure on suppliers of towers, blades, gearboxes, bearings, and generators. Wind farm developers were forced to wait from twelve to twenty-four months for wind turbines. It took a couple of years for manufacturers and their component vendors to ramp up their manufacturing capability to meet growing global demand.

Competition between international turbine manufacturers began to intensify as each fought for their own share of the growing global pie. For example—in 2003—Denmark's Vestas and NEG Micon merged to form the world's largest wind turbine manufacturer. The growing wind market also caught the attention of traditional industrial manufacturers, who wanted a piece of the action. After watching American industrial conglomerate GE enter the wind power market with its purchase of Enron Wind in 2002—two years later—German industrial conglomerate Siemens purchased Bonus Energy of Denmark, the world's fifth largest wind turbine manufacturer at the time.

Siemen's had formed Siemens Wind Power with its purchase of Denmark's Bonus Energy in 2004. With the purchase they inherited the legacy of Danish wind power innovation, design, and manufacturing. Recall, Henrik Stiesdal partnered with blacksmith Karl-Erik Jørgenson to develop a wind turbine prototype that was purchased by Vestas and became Vestas's first production model. When Siemens purchased Bonus, Stiesdal was the CTO. Siemens kept him as the CTO of Siemens Wind Power when the purchase was complete. Siemens Wind Power thrived as the European wind market burgeoned and quickly became one of the top wind turbine manufactures in the world. Later—in 2016—Siemens Renewable Energy joined forces with Spanish wind turbine manufacturer Gamesa to challenge Vestas as the world's largest wind turbine manufacturer. Interestingly, Gamesa itself had acquired its original wind turbine technology from Vestas in 1994. It seems that all wind power technology roads led back to Denmark's reinvention of wind power in the 1970s.

In his role as CTO—first at Bonus and then later at Siemens—Stiesdal was instrumental in advancing wind power technology. One of his most important contributions was the development of the world's first offshore wind farm in 1991, the 4.95-MW Vindeby Offshore Wind Farm in Denmark. Danish electric utility Elkraft had been eyeing wind power since 1987. They approached Bonus about purchasing wind turbines for the offshore environment and—instead of laughing them out of the room—Stiesdal was enthusiastic. He understood the potential of offshore wind power and he wanted Bonus to be the first wind turbine manufacturer to provide products for the offshore environment. Bonus later supplied eleven 450-kW wind turbines for the project. To ensure that they could withstand the offshore environment, they painted the wind turbines using same paint used on the North Sea drilling platforms. Vindeby was decommissioned in 2007 by new owner DONG Energy. It is still recognized as the "cradle of the offshore wind industry."[3]

By 2010, the European wind power market experienced a slowdown because of the global economic recession, which started in the United States in 2008 and spread to the rest of the world in 2009. Nonetheless, nearly ten gigawatts of new offshore capacity was installed. Spain, Germany, France, the United Kingdom, and Italy led the way. Before the year was over, European wind power capacity would reach 85 GW. The offshore market grew by over 50 percent, and offshore wind power additions accounted for nearly 10 percent of total European wind power additions.[4]

The EC keep its foot on the accelerator by implementing a new Renewable Energy Directive (2009/28/EC). The Directive went into effect in December of 2009 and it set an EU renewable energy target of 20 percent by 2020. Each member state had a legally binding target that must be achieved by 2020. The member states were required to submit National Renewable Energy Action Plans (NREAPs) detailing their sector-level targets and the policies needed to reach them. Fifteen EU countries immediately indicated that they will exceed the target. With these higher targets in place, Europe continued to press ahead with wind power. By 2016, 153 GW of wind power had been installed, including 12.6 GW offshore.[5]

On the policy front, many European countries began following the lead of Brazil, which had implemented competitive auctions for wind power projects starting in 2009. Brazil's auctions have led to the development of wind farms that were well below previous wind power price expectations. To manage the cost of wind power projects in Europe, competitive auctions were held in EU Member States, Denmark, Italy, Spain, the United Kingdom, Germany, France, and the Netherlands. The introduction of wind auctions has complicated the policy picture across Europe, with some countries immediately transferring to auctions and other countries keeping other policy mechanisms in place. The shift to auctions has also led to a surge in wind power development as market participants raced to install projects under existing feed-in tariff policies. This led to a 20 percent increase in European wind power capacity in 2017 as total installed capacity jumped to 169 GW.[6] Record wind power growth occurred in Germany, the United Kingdom, France, Finland, Belgium, Ireland, and Croatia.

A big challenge for wind power in Europe is the uncertainty about what happens after the Renewable Energy Directive's 2020 target is met. Only seven out of the current twenty-eight members of the EU have a post-2020 policy in place. The European Commission released its Clean Energy Package in 2016, which proposed a 27 percent renewable energy target by 2030.[7] Outside renewable energy groups have proposed a higher target of 35 percent. As of early 2018, the package had not yet been adopted by the European Council and the European Parliament. Wind power's future in Europe hangs in the balance.

20.2   The Final Frontier

In 1930s, German engineer Hermann Honnef was the first to advance the possibility of offshore wind power. Later, American William Heronemus proposed the development of a large network of offshore wind turbines to help solve the American energy crisis in the 1970s. Although the first offshore wind farm can be traced back to Vindeby in 1991, offshore wind didn't really take off until the twenty-first century. Thus, after nearly eighty years of possibilities and promise— starting around the turn of the twenty-first century— offshore wind power gradually started making a splash.[8]

There are very good reasons to place a wind turbine offshore. The winds are faster and the overall resource base is larger. For example, a recent European study concluded that offshore wind could—in theory—generate between 2,600 and 6,000 billion kWh per year. That represents between 80 and 180 percent of the European Union's total electricity demand. Furthermore, faster wind speeds mean higher production levels for each wind turbine. Also, because some of the best onshore sites become saturated the movement offshore represents a new frontier of possibilities. Finally, in some countries—like China and the United Kingdom—favorable offshore wind resources are closer to major population centers than the best onshore locations.

But there are some challenges associated with offshore development too. First, building turbines that can withstand the extreme wind, waves, and corrosion is a challenge. They must be extremely durable to handle the maritime environment. They also need to have low maintenance requirements, because maintaining machines out in the sea is costly. A few broken wind turbines can reduce profit from an offshore wind farm very quickly. This has tilted the scales toward lower-maintenance direct drive wind turbines for offshore environments. Beyond the turbines themselves, installing the foundations in rough waters is challenging. Then there's the trouble of laying the transmission cables across the seabed to connect with the land-based network.

Given the time and cost required to install the wind turbines out at sea, technology improvements to reduce the cost and improve the efficiency of the wind turbine are more impactful when they are onshore. As a result, the size of offshore wind turbines is larger than their onshore brethren. If you want to get as many kilowatt-hours as possible out of a single machine, then it needs to be large. As of 2018, offshore wind turbines are being built with new generation wind turbines that are greater than 10 MW each.

Among the large international wind turbine manufacturers, Siemens Gamesa has the largest offshore market share having supplied over two-thirds of the world's offshore capacity. This is the legacy of Henrik Stiesdal's decision as CTO of Bonus to supply wind turbines for Vindeby back in 1991. Vestas and

Figure 20.1 Built in 1991, the Vindeby Offshore Wind Farm in Denmark was the world's first. The wind farm used 11 Bonus B35/450 kW wind turbines for a total installed capacity of 4.95 MW. The turbines were modified for offshore use by sealing the towers.

Mitsubishi Heavy Industries (MHI) formed a joint venture in 2013 to develop an 8-MW wind turbine for offshore applications.[9]

With over 5.1 GW of installed offshore capacity, the United Kingdom is the world's offshore wind power leader.[10] That makes sense because the United Kingdom has some of the world's best offshore wind resources. The United Kingdom's offshore leadership position also reflects Europe's role in the world's offshore innovator—much the same role that European countries played as the vanguard of onshore wind farms in the 1990s. Total global offshore capacity stood at 12.6 GW at the end of 2016 and nearly 90 percent of it was installed in the waters of the coasts of the

United Kingdom, Denmark, and Germany.[11] Beyond Europe, the rest of the world's offshore wind projects are scattered across China, Japan, South Korea, and the United States.

In the late 1990s, the British Wind Energy Association issued a set of guidelines to help wind power developers gain access to the UK coastline and seabed. Over the subsequent decade, the UK government conducted three rounds of bidding to provide offshore wind farm developers with permits to develop large projects off the coast of the United Kingdom. As a result, twenty-seven wind farms with a capacity of 5.1 GW have been installed since 2000. The largest wind farm is the colossal 630-MW London Array project, followed by the Welsh wind farm Gwynt-y-Mor at 576 MW, and the Greater Gabbard 504-MW project. Furthermore, there is an existing pipeline of 12 GW of offshore projects.[12]

The maturation of the offshore wind industry coupled with technology and management improvements has brought costs down to levels that are approaching onshore costs in some cases. In addition, the introduction of larger wind turbines has also put downward pressure on the cost of electricity. In the United Kingdom, 2016 auction prices were as low as £65–70/MWh (US$82–88/MWh). In 2016, the Dutch tender for Borssele 1 and 2 came in at €72/MWh (US$80/MWh). This was followed by a Danish nearshore tender at €64/MWh (US$71/MWh). In the United Kingdom's 2017 auction, offshore wind developer Ørsted secured a contract worth £57.50/MWh ($73/MWh) for the 1.3-GW Hornsea Project Two project—a price that was 50 percent lower than those reached in a round two years earlier.[13]

Ørsted's Hornsea Project Two will feature turbines from Siemens Gamesa. The company's SG-8.0-167 DD is an 8-MW direct drive wind turbine built for the offshore environment. The Hornsea Project will contain 170 of these wind turbines upon completion in 2022.[14] With its 167-m rotor diameter, it looks a lot like the wind turbine of the future. At the same time, one cannot help but think that this turbine was created by a ghost from the past. It was built in the image of Herman Honnef's High Altitude Wind Power Generator from 1938. Honnef's designs have now officially moved from science fiction to science fact.

In 2018 GE announced the release of it's new Haliade-X wind turbine. GE announced the release of its new Haliade-X 12-MW wind turbine. In addition to being the largest offshore wind turbine, the Haliade-X will also be the world's most efficient offshore wind turbine.

GE's 12-MW giant has a very big rotor but a modest "specific power rating." The specific power rating is the ratio of the generator size to the rotor size. Large wind turbines with low specific power ratings are a growing trend in the wind industry. When the generator is small relative to the rotor size, the utilization increases and the wind turbine's capacity factor is higher.

This means the wind turbine will generate at—or close to—full capacity more frequently. This design approach is becoming increasingly attractive because wind turbines with lower specific power produce electricity for more hours over the course of a year, which further lowers the cost of wind power.

20.3 A Stable Sunset

The wind power story in the United States was like that of Europe in the first decade of the twenty-first century. Wind power additions and total capacity additions rose to levels that were hitherto unexpected. In 2000, just 2.6 GW of wind power capacity was installed across the country. By 2016, the level of installed wind power capacity had jumped to 82 GW.[15] Despite never being a leader in wind power globally, the United States continues to play a dominant role in the global wind power market. It is currently second only to China in installed wind power.

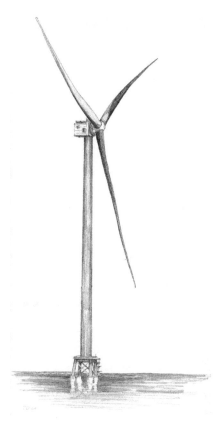

Figure 20.2 In 2018, GE announced the 12-MW Haliade-X offshore wind turbine. The race to create the biggest and best offshore wind turbines is on, a clear sign that William Heronemus's 1972 dream of a future with a large network of offshore wind turbines is now upon us.

The growth of wind power in America was driven by a combination of state and federal policies. At the state level, a growing number of states enacted renewable portfolio standards. RPS programs can have different configurations. Utilities can be required to own or purchase a specific quantity of wind power, or they can be required to ensure that a certain percentage of their sales to customers comes from renewable energy. This is a different approach from tax incentives which attempt to make wind power more economically competitive by providing direct incentives that lower the cost of wind power. RPS programs can be used in combination with tax incentives. When this occurs, the tax incentives reduce the cost to meet RPS requirements.

In 1983, Iowa became the first state in the United States to adopt a renewable portfolio standard by enacting the Alternative Energy Production law. The Iowa RPS requires its investor-owned utilities to own or to contract 1.5 percent of their power from renewable energy sources. Minnesota followed suit with its own program in 1994. The enactment of Minnesota's policy and the development of wind power projects in response demonstrated how effective state programs could be in jumpstarting wind power.

Recognizing the potential of state-level policies, AWEA's Executive Director—Randy Swisher—hired Nancy Rader to develop new wind policies that could be presented to national- and state-level policy makers. Rader began her renewable energy career at Public Citizen in Washington, DC. She then began advising renewable energy companies and trade associations in the United States and Canada. After months of consultation and debate and an evaluation of an entire array of alternative approaches, AWEA fleshed out an RPS proposal and engaged the Union of Concerned Scientists (UCS) to help with political outreach.

After a proposed national RPS bill failed in Congress, UCS focused its outreach on the states. Arizona adopted an RPS in 1996; Massachusetts, Maine, and Nevada in 1997; Wisconsin, Connecticut, and Pennsylvania in 1998; New Jersey and Texas in 1999; and New Mexico in 2000. By the end of the 1990s, twelve states had implemented their own RPS program. By 2007, twenty-one states and the District of Columbia had implemented mandatory RPS obligations. These policies covered roughly 40 percent of total US electrical load.[16]

At the time of their enactment, state electricity ratepayers would have had to pay a premium to meet the RPS obligations because wind power wasn't quite economic with conventional electricity generators yet. However, the presence of the federal PTC reduced the cost of wind power in RPS states by shifting some of the costs of wind power to federal income tax ratepayers. The PTC was initially created by the Energy Policy Act of 1992. It provided a ten-year, 1.5- ¢/kWh inflation-adjusted production tax credit for wind power and other renewable energy sources. When accelerated depreciation was coupled with the PTC, wind farm investors could achieve up to 50 percent of their return from these federal tax benefits.

The combination of the PTC and state-level RPS policies reignited wind power in America starting in 1998. Momentum was restored. In 1998 and 1999, US wind power capacity grew by approximately one gigawatt. The biggest challenge for the wind power industry at that time was dealing with the the periodic lapses in the PTC, the first of which occurred at the end of 1999. The problem with the PTC lapse was not just the negative impact on project economics, but it was the gyrations it caused in the marketplace. Wind farm developers and investors would rush to complete projects in advance of an expected PTC expiration, and they would stop building anything after an expiration in case the credit wasn't renewed and wasn't retroactive. It created an unhealthy boom and bust cycle driven solely by federal policy uncertainty.

The PTC was extended for what may be the final time in 2016. The 2015 extension pushed the credit expiration to the end of 2019. With the final PTC extension in hand, many states began expanding their RPS obligations starting in 2015. By 2016, five states and the District of Columbia had RPS targets of 50 percent or more. Four states—New Mexico, Washington, Nevada, and Maryland—and the District of Columbia have updated their RPS programs since the start of 2019.[17] State-level RPS programs continue to remain an important driver of wind power in the United States and will become increasingly important as the federal PTC phases out.

The importance of both federal tax credits and state RPS programs as a wind power accelerant is heightened in the United States because of the emergence of a new Presidential Administration in 2016 that is antagonistic towards wind power. Although the June 2017 announcement that the United States would cease participation in the 2015 Paris Agreement garnered the most headlines, the EPA plan to scrap the Clean Power Plan is arguably more impactful. The policy was aimed at cutting Greenhouse Gases (GHG) emissions from electric power generation. Under the rule, US states would have been required to draw up their own plans to cut emissions by 32 percent by 2030. This would have inevitably translated into more wind power installations. However, without the Plan, States and utilities will be left to their own devices on how to meet their power generation needs.

However, economic considerations are likely to override recent political decisions and continue to drive the United States toward a low-carbon energy future. Given the increasingly low cost of wind and solar power and the abundance of natural gas in the United States, the future of electric power in America will continue to move toward renewable energy generation and natural-gas fired power plants. As a result, US electric sector carbon dioxide emissions are on a downward trajectory independent of America's participation in the Paris Agreement or the presence of Clean Power Plan.

In fact, forecasts assembled by the US Department of Energy show United States' electric sector carbon dioxide emissions declining by over 6 percent through 2025 even as electricity use climbs by 8 percent over the same period.[18] Wind power is expected to continue to grow, even as the PTC is phased out. The Department of Energy now believes that wind power capacity will climb to 118 GW by 2025. It seems that, in America, wind power has found stability in a sea of conflicting signals and opposing interests.

20.4   Around the World

The rise of wind power in Europe, the United States, India, and China provides an illustration of how the combination of improving technology economics and supportive policies can help launch a multi-billion-dollar energy industry and create a new set of domestic and international equipment manufacturers. Wind power's journey across the course of the twentieth century was essentially a prelude to the explosive growth that started at the dawn of the twenty-first century.

Another interesting aspect of wind power's twenty-first century rise is the extent that it has been made more available across the globe. Wind power policies have been exported from Western Europe and the United States, first to India and China, and then later to Eastern Europe, Latin America, Africa, and the Middle East. As a result, wind power is now installed in ninety countries around the world.[19]

Morocco is an example of how wind power's reach has spread beyond the core markets in Europe, Asia, and North America. The country has set a target of installing 2 GW of wind capacity by 2020. Morocco has 787 MW of installed wind power capacity and recently issued a competitive tender for 850 MW of additional capacity spread over five sites. A consortium consisting of Enel Green Power (EGP), Nareva and Siemens Gamesa was officially named as preferred bidders in March of 2016.[20] The bidders were required to include plans for local content and the projects will be built under a public–private partnership. Siemens Gamesa's local content plans included the supply of local blades. In response, the company officially opened a new blade manufacturing plant in Tangier, northern Morocco, in October 2017.[21] Siemens already has roughly 70 percent of Morocco's installed wind market and expects to provide more wind turbines as the supplier for Nareva.

Mexico is another country that has caught the wind power bug. The country is in the midst of a transformation from a fully regulated, state-owned electricity system to a competitive system with a functioning wholesale electricity market, independent generation companies, and independent system operators. The country has also decided to facilitate a transition to cleaner energy sources, including wind power. As a result, Mexico had held three clean energy auctions that have resulted in contracts for more than 2 GW of wind power that will come online before 2020. This will place the country's installed capacity at around 12 GW by 2020.[22]

Outside of Europe, Asia, and North America, Brazil has the best track record of recent wind power development. With less than 1 GW of installed wind power as late as 2010, the country raced forward and boasted an installed capacity of approximately eleven gigawatts by the end of 2016. Traditionally reliant on hydropower generation for the lion's share of its electricity needs, in the last decade Brazil started holding auctions for wind power as part of a diversification strategy.

Brazil's wind power journey began in 1992 when the first Brazilian wind turbine was installed on the archipelago of Fernando de Noronha. The 75-kW wind turbine was the result of a collaboration between the Danish Folkecenter and the Brazilian Wind Energy Centre. At the time, the archipelago used diesel power to meet its electricity needs. This was a rare exception in Brazil where 92 percent of electricity generation came from hydropower.[23] A decade later in 2001, Brazil experienced an electricity shortage due to a lack of rainfall. The government realized that it needed to diversify its electricity sources and instituted a Program of Incentives for Alternative Energy Sources (PROINFA). PROINFA had a mixed track record and the Brazilian government moved on to an auction-based approach in 2009.

Brazilian wind energy auctions have led to the development of ten gigawatts of wind power since 2009. The most recent award for 1.4 GW of power purchase agreements from energy regulator Aneel came in at an average price less than US$30/MWh, which shows that Brazil's efforts to diversify away from hydropower make economic sense as well.[24] However, the country's recent economic slowdown has created some headwinds for wind power. There is currently a problem with excess supply in some part of the electric power system. This caused Brazil to cancel its auctions in 2016. Furthermore—as in other countries—transmission constraints are starting to become an issue for wind farms that are not located near demand centers. The states of Rio Grande do Norte, Rio Grande do Sul, and Bahia have all experienced transmission constraints.

20.5 The Tipping Point

To the casual observer, wind power exploded on the global electricity scene in the twenty-first century. From out of nowhere, suddenly wind power was accounting for overall half of global electric power additions in some countries. Industrial giants like GE and Siemens suddenly had entire businesses devoted to wind energy. International wind turbine manufacturers like Vestas became household names. Countries like China, India, Germany, and the United States began vying for wind power supremacy. It appeared to have happened so quickly.

On the contrary, a look back at the historical record reveals a long road littered with generous amounts of both success and failure. The goal of wind power R&D has always been to build reliable wind turbines that produced electricity at prices at or below conventional generation options. This goal was elusive throughout the twentieth century as the cost of conventional options generally moved downward in response to fuel price changes and technology improvements. In fact—well into the second decade of the twenty-first century—wind power struggled to achieve cost parity with conventional electricity sources such as coal, natural gas, and nuclear power.

However—in the past decade—a tipping point has been reached. One a pure energy basis, wind power is often the least cost source of power in many instances across the globe. This fact is difficult to digest—particularly for long-term energy market observers and those who have political or philosophical biases against wind power. Wind power is economically competitive today for several reasons. First, the size of wind turbines has increased. Wind turbines exhibit economies of scale, as they grow larger, they produce greater amounts of electricity than can be spread over the one-time installation cost. Onshore wind turbines are typically 3 MW or larger; as we have seen, new offshore wind farms have wind turbines that are 10 MW or more. Second, the productivity of wind turbines has increased as control systems have gotten smarter, wind turbine blades have been lengthened to enable wind turbines and scoop up more wind, and the specific power of wind turbines has been reduced. Modern wind turbines now have capacity factors that commonly exceed 50 percent—a figure that would have been unheard of just a decade ago. Third, industry supply chains have matured and globalized. Wind turbine component designs have advanced and suppliers are realizing economies of scale. Components are constructed across the globe in the least-cost locations. Fourth, the new approach to procuring large amounts of wind power through competitive auctions has also driven down costs and increased competition among bidders.

New wind power projects in markets in such diverse locations as Morocco, India, Mexico, and Canada are all coming in with prices in the range of US\$30/MWh or below, with a recent Mexican tender coming in with prices below US\$20/MWh.[25] Updated cost of energy estimates by industry organization such as the International Energy Agency and the International Renewable Energy Agency (IRENA) confirm that the costs of electricity from onshore wind are now at the lower end of the fossil fuel cost range.

Today's low cost of wind power is most evident in the US state of Colorado, home of the legendary Telluride AC electric power network that helped turn the tide in favor of AC during the War of the Currents near the turn of the twentieth century. In 2017, Xcel Energy solicited bids for projects to help meet its goal of going completely carbon-free by 2050.[26] Xcel has reported that they received bids for forty-two wind projects totaling over 17 GW with an average price of US\$18.10/MWh.[27]

The tipping point is also evident in IEA's World Energy Outlook, which reflects a shift from fossil fuels to renewables. Between 2010 and 2016, global coal-fired capacity increased at a rate of 65 GW a year, the highest of any electricity-generating source. However, looking forward—for the 2017–2040 period—anticipated annual additions of coal-fired generation are less than 20 GW. Based on their attractive economics, wind farms are expected to be built at a rate of approximately 50 GW per year.[28]

The twenty-first century rise in wind power has created electricity networks with high levels of variable wind generation. In Denmark, wind power already accounts for around 40 percent of annual electricity generation. In Ireland, wind power accounts for about 20 percent of annual electricity generation. Spain, Germany, and the United Kingdom are close behind.[29] That's on an annual basis, but on the windiest days, wind power's share of total generation will reach two to three times these levels. That means that on some days, these countries are already experiencing days when wind power accounts for all of the electricity generation. In July 2015, Denmark generated 140 percent of its electricity from wind.[30] On a recent breezy Sunday, wind power and solar combined accounted for 85 percent of Germany's electricity generation.[31] In Texas, 22.6 GW of wind power has been installed partially in response to the state's RPS program and because the Texas winds make wind power an economic choice. That's triple the installed capacity of the next nearest state Iowa. As a result, in the Lone Star State wind power can account for nearly half of total electricity generation on any given day.[32]

Remember though, electric power systems must remain in balance at all times. High levels of wind power are fine if electricity demand is also high, transmission capacity is available to transport the wind power, and other generators are not producing electricity at the same time. If one or more of these conditions doesn't hold, then electric grid operators need to figure out how to accommodate the wind. In these cases, the flexibility of other generators and even electricity customers becomes important as electricity demand and supply vary. This means that the operating patterns of all power plants on the system will vary as they accommodate the wind. High wind levels can also yield significant changes in power flow patterns across the grid.[33]

What happens when the amount of wind power generated in a power system is greater than the level of electricity demand? The first response is to curtail the wind power so that the wind farm produces less electricity than it can produce given the prevailing wind speed. But in some power markets that have real-time competitive pricing and are relatively isolated, like the Electricity Reliability Council of Texas (ERCOT), there's another possibility: negative electricity prices. If there's too much wind power for the electric system to absorb, market prices will turn negative. This means that wind power producers will pay the electricity system to take the electricity off their hands.

In ERCOT—which will have nearly 30 GW of installed wind capacity by 2019—surplus wind energy continues to create negative prices into the real-time wholesale electricity market.

The challenges associated with increasing levels of wind power are being addressed. However, dealing with these issues requires flexibility and careful consideration, and there are costs involved. Transmission capacity upgrades and grid reconfigurations are needed to accommodate new and different power flows. Existing and new nonvariable generators need to be increasingly flexible to ramp up and down in response to variable wind power. And—in some cases—older, inflexible generators will need to be phased out of the power system to accommodate new system dynamics. In short, the introduction of variable wind power creates a need to transform inflexible, balkanized power systems into more expansive, flexible networks. It is happening, but accommodating variable resources is a gradual transformational process, not a singular event.

At high levels of wind penetration, electric system stability also becomes a consideration. Looking forward, some power networks are going to begin grappling with power stability issues. The electric power system must be able to maintain stable operating conditions following disturbances. Typically, grid stability is maintained by the inertia in the system. This refers to the kinetic energy stored in the rotating masses of the generators connected to the system. This serves as a short-term energy storage. Any shortfall in power will be experienced as a force acting against their rotation. However, when lots of wind power is present, there are fewer synchronous generators on the system and so less inertia is available to respond to disturbances. For example, ERCOT has reported a continuous decline in the inertial response of its system and recommends additional inertial response.[34]

One way to address this issue is to equip wind turbines with synthetic or virtual inertia to overcome this problem. Wind turbines can be programmed to provide synthetic or emulated inertia by increasing the active power in the face of a drop in the grid frequency. The energy for this increase is drawn from the rotating masses of the wind turbine blades. Enercon introduced its "Inertia Emulation" control feature in 2008.[35] Other wind turbine manufacturers have followed suit. The importance of virtual inertia and sophisticated wind turbine power control systems will increase over time as wind penetration levels grow. Advanced wind turbine power control features will become increasingly important within the transformed energy systems of the future.[36]

20.6 Back to the Future

When it's all said and done, two important themes emerge from recent events on the global wind power stage. First, because of recent innovations, wind power has now become the least-cost source of electricity in many cases.

Second, increasing amounts of wind power are being added to power networks across the globe. Dealing with growing amounts of wind power is an emerging challenge that is currently being addressed. New innovations will be required to transform power systems around the world, and emerging technology solutions are already providing a glimpse into the future of wind power.

Although batteries are not a new technology—when integrated with wind power either at the wind turbine or dispersed in critical locations throughout the grid—batteries provide the ability to smooth out power fluctuations and provide a steadier stream of electricity to the surrounding network. In addition, battery-integrated wind power can enable wind farms to deliver electricity when it is needed and can be effectively used by the grid. If a congested transmission line doesn't have room for wind power as soon as it is generated, a battery can be used to enable the electricity to "wait" at the wind farm and then be delivered later when the constraint has eased. That will help solve some of the challenges associated with high-penetration variable wind energy.

As battery costs have fallen, an increasing number of wind turbine manufacturers are eyeing batteries as a solution, either directly integrated into wind turbines or placed strategically throughout the distribution system to provide support. In 2012, Vestas began a battery-integrated wind power initiative at its Lem Kaer wind farm in Denmark. Vestas—equipped the facility with two lithium-ion batteries, which successfully stored excess energy for future use. GE was the first out of the gate with a commercial offering with its "Brilliant Wind Turbine" in 2013. Three units went into operation at Invenergy's Goldthwaite Wind Energy Center in Mills County, Texas in 2014.

More recently, Vestas has partnered with KK Wind Solutions, a Danish wind systems developer; PowerCon, Danish engineering firm; and Denmark's Aalborg University to develop a battery-integrated wind farm that reduces power output fluctuations by up to 90 percent using batteries that are sized to just 80 percent of total wind farm capacity. In 2017, Toshiba announced the installation of a 2-MW lithium-ion battery system near NRG's Elbow Creek Wind Farm in Howard County, Texas, to correct short-term grid imbalances from variable wind generation on the ERCOT grid.[37] Spanish wind power developer Acciona recently equipped one of its 3-MW wind turbines in Barásoain, Navarre, with two Samsung lithium-ion batteries.[38]

In 2017, Vestas announced that it was working with different energy storage technologies with specialized companies—including Tesla Inc.—to explore and test how wind turbines and energy storage can work together in sustainable energy solutions that can lower the cost of energy. Tesla supplied batteries for the first phase of the Kennedy Energy Park in Australia. The project will deploy twelve of Vestas's V136 3.6-MW wind turbines as well as 15 MW of solar panel capacity and a 4-MWh Tesla lithium ion battery, all managed by Vestas's control system.[39]

GE too has continued to innovate in this area by releasing a new battery solution called the GE Reservoir that delivers a suite of customized battery solutions to help customers address new challenges and seek new opportunities in a rapidly transforming power grid.[40]

Hydrogen is another emerging energy source that can assist with the transformation of the global power system. We know that electrolysis can be used to turn excess wind-generated electricity into hydrogen and oxygen. The hydrogen can then be injected into pipelines or compressed and stored for later use. The stored hydrogen can be used to generate electricity, either in a gas turbine through a hydrogen-natural gas mixture, or via a fuel cell, or used as a fuel in transportation networks. If wind power is inexpensive enough, it's not difficult to envision a scenario where wind turbines supply electric power to the grid when it is needed, and, instead of curtailing wind when it is not needed, it is used to produce hydrogen for a variety of uses.

There are several wind-hydrogen demonstration projects across Europe, Asia, and North America. In Yokohama City, Japan, a Vestas V80 2-MW turbine is being used to produce hydrogen which is then stored in an onsite tank before being shipped by truck to neighboring Kawasaki City and used to power fuel-cell driven forklifts at a fruit and vegetable market, a factory, and warehouses.[41] In Denmark, a hydrogen system supplier is supplying a 1.2-MW electrolyzer that will produce hydrogen from excess wind energy and will be used to enable grid balancing services.[42]

In Germany, utility E.ON has partnered with Swissgas AC to operate a 2-MW electrolyzer in Falkenhagen that is connected to a nearby wind farm. The produced hydrogen is fed directly into a nearby natural gas pipeline.[43] Recent studies have shown that hydrogen can be safety blended into natural gas pipelines. Many existing gas turbines are equipped to accept a 50 percent natural gas-hydrogen mixture already. Furthermore—in 2018—the chemical arm of AkzoNobel and a research unit of natural gas company Gasunie joined forces to investigate the potential of a 20-MW electrolysis plant powered by renewable energy to produce hydrogen in the northeast of the Netherlands. The produced hydrogen will be used as fuel for Netherland's bus fleet.[44] The possibilities are endless; and many observers believe that the wind-hydrogen alliances represents the future of energy.

If these new innovations sound familiar, they should.

Back in 1887, the electrical output from Charles F. Brush's wind turbine in the backyard of his Cleveland mansion was used to charge twelve batteries before they were discharged to power 350 incandescent lamps, two arc lamps, and three motors. Brush used batteries as the intermediary between his wind turbine and the electrical needs of his home.

And in Denmark, in 1891, Poul la Cour directed the electricity generated by his wind turbine into a tub of water where oxygen was separated from hydrogen through electrolysis. The hydrogen and oxygen gas was then collected in tanks and burned later for lighting. Both Brush and la Cour knew a thing or two about wind turbines; and, as the history of wind power has demonstrated, there are no bad ideas—just innovations whose time has yet to come.

These nineteenth century wind power ideas are likely to continue to play an increasing role in twenty-first century power systems. Thus, we find ourselves back at the beginning of the wind power journey, still searching for better ways to tame Mother Nature in order to help meet humankind's energy needs. Of course, as we have seen, technology innovation is a continuous process. Indeed, it appears that the wind power story is a journey without end.

Notes

1 GWEC, 2007, 8.
2 Ibid.
3 Neilson, 2016.
4 GWEC, 2011, 39.
5 GWEC, 2017, 18.
6 Ibid.
7 Ibid., 42.
8 For background information on the rise of offshore wind power see Kaldellis and Kapsali (2013).
9 Scott and Reid, 2013.
10 GWEC, 2017, 61.
11 Ibid.
12 Ibid., 60.
13 Ibid, 59.
14 WindPower Offshore, 2018.
15 GWEC, 2017, 71.
16 Wiser et al., 2007, 8.
17 GWEC, 2017, 70.
18 EIA, 2018.
19 GWEC, 2017, 13.
20 SeeNews, 2015.
21 Richard and Weston, 2017.
22 Hill, *Third Mexican Auction Awards Enel 593 Megawatts of Wind, Canadian Solar Awarded 367 Megawatts Solar*, 2017.
23 Feitosa, 2014, 320.
24 Azzopardi, 2017.
25 GWEC, 2017, 4.
26 Gillis and Harvey, 2018.
27 Arcus, 2018.

28 IEA, 2017a.

29 IEA, 2017b, 32.

30 Hill, *Wind Power Generates 140% of Denmark's Power Demand*, 2015.

31 Parkinson, 2017.

32 Hill, *Texas Hits New Wind Generation Record of 45% Total Electric Demand*, 2016.

33 See IEA (2017c) for a good discussion of these factors.

34 ERCOT, 2017.

35 ENERCON, 2015.

36 See Tamrakar et al. (2017) for more information on virtual inertia.

37 Deign, *Danish Tech Firm Looks to Batteries for Integrated Wind-Plus-Storage Solution*, 2017b.

38 Deign, *Battery Storage Takes Hold in the Wind Industry*, 2017a.

39 Shankleman, 2017.

40 Battery Power, 2018.

41 Foster, 2017.

42 Ver Bruggen, 2016.

43 Danko, 2013.

44 Weston, 2018.

Bibliography

Arcus, Christopher. 2018. "Wind & Solar + Storage Prices Smash Records." *Clean Technia*, Janurary 18. https://cleantechnica.com/2018/01/11/wind-solar-storage-prices-smash-records.

Azzopardi, Tom. 2017. "Brazil Tenders 1.4GW of Wind Power." *WindPower Monthly*, December 22.

Battery Power. 2018. "GE Announces Energy Storage Platform Called the Reservoir." *Battery Power*, March 11. http://www.batterypoweronline.com/news/ge-announces-energy-storage-platform-called-the-reservoir.

Danko, Pete. 2013. "Wind Power Makes Hydrogen for German Gas Grid." *Greentech Media*, September 9. https://www.greentechmedia.com/articles/read/wind-power-makes-hydrogen-for-german-gas-grid#gs.Gh=yje8.

Deign, Jason. 2017a. "Battery Storage Takes Hold in the Wind Industry." *Greentech Media*, June 9. https://www.greentechmedia.com/articles/read/battery-storage-takes-a-hold-in-the-wind-industry#gs.E3FBLQY.

———. 2017b. "Danish Tech Firm Looks to Batteries for Integrated Wind-Plus-Storage Solution." *Greentech Media*, November 13. https://www.greentechmedia.com/articles/read/danish-tech-firm-builds-wind-plus-storage-solution#gs.ghGBAbk.

EIA. *Annual Energy Outlook 2018*. Washington, DC: Energy Information Administration (EIA), 2018.

ENERCON. 2015. ENERCON Wind Energy Converters: Technology and Service. *ENERCON.* https://www.enercon.de/fileadmin/Redakteur/Medien-Portal/broschueren/pdf/en/ENERCON_TuS_en_06_2015.pdf.

ERCOT. 2017. Dynamic Stability Assessment of High Penetration Renewable Generation in the ERCOT Grid. *Electric Reliability Council of Texas.* http://www.ercot.com/content/wcm/key_documents_lists/108892/Nov_2017_RPG_Update_Dynamic_Stability_Assessment_of_High_Penetration_of_Renewable_Generation_in_the_ERCOT_Grid-Year2031_Preliminary.pdf.

Feitosa, Everaldo Alencar. "The Emergence of Wind Energy in Brazil." In *Wind Power for the World: International Reviews and Development,* edited by Preben Maegaard, Anna Krenz, and Wolfgang Palz. 319–30. Singapore: Pan Stanford Publishing, 2014.

Foster, Martin. 2017. "Japan Plans Wind-Driven Hydrogen Project." *WindPower Monthly,* July 21. https://www.windpowermonthly.com/article/1440164/japan-plans-wind-driven-hydrogen-project.

Gillis, Justin, and Hal Harvey. 2018. "Why a Big Utility Is Embracing Wind and Solar." *The New York Times,* February 6.

GWEC. *Global Wind Energy Report 2006.* Brussels, Belgium: Global Wind Energy Council, 2007.

——. *Global Wind Report: Annual Market Update 2010.* Brussels, Belgium: Global Wind Energy Council, 2011.

——. *Global Wind Report 2016.* Brussels, Belgium: Global Wind Energy Council, 2017.

Hill, Joshua S. 2015. "Wind Power Generates 140% of Denmark's Power Demand." *Clean Technia,* July 13. https://cleantechnica.com/2015/07/13/wind-power-generates-140-denmarks-power-demand.

——. 2016. "Texas Hits New Wind Generation Record of 45% Total Electric Demand." *Clean Technia,* December 2, 2016. https://cleantechnica.com/2016/12/02/texas-hit-new-wind-generation-record-45-total-electric-demand.

——. 2017. "Third Mexican Auction Awards Enel 593 Megawatts of Wind, Canadian Solar Awarded 367 Megawatts Solar." *Clean Technia,* November 28. https://cleantechnica.com/2017/11/28/third-mexican-auction-awards-enel-593-mw-wind-canadian-solar-awarded-367-mw-solar.

IEA. *World Energy Outlook 2017.* Paris, France: International Energy Agency, 2017a.

——. *Renewables 2017: Analysis and Forecasts to 2022.* Paris, France: International Energy Agency, 2017b.

——. *Getting Sun and Wind onto the Grid: A Manual for Policy Makers.* Paris, France: International Energy Agency, 2017c.

Kaldellis, J.K., and M. Kapsali. "Shifting Toward Offshore Wind Energy: Recent Activity and Future Development." *Energy Policy* 53 (2013): 136–48.

Maegaard, Preben. "Introduction." In *Wind Power for the World: International Reviews and Developments,* edited by Preben Maegaard, Anna Krenz, and Wolfgang Palz, xxiii–xxv. Singapore: Pan Stanford Publishing, 2014.

Neilson, Michael Korsgaard. 2016. "Fra Energiens Originaler Til Bølgernes Gulddrenge." *Berlingske Business*, February 8. https://www.business.dk/energi/fra-energiens-originaler-til-boelgernes-gulddrenge.

Parkinson, Giles. 2017. "Graph of the Day: Germany's Record 85% Renewables Over the Weekend." *REneweconomy*, May 4. http://reneweconomy.com.au/graph-of-the-day-germanys-record-85-renewables-over-weekend-60743.

Rader, Nancy, and Richard Norgaard. "Efficiency and Sustainability in Restructured Electric Markets: The Renewables Portfolio Standard." *The Electricity Journal*, 9, no. 6 (1996): 37–49.

Richard, Craig, and David Weston. 2017. "Siemens Gamesa Opens First Blade Factory in MEA Region." *WindPower Monthly*, November 1. https://www.windpowermonthly.com/article/1448417/siemens-gamesa-opens-first-blade-factory-mea-region.

Scott, Mark, and Stanley Reid. 2013. "Vestas Joins With Mitsubishi for Offshore Turbines." *The New York Times*, September 27. http://www.nytimes.com/2013/09/28/business/international/vestas-joins-with-mitsubishi-for-offshore-turbines.html.

SeeNews. 2015. "Enel Green Placed Lowest Bid in Morocco's 850-MW Wind Tender: Report." *Renewables Now*, December 10. https://renewablesnow.com/news/enel-green-placed-lowest-bid-in-moroccos-850-mw-wind-tender-report-505134.

Shankleman, Jess. 2017. "Tesla and Vestas Partner in $160 Million Australian Project." *Bloomberg Markets*, October 9. https://www.bloomberg.com/news/articles/2017-10-19/tesla-and-vestas-partner-in-160-million-australian-project.

Tamrakar, Ujjwol, Dipesh Shrestha, Manisha Maharjan, Bishnu P. Bhattarai, Timothy M. Hansen, and Reinaldo Tonkoski. "Virtual Inertia: Current Trends and Future Directions." *Applied Sciences* 7 (2017): 654.

Ver Bruggen, Sara. 2016. "Industrial-Scale Hydrogen Storage on Trial." *WindPower Monthly*, October 13. https://www.windpowermonthly.com/article/1412122/industrial-scale-hydrogen-storage-trial.

Weston, David. 2018. "Largest Hydrogen-from-Wind Project Proposed." *WindPower Monthly*, Janurary 10. https://www.windpowermonthly.com/article/1454127/largest-hydrogen-from-wind-project-proposed.

WindPower Offshore. 2018. "Siemens Gamesa Wins Hornsea Project Two Deal." *WindPower Offshore*, February 14. https://www.windpoweroffshore.com/article/1457120/siemens-gamesa-wins-hornsea-project-two-deal.

Wiser, Ryan, Christopher Namovicz, Mark Gielecki, and Robert Smith. *Renewables Portfolio Standards: Factual Introduction to Experience from the United States*. Berkeley, CA: Lawrence Berkeley Nattional Laboratory, 2007.

Index

The Wind Power Story: A Century of Innovation that Reshaped the Global Energy Landscape,
First Edition. Brandon N. Owens.
© 2019 by The Institute of Electrical and Electronics Engineers, Inc.
Published 2019 by John Wiley & Sons, Inc.